MICRO- AND NANO-STRUCTURED INTERPENETRATING POLYMER NETWORKS

MICRO- AND NANO-STRUCTURED INTERPENETRATING POLYMER NETWORKS

From Design to Applications

Edited by

PROF. DR. SABU THOMAS

DR. DANIEL GRANDE

DR. UROŠ CVELBAR

DR. K.V.S.N. RAJU

DR. RAMANUJ NARAYAN

DR. SELVIN P. THOMAS

AKHINA H.

WILEY

Published by John Wiley & Sons, Inc., Hoboken, New Jersey
Published simultaneously in Canada

Library of Congress Cataloging-in-Publication Data:

Names: Thomas, Sabu, editor.
Title: Micro- and nano-structured interpenetrating polymer networks : from design to applications / edited by Prof. Dr. Sabu Thomas, Dr. Daniel Grande, Dr. Uroš Cvelbar, Dr. K.V.S.N. Raju, Dr. Ramanuj Narayan, Dr. Selvin P. Thomas, Akhina H.
Description: Hoboken : Wiley, 2016. | Includes bibliographical references and index.
Identifiers: LCCN 2015047249 (print) | LCCN 2015050095 (ebook) | ISBN 9781118138175 (hardback) | ISBN 9781119138952 (Adobe PDF) | ISBN 9781119138969 (ePub)
Subjects: LCSH: Polymer networks. | Polymers–Industrial applications. | BISAC: TECHNOLOGY & ENGINEERING / Chemical & Biochemical.
Classification: LCC QD382.P67 M53 2016 (print) | LCC QD382.P67 (ebook) | DDC 620.1/92–dc23
LC record available at http://lccn.loc.gov/2015047249

Cover image courtesy of Getty Images/ Milanares

Set in 10.5/13.5pt Times by SPi Global, Pondicherry, India

Printed in the United States of America

10 9 8 7 6 5 4 3 2 1

CONTENTS

LIST OF CONTRIBUTORS

Rameshwar Adhikari, Central Department of Chemistry, Tribhuvan University, Kathmandu, Nepal

Akhina H., International and Interuniversity Centre for Nanoscience and Nanotechnology, Mahatma Gandhi University, Kottayam, Kerala, India

Mohammed N. Alghamdi, Mechanical Engineering Technology Department, Yanbu Industrial College, Royal Commission for Yanbu Colleges and Institutes, Yanbu Al-sinaiah, Saudi Arabia

Mladen Andreis, Physical Chemistry Division, Rudjer Bošković Institute, Zagreb, Croatia

S. Anil Kumar, Department of Chemistry, N.S.S. College, Ottapalam, Kerala, India

Prathab Baskar, Materials Modeling and Product Design, Research & Development, Tata Steel, Jamshedpur, Jharkhand, India

Emilio Bucio, Departamento de Química de Radiaciones y Radioquímica, Instituto de Ciencias Nucleares, Universidad Nacional Autónoma de México, Mexico City, México

Guillermina Burillo, Departamento de Química de Radiaciones y Radioquímica, Instituto de Ciencias Nucleares, Universidad Nacional Autónoma de México, México City, Mexico

Shoubing Chen, State Key Laboratory of Solid Lubrication, Lanzhou Institute of Chemical Physics, Chinese Academy of Sciences, Lanzhou, China

Uroš Cvelbar, Department of Surface Engineering, Jozef Stefan Institute, Ljubljana, Slovenia

Branko Dunjić, Faculty of Technology and Metallurgy, University of Belgrade, Belgrade, Serbia

Lorena Garcia-Uriostegui, Departamento de Química de Radiaciones y Radioquímica, Instituto de Ciencias Nucleares, Universidad Nacional Autónoma de México, México City, Mexico

Sudipta Goswami, Chemical Engineering and Technology, Birla Institute of Technology, Mesra, Ranchi, Jharkhand, India

Daniel Grande, Institut de Chimie et des Matériaux Paris-Est (ICMPE), UMR 7182 CNRS-Université Paris-Est Créteil, Thiais, France

Gaohong He, State Key Laboratory of Fine Chemicals, Research and Development Center of Membrane Science and Technology, Dalian University of Technology, Dalian, China

Jose James, Research and Post-Graduate Department of Chemistry, St. Joseph's College, Moolamattom, Kerala, India

Xiaobin Jiang, State Key Laboratory of Fine Chemicals, Research and Development Center of Membrane Science and Technology, Dalian University of Technology, Dalian, China

Damir Klepac, Department of Chemistry and Biochemistry, School of Medicine, University of Rijeka, Rijeka, Croatia

Xiangcun Li, State Key Laboratory of Fine Chemicals, Research and Development Center of Membrane Science and Technology, Dalian University of Technology, Dalian, China

Sanja Marinović, Department of Catalysis and Chemical Engineering, Institute of Chemistry, Technology and Metallurgy, University of Belgrade, Belgrade, Serbia

Miran Mozetič, Department of Surface Engineering, Jozef Stefan Institute, Ljubljana, Slovenia

Ramanuj Narayan, Polymers and Functional Materials Division, CSIR-Indian Institute of Chemical Technology (IICT), Hyderabad, Telangana, India

Ivanka Popović, Faculty of Technology and Metallurgy, University of Belgrade, Belgrade, Serbia

K.V.S.N. Raju, Polymers and Functional Materials Division, CSIR-Indian Institute of Chemical Technology (IICT), Hyderabad, Telangana, India

Chepuri R.K. Rao, Polymers and Functional Materials Division, CSIR-Indian Institute of Chemical Technology (IICT), Hyderabad, Telangana, India

Radhika Raveendran, Division of Biosurface Technology, Biomedical Technology Wing, Sree Chitra Tirunal Institute for Medical Science and Technology, Thiruvananthapuram, Kerala, India

Sa-Ad Riyajan, Department of Chemistry, Faculty of Science and Technology, Thammasat University, Pathum Thani, Thailand

Chandra P. Sharma, Division of Biosurface Technology, Biomedical Technology Wing, Sree Chitra Tirunal Institute for Medical Science and Technology, Thiruvananthapuram, Kerala, India

George V. Thomas, Research and Post-Graduate Department of Chemistry, St. Joseph's College, Moolamattom, Kerala, India

Sabu Thomas, International and Interuniversity Centre for Nanoscience and Nanotechnology and School of Chemical Sciences, Mahatma Gandhi University, Kottayam, Kerala, India

Selvin P. Thomas, Yanbu Research Center, Royal Commission for Yanbu Colleges and Institutes, Yanbu Al-sinaiah, Saudi Arabia

Srećko Valić, Department of Chemistry and Biochemistry, School of Medicine, University of Rijeka, Rijeka and Physical Chemistry Division, Rudjer Bošković Institute, Zagreb, Croatia

Qihua Wang, State Key Laboratory of Solid Lubrication, Lanzhou Institute of Chemical Physics, Chinese Academy of Sciences, Lanzhou, China

Tingmei Wang, State Key Laboratory of Solid Lubrication, Lanzhou Institute of Chemical Physics, Chinese Academy of Sciences, Lanzhou, China

Runcy Wilson, School of Chemical Sciences, Mahatma Gandhi University, Kottayam, Kerala, India

Xuemei Wu, State Key Laboratory of Fine Chemicals, Research and Development Center of Membrane Science and Technology, Dalian University of Technology, Dalian, China

Wu Xiao, State Key Laboratory of Fine Chemicals, Research and Development Center of Membrane Science and Technology, Dalian University of Technology, Dalian, China

Xiaoming Yan, State Key Laboratory of Fine Chemicals, Research and Development Center of Membrane Science and Technology, Dalian University of Technology, Dalian, China

1

MICRO- AND NANO-STRUCTURED INTERPENETRATING POLYMER NETWORKS: STATE OF THE ART, NEW CHALLENGES, AND OPPORTUNITIES

JOSE JAMES[1], GEORGE V. THOMAS[1], AKHINA H.[2] AND SABU THOMAS[2,3]

[1] *Research and Post-Graduate Department of Chemistry, St. Joseph's College, Moolamattom, Kerala, India*
[2] *International and Interuniversity Centre for Nanoscience and Nanotechnology, Mahatma Gandhi University, Kottayam, Kerala, India*
[3] *School of Chemical Sciences, Mahatma Gandhi University, Kottayam, Kerala, India*

1.1 INTRODUCTION

Polymer mixtures are materials that play a significant role in modern industry. The preparation and properties of multicomponent polymeric systems are of great practical and academic interest. They provide a convenient route for the modifications of properties to meet specific needs. Interpenetrating polymer networks (IPNs) are one of the most rapidly growing areas in polymer material science. The golden history of IPN had begun with its discovery by Aylsworth in 1914 [1].

Micro- and Nano-structured Interpenetrating Polymer Networks: From Design to Applications, First Edition. Edited by Sabu Thomas, Daniel Grande, Uroš Cvelbar, K.V.S.N. Raju, Ramanuj Narayan, Selvin P. Thomas and Akhina H.

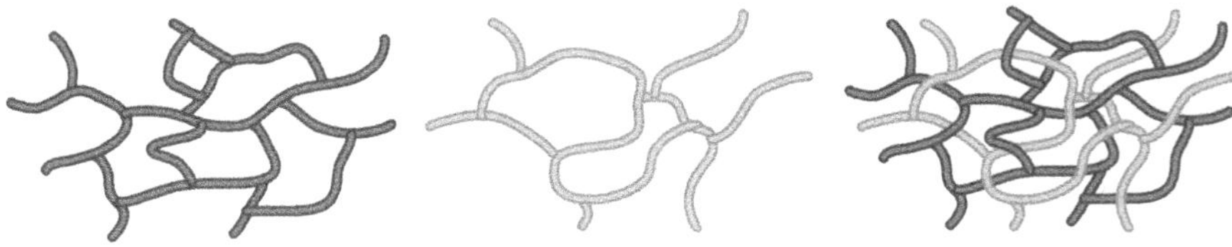

FIGURE 1.1 Schematic representation of component networks and IPN. Source: Klempner *et al.* [3]. Reproduced with permission of the American Chemical Society

IUPAC Compendium of chemical terminology defines IPNs as polymers comprising two or more networks that are at least partially interlaced on a molecular scale but not covalently bonded to each other and cannot be separated unless chemical bonds are broken [1, 2]. IPNs are a combination of incompatible polymer networks, at least one of which is synthesized and/or crosslinked in the presence of the other [3, 4] (Figure 1.1).

Figure 1.2 shows a schematic representation of mechanical blends, graft copolymers, block copolymers, semi-IPNs, and full-IPNs.

IPNs are in fact a special class of polymer blends. The two characteristic features of IPNs that distinguishes itself from the other types of multiphase polymer systems are as follows: (1) IPNs swell but do not dissolve in solvents and (2) creep and flow are suppressed in IPNs [5].

While the science of IPNs began with the work of Millar in 1960 [6], the first publication on the subject came through a patent by Aylsworth in 1914 [1]. Since then, IPNs have been the subject of extensive study by investigators looking into the synthesis, morphology, properties, and applications of these materials. Table 1.1 summarizes the history of IPNs and related materials [7, 8].

1.2 TYPES OF INTERPENETRATING POLYMER NETWORKS

Figure 1.3 classifies the following types of IPNs [1, 9]: semi-IPN, sequential IPN, simultaneous IPN, and full-IPN. They are grouped based on chemical bonding and rearrangement pattern.

1.2.1 Full-Interpenetrating Polymer Networks

They comprise two or more polymer networks, which are at least partially interlocked on a molecular scale but not covalently bonded to each other and cannot be separated unless chemical bonds are broken. It can be represented as in Figure 1.4a.

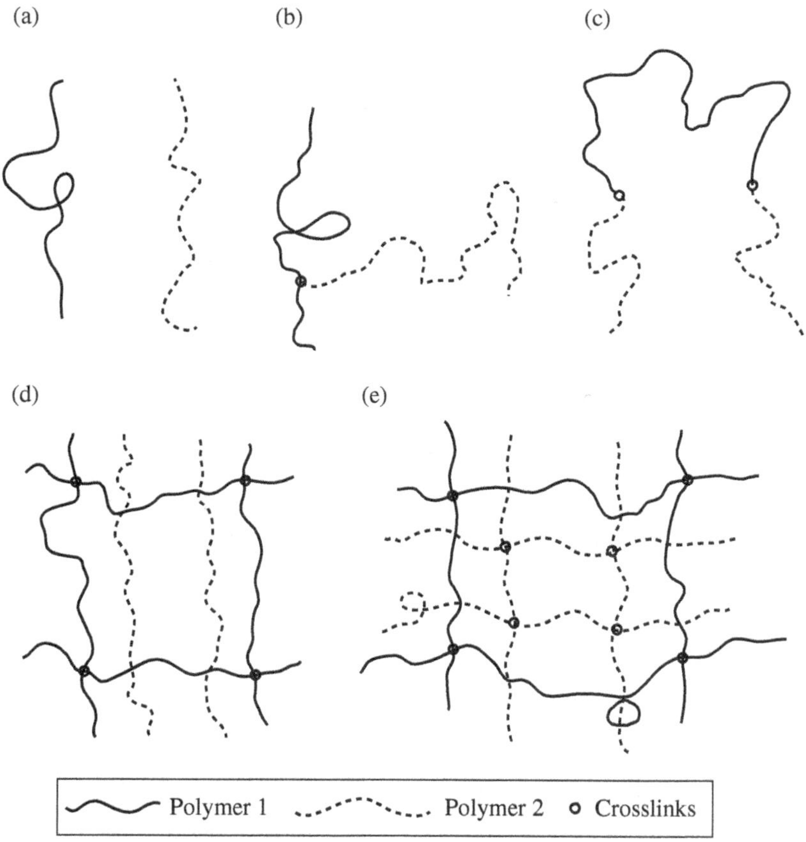

FIGURE 1.2 Schematic representation of (a) mechanical blends, (b) graft copolymers, (c) block copolymers, (d) semi-IPNs, and (e) full-IPNs. Source: Sperling and Mishra [5]. Reproduced with permission of John Wiley & Sons

TABLE 1.1 History of IPN

Year	Event	First Investigators
1914	IPN-type structure	J. W. Aylsworth
1927	Graft copolymers	I. Ostromislensky
1951	IPNs	J. J. Staudinger and H. M. Hutchinson
1952	Block copolymers	A. S. Dunn and H. W. Melville
1960	Homo-IPNs	J. R. Millar
1969	Sequential IPNs	L. H. Sperling
1969	Latex IPNs	K. C. Frisch, D. Klempner, and H. L. Frisch
1971	Simultaneous IPNs	L. H. Sperling and R. R. Arnts
1974	IPN nomenclature	L. H. Sperling
1977	Thermoplastic IPNs	S. Davison and W. P. Gergen

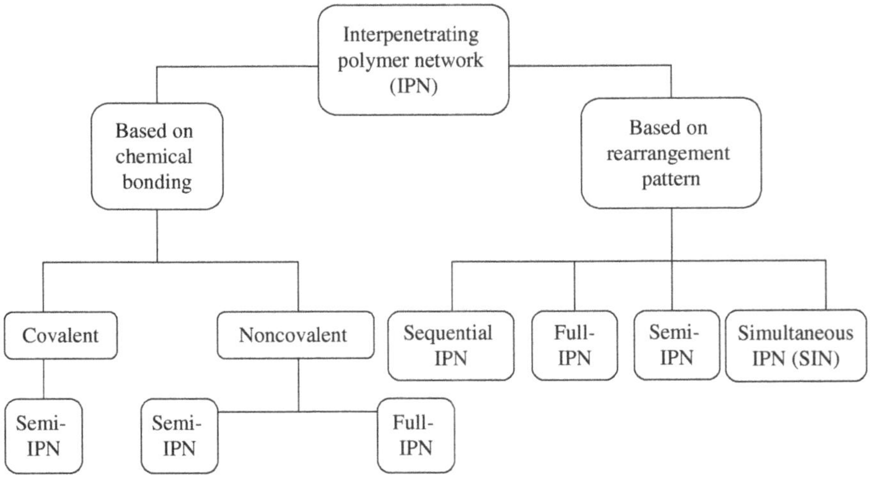

FIGURE 1.3 Classifications of IPNs. Source: Shivashankar, http://www.ijppsjournal.com/. Used under CC-BY-4.0 http://creativecommons.org/licenses/by/4.0/

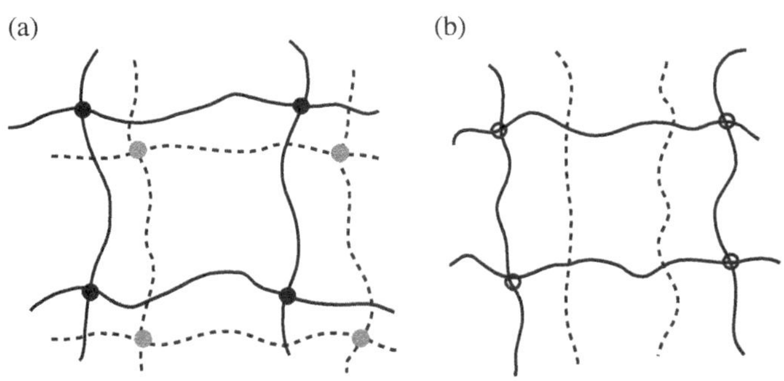

FIGURE 1.4 (a) Full-IPN. (b) Semi-IPN. Source: Sperling and Mishra [6]. Reproduced with permission of Taylor & Francis

1.2.2 Sequential Interpenetrating Polymer Networks

This type of IPN involves the preparation of a polymer network I followed by the *in situ* polymerization of monomer II along with crosslinker and activator, which are then swollen into the network I.

1.2.3 Simultaneous Interpenetrating Networks

In simultaneous interpenetrating networks (SINs), the monomers along with the crosslinkers and activators of both networks are mixed. The reactions are carried out simultaneously; however, they are noninterfering types of reactions such as chain and step polymerization reactions.

1.2.4 Latex Interpenetrating Polymer Networks

These IPNs are made in the form of lattices, frequently with a core and shell structure.

1.2.5 Gradient Interpenetrating Polymer Networks

Gradient IPNs are materials in which the overall composition or crosslink density of the material varies from location to location on the macroscopic level. For example, a film can be made with network I predominantly on one surface, network II on the other surface, and a gradient in composition throughout the interior.

1.2.6 Thermoplastic Interpenetrating Polymer Networks

Thermoplastic IPN materials are hybrids between polymer blends and IPNs that involve physical crosslinks rather than chemical crosslinks. Thus, these materials flow at elevated temperatures, similar to the thermoplastic elastomers, and at use temperature, they are crosslinked and behave like IPNs.

1.2.7 Semi-Interpenetrating Polymer Networks

In semi-IPNs, only one component of the assembly is crosslinked leaving the other in linear form. They are also called pseudo-IPNs. It can be depicted as in Figure 1.4b.

Mishra and Sperling [6] investigated semi-IPNs composed of poly(ethylene terephthalate) and castor oil. They found that bond interchange between these two materials played a major role in initial miscibility and morphology. The semi-IPNs displayed much better mechanical properties than the individual component materials did.

1.3 SYNTHESIS OF INTERPENETRATING POLYMER NETWORKS

The IPN synthesis techniques can be summarized as follows.

1.3.1 Sequential Interpenetrating Polymer Networks

This technique involves the sequential addition of selective crosslinkers to a homogenous mixture of two polymers in solution or in melt form [6]. They are in fact synthesized by a two-step process. In the first step, polymerization

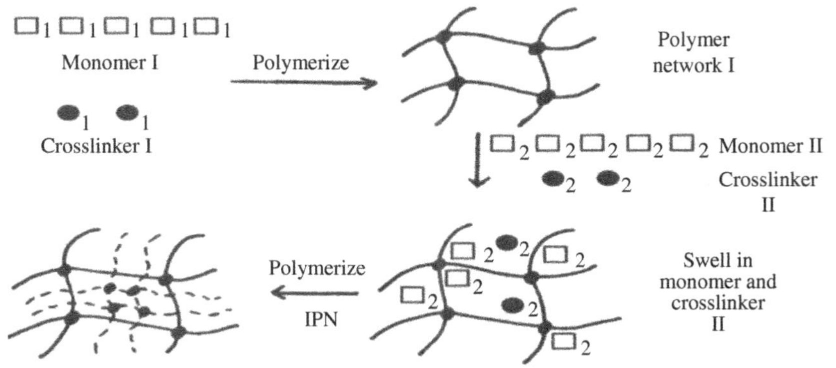

FIGURE 1.5 Schematic representation of sequential IPN synthesis. Source: Sperling [1]. Reproduced with permission of Springer

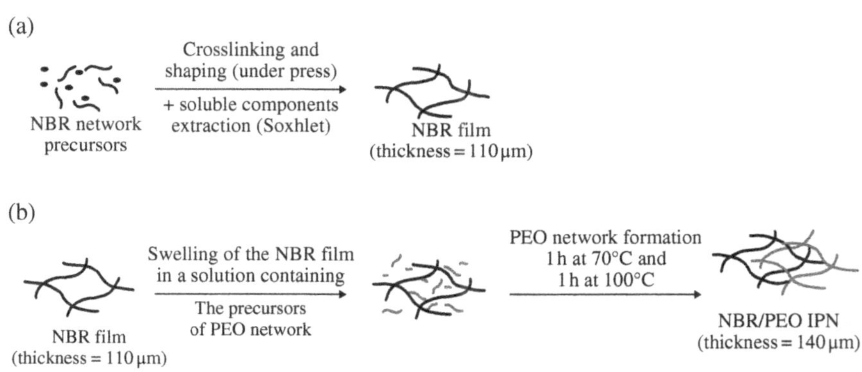

FIGURE 1.6 Outline of NBR/PEO IPN synthesis. (a) NBR network synthesis and (b) PEO network synthesis within NBR network. Source: Goujon *et al.* [10]. Reprinted with permission of the American Chemical Society

of first mixture (consisting of monomer, crosslinking agent, and initiator or catalyst) forms a network I. This network is swollen with the second combination of monomer and crosslinking agent and polymerized to form an IPN, that is, the polymer-2 is polymerized and crosslinked *in situ* in network I [6]. The route can be represented as in Figure 1.5.

Nitrile butadiene rubber/poly(ethylene oxide) (NBR/PEO) IPNs prepared by Goujon *et al.* successfully employed the sequential route for IPN synthesis [10]. Here, IPNs were prepared from NBR and PEO using a two-step process. The NBR network was obtained by dicumyl peroxide crosslinking at high temperature and pressure. A free radical copolymerization of poly(ethylene glycol) methacrylate and dimethacrylate led to the formation of the PEO network within the NBR network. It can be depicted as in Figure 1.6.

Sequential IPNs based on a nitrile–phenolic blend and poly(alkyl methacrylate) were prepared and characterized by Samui *et al.* [11]. The IPNs were not fully compatible but exhibited higher tensile strength compared to corresponding nitrile–phenolic blends. The strength increased with the increase in the concentration of poly(alkyl methacrylate).

1.3.2 Simultaneous Interpenetrating Networks

In SINs, a polymer is synthesized (from the monomer) and simultaneously crosslinked within the network of another polymer to give rise to an interpenetrating network [12]. Here, an IPN is formed by polymerizing two different monomers and crosslinking agent pairs together in one step. The key to the success of this process is that the two components must polymerize by reactions that will not interfere with one another. This is often accomplished by polymerizing one network by a condensation reaction, while the other network is formed by a free radical reaction. The route can be represented as in Figure 1.7.

A notable IPN synthesis by adopting sequential method was reported by M. Ivankovic *et al.* [13]. Here, IPN consists of methyl methacrylate (MMA) and an organically modified silicon alkoxide, 3-glycidyl oxy propyltrimethoxysilane (GLYMO), with varying MMA/GLYMO molar ratios.

SINs were first reported by Frisch by blending a urethane prepolymer with a low-molecular-weight epoxy resin [7]. In SINs, the relative rates of formation of each network would be the controlling factor in determining the morphology. It was also observed that the morphology of IPNs was slightly finer than the corresponding pseudo-IPN [8].

Classical works on IPN include damping applications [14], photodiodes from IPN [15], IPN membranes [16], complex-forming IPNs [17], fire-retarding IPNs [18], conducting IPNs [19], encapsulating IPNs, conducting hydrogel IPNs [20], elastomeric IPNs [21], and polymer solar cell based on IPNs [22].

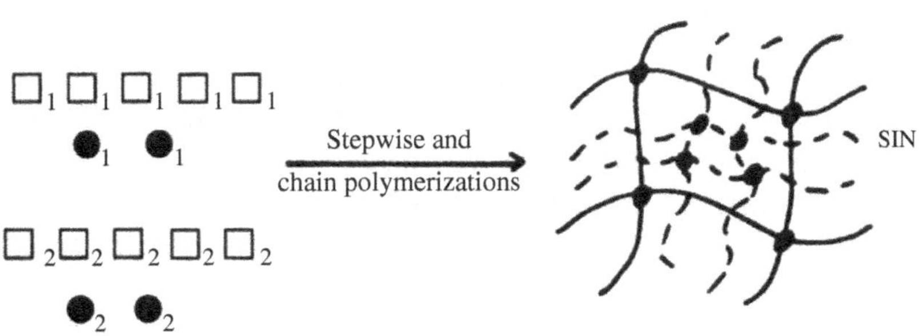

FIGURE 1.7 Schematic representation of simultaneous IPN synthesis. Source: Sperling *et al.* [1]. Reproduced with permission of Springer

1.4 CHARACTERIZATION OF INTERPENETRATING POLYMER NETWORKS

Modern instrumental techniques are used to characterize the interpenetration and ultimate properties of IPNs. Various methods are there to determine the morphology, thermal properties, physical properties, and other characteristics of IPNs. The interpenetration in IPN may be identified by:

1. Comparing the shift of dynamic glass transition temperatures of IPNs with their homopolymers.
2. Comparing the morphologies (micro- or nanoscale) and distribution of phase domains, shape, and sizes.

The existence of interpenetration is generally better judged through the combination of aforementioned inferences [23].

1.4.1 Morphology

The morphology of IPNs depends on the method of synthesis, on the compatibility of the polymer systems employed, and on the relative rates of formation. In sequential IPNs, the network first formed is most likely to be the continuous network. Its crosslink density is the controlling factor in determining the morphology of the system of each network [24]. The morphological properties of IPNs are best characterized by using the scanning electron microscopy (SEM) and transmission electron microscopy (TEM).

The micrographs obtained using the SEM and TEM appears to be useful in detecting the interpenetration of phase domains, shape, and structures of the order of magnitude of a micron to a tenth of a micron and the degree of mixing [25].

TEM is another powerful technique for the elucidation of IPN architecture. TEM analysis proves the fine distribution of PEO phase in continuous NBR phase for the IPN of the PEO–NBR system [10] (Figure 1.8).

Here, the black domains correspond to NBR-rich phases, and the white domains represent the PEO-rich phases. Since the TEM pictures are grayish, the boundary between the dark and clear gray regions is not obvious. This reinforces the fact that the IPN architecture is a very finely distributed PEO phase in a continuous NBR matrix.

A recent work on magnetically doped multistimuli-responsive hydrogel microspheres with IPN structure reported by Ahmad *et al.* [26] successfully employed SEM and TEM images for morphology characterization. In this

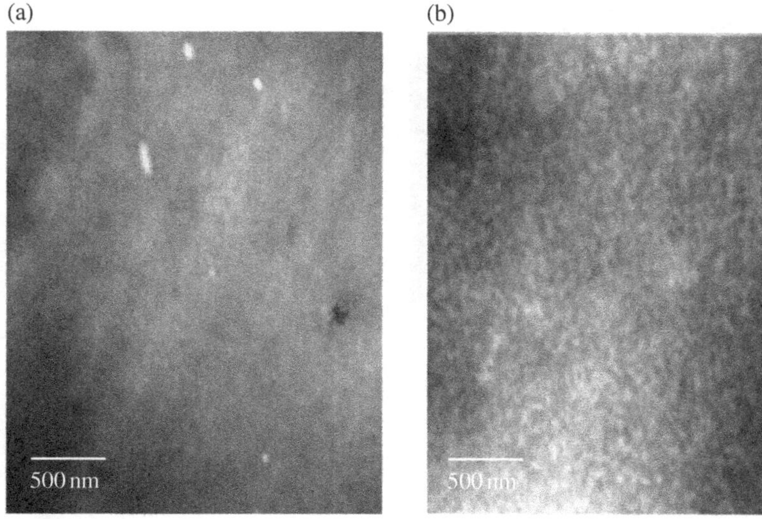

FIGURE 1.8 TEM pictures of NBR/PEO (50% PEGDM) IPN. Source: Goujon *et al.* [10]. Reprinted with permission of the American Chemical Society

study, it is found that the presence of Fe_3O_4 nanoparticles on the surface of P(NIPAM-MBAM)/P(MAA-EGDM) IPN hydrogel microspheres is clearly observed from microscope images. Both TEM and SEM images suggest that the deposition of Fe_3O_4 nanoparticles on the surface of P (NIPAM-MBAM)/P (MAA-EGDM) IPN hydrogel microspheres most likely improved the stability during sample preparation as the microspheres remained mostly spherical (Figures 1.9 and 1.10).

IPN based on polyisoprene and PMMA by John *et al.* [27] also employed SEM images as a powerful tool for characterization. SEM images of IPN (poly isoprene–PMMA IPN) at different compositions are given in the succeeding text. These are SEM images of IPNs with same crosslink density and varying compositions.

Figure 1.11a shows the sample with 20 wt% PMMA. It has a sea-island morphology. Figure 1.11b shows the sample with 35% PMMA. It shows a discrete state with larger domain dimension. In Figure 1.11c, the 50% PMMA composition shows a dual-phase morphology with polyisoprene as the matrix. The sample with 60% PMMA composition shown in Figure 1.11d shows a continuous phase morphology.

An unusual semi-IPN nanoencapsulating layer composed of thermally cured polyimide (PI) and polyvinyl pyrrolidone (PVP) with lithium cathode materials (here $LiCoO_2$ (LCO)) reported by Kim *et al.* utilized the potential application of SEM and TEM images for morphology characterization [29]. The surface

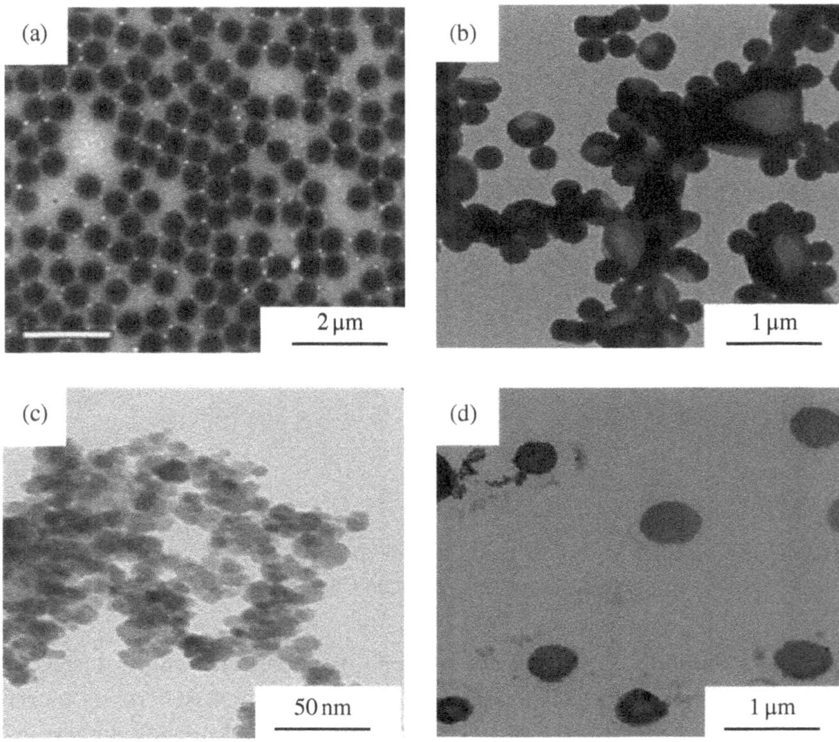

FIGURE 1.9 TEM images of (a) P(NIPAM-MBAM) hydrogel microspheres, (b) P(NIPAM-MBAM)/P(MAA-EGDM) IPN hydrogel microspheres, (c) Fe_3O_4 particles, and (d) P(NIPAM-MBAM)/P(MAA-EGDM)/Fe_3O_4 IPN hydrogel microspheres. Source: Ahmad *et al.* [26]. Reprinted with permission of Elsevier

FIGURE 1.10 SEM image of washed P (NIPAM-MBAM)/P (MAA-EGDM)/ Fe_3O_4IPN hydrogel microspheres. Source: Ahmad *et al.* [26]. Reprinted with permission of Elsevier

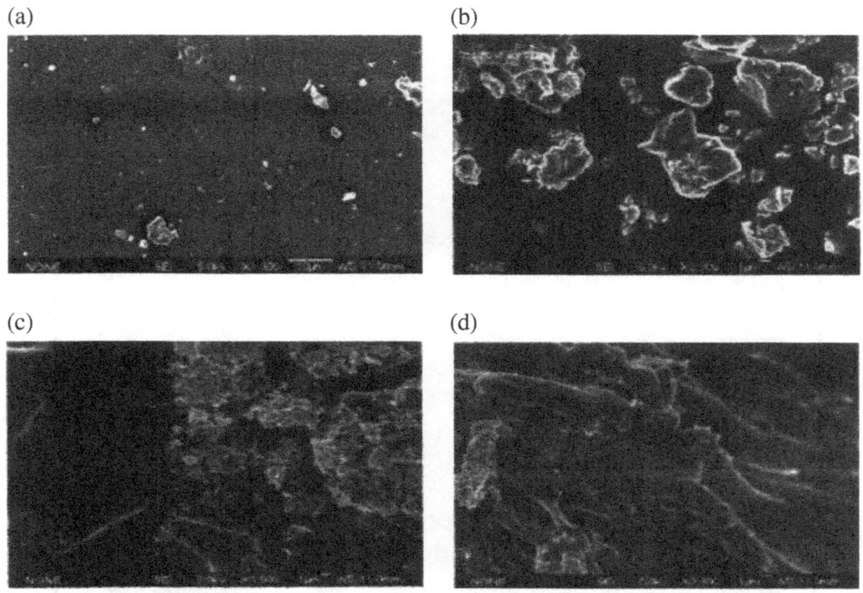

FIGURE 1.11 SEM images of PI-PMMA (a) with 20 wt% PMMA, (b) with 35% PMMA, (c) with 50% PMMA, and (d) with 60% PMMA. Source: John *et al.* [28]. Reproduced with permission of Elsevier

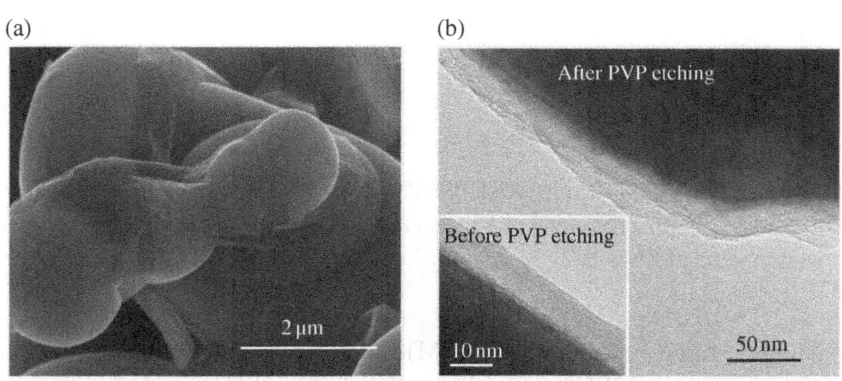

FIGURE 1.12 Structural characterization of PI/PVP-LCO. (a) FE-SEM photograph; (b) TEM photographs before/after selective etching of PVP phase. Source: Kim *et al.* [29]. Reproduced with permission of Nature Publishing Group

morphology of PI/PVP-LCO was characterized with a focus on coverage area of the LCO surface. Figure 1.12a shows that the PI/PVP-LCO has highly continuous and conformal polymeric layers on the LCO surface. The highly continuous conformal morphology of the PI/PVP nanoencapsulating layer was further elucidated by conducting TEM studies (Figure 1.12b).

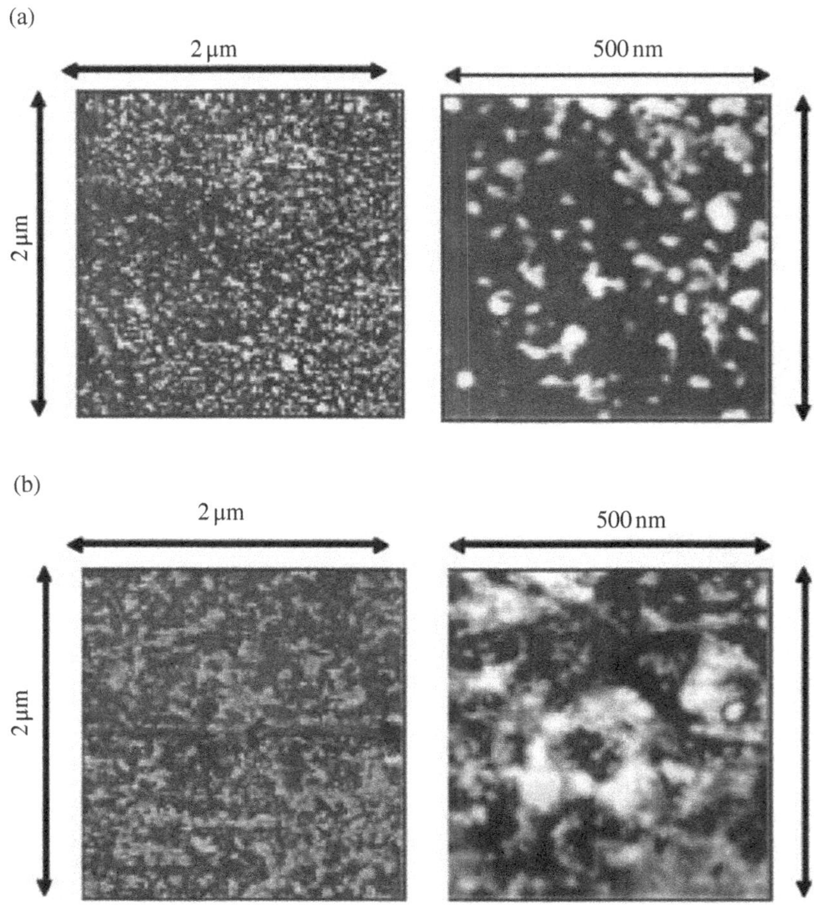

FIGURE 1.13 (a) AFM images of PIB/PMMA (60/40) IPN. (b) AFM images of PIB/PMMA (30/70) IPN. Source: Vancaeyzeele *et al.* [30], figures 6, 7. Reprinted with permission of Elsevier

Atomic force microscopy (AFM) analysis also provides valuable information on the IPN morphology. It helps to differentiate continuous phase and dispersed phase in IPN. AFM is also used to identify the domain size of the components in IPN. In the characterization of polyisobutene (PIB)–PMMA IPN, AFM was used successfully to elucidate the morphology as shown in Figure 1.13a and b [30].

Confocal laser-scanning microscopy (CLSM) also provides some useful insight into IPN morphology [31]. It helps to identify the hydrophilic and hydrophobic regions of hydrogel type IPN. X-ray scattering experiments are widely used for enlightening the morphology of IPN [32]. It can be used for

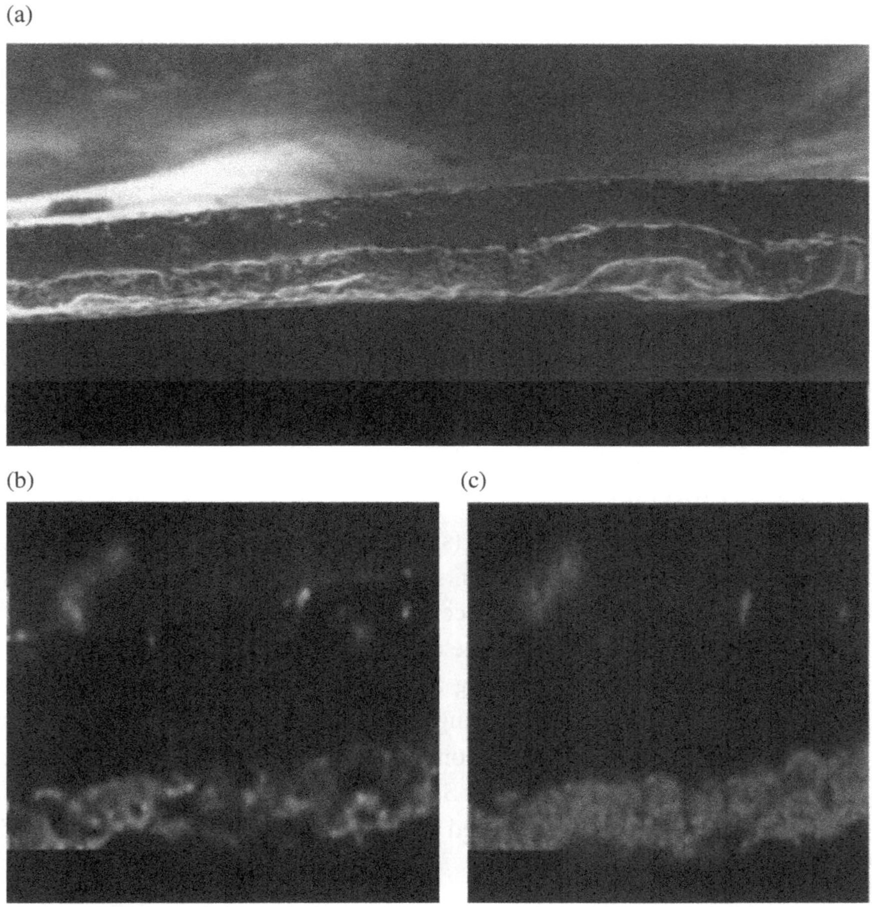

FIGURE 1.14 (a) SEM images of NBR-PEDOT network. (b) EDAX analysis showing the repartition of sulfur, (c) that of oxygen in the NBR-PEDOT semi-IPN. Source: Francke *et al.* [33]. Reprinted with permission of Elsevier

the visualization of interpenetrating channels in IPN. Energy-dispersive X-ray analysis (EDAX) explains the distribution of dispersed phase in the IPN matrix [33]. SEM pictures and EDAX analysis give immediate information about the repartition of polyethylenedioxythiophene (PEDOT) within the NBR matrix. The scanning electron micrograph of a cross section of NBR/PEDOT semi-IPN obtained at 10 kV is shown in Figure 1.14a.

As expected, this picture clearly shows the formation of a heterogeneous system: the two different phases observed correspond to pure NBR (phase 1) and PEDOT/NBR (phase 2). We can deduce that PEDOT interpenetrates within the NBR matrix. This observation corroborates the fact that PEDOT cannot be removed from the surface of the matrix by scratching. From the SEM picture, the depth of interpenetration can be estimated around 35 μm. Moreover, the distribution of the characteristic elements S and O of PEDOT in the cross-sectional area observed in the EDAX mapping (Figure 1.14b and c) clearly indicated that PEDOT was present inside the NBR in its doping state as expected.

Laser scanning confocal microscopy (LSCM) is a very good technique to provide the morphology of IPN as a function of depth [34].

1.4.2 Thermal Properties

The collective observations from electron microscopy and thermal measurements provide the real morphology of an IPN. Differential scanning calorimetry (DSC), thermogravimetric analysis (TGA), differential thermal analysis (DTA), and so on have been extensively used to study the thermal properties of IPN [35].

DSC shows one or two T_g values (shifted or unshifted) based on the extent of interaction between the components. The T_gs often lie at or between the T_gs of the pure networks. The appearance of single T_g indicates the formation of an IPN, and in this case, one can rule out the formation of any macro-/micro-phase separation. The slight shifting and broadening of the peaks indicate a moderate degree of molecular mixing [35]. In many cases, two distinct T_gs corresponding to those individual component networks have been observed. Such behavior has been reported by Sperling and others [36]. Merging of T_gs or an inward shift can be considered as evidence for the interpenetration of phases in IPN [37] (Figure 1.15).

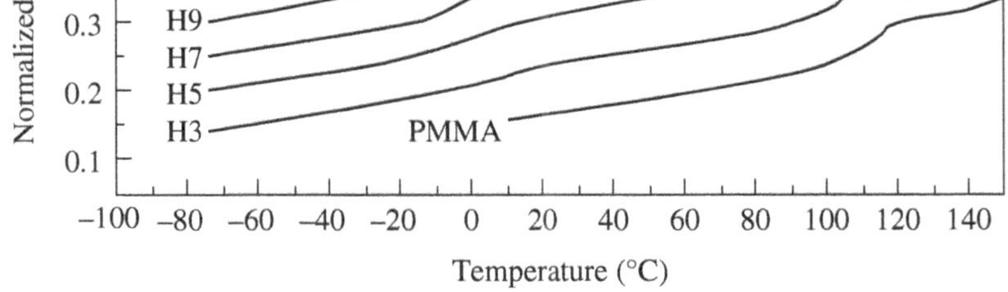

FIGURE 1.15 DSC thermograms of PMMA and hybrid systems. Source: Ivankovic *et al.* [13]. Reproduced with permission of Elsevier

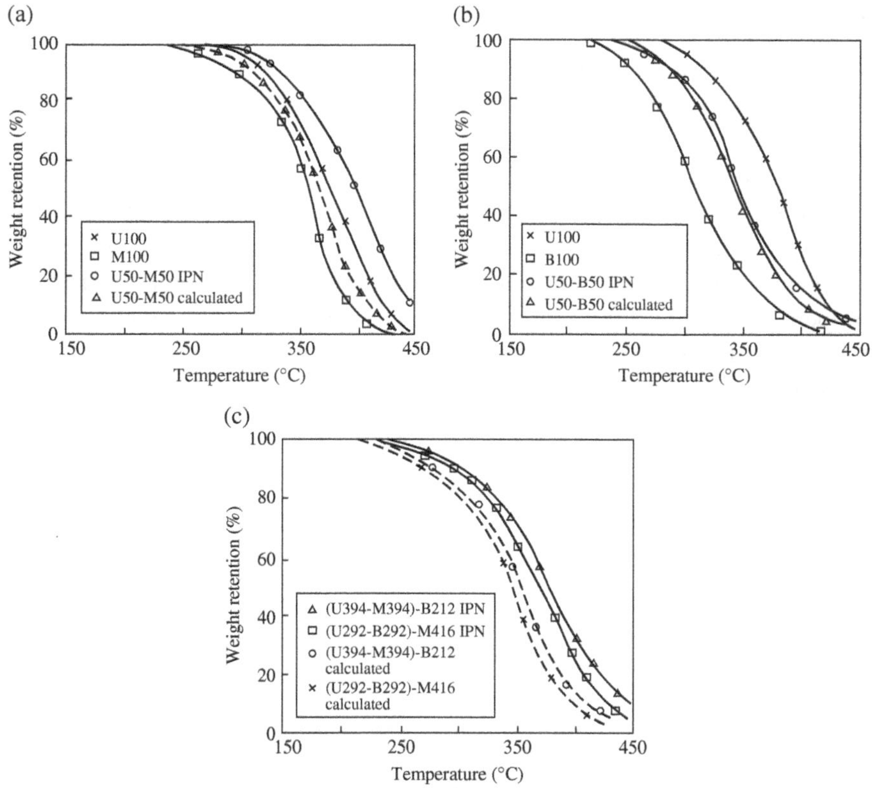

FIGURE 1.16 TGA thermograms of PU-PMMA SIN (a), PU-PBMA SIN (b), and three-component IPN homopolymers (c). Source: Lee and Kim [32]. Reproduced with permission of Nature Publishing Group

Ivankovic *et al.* [13] has used the DSC technique for the characterization of organic–inorganic hybrids based on PMMA and sol–gel-polymerized glycidyloxypropyl trimethoxysilane (GLYMO). The glass-transition temperatures of the GLYMO and PMMA were −17 and 114°C, respectively. The hybrids showed two T_g values suggesting heterophase morphology.

The weight loss of IPNs indicates the thermal stability of IPNs. In most reports, the thermal stability of the IPN has been found to be higher than their homopolymers [38]. The enhancement or depression in thermal stability during IPN formation can be best explained in terms of TGA [39]. Thermal stability characteristics of three-component IPNs have been compared with two-component IPNs with the aid of TGA analysis by Lee and Kim [32].

By comparing Figure 1.16a and b, it can concluded that PU-PMMA SIN posses better thermal stability than PU-PBMA SIN. From Figure 1.16c, the enhanced thermal stability of three-component IPNs can also be understood.

1.4.3 Mechanical Properties

The composition and nature of the constituent polymeric network defines the mechanical properties of IPN. Common mechanical properties used to characterize IPN are tensile strength, elongation at break, Young's modulus, and hardness [39]. Mechanical properties (tensile stress–strain behavior, tensile stress relaxation, and tensile creep) were extensively used to elucidate the composition of IPN of the PEA–PMMA system [40]. From the following tensile stress relaxation master curves and dynamic mechanical properties (Figures 1.17 and 1.18), one can get the idea of kinetics of the formation of IPNs and the extent of interpenetration in sequential IPN of the PEA–PMMA system.

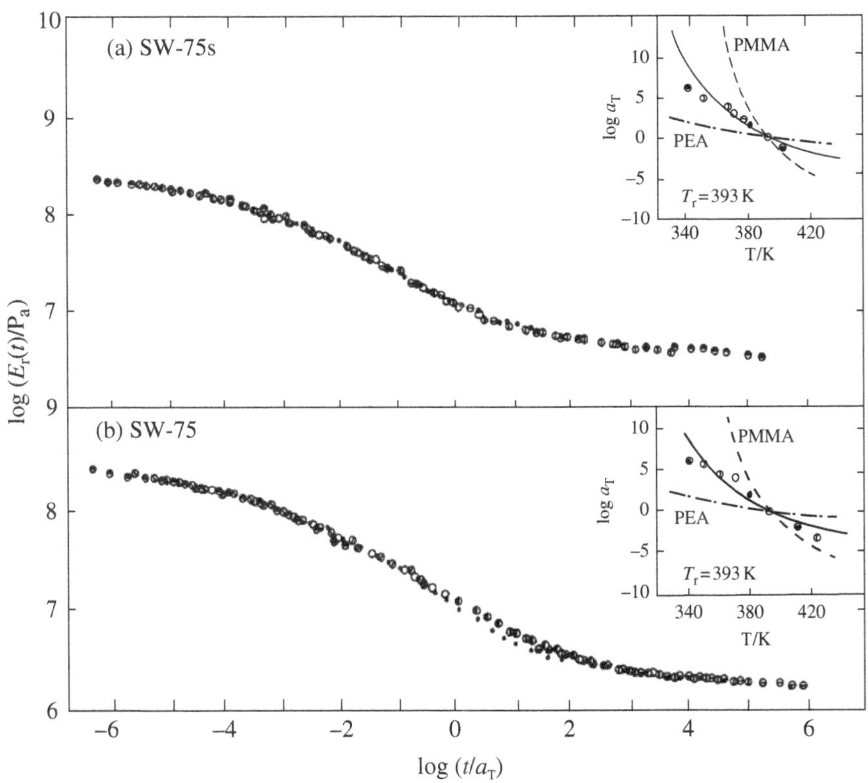

FIGURE 1.17 Tensile stress relaxation master curves for (a) semi-IPN and (b) full-IPN. Curves in the inset represent the WLF equation for PEA–PMMA and IPN. Source: Adachi and Kotaka [40]. Reproduced with permission of Nature Publishing Group

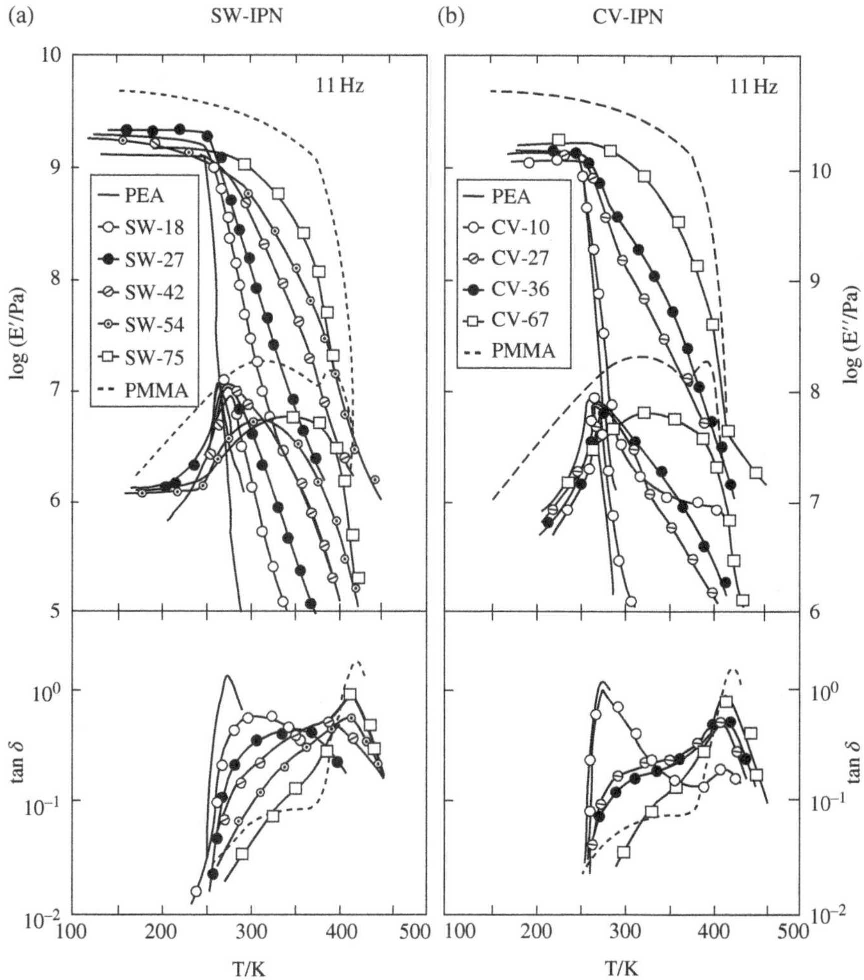

FIGURE 1.18 Storage (E′) and loss (E″) modulus and loss strength (tan δ) plotted against temperature for (a) SW- and (b) CV-full-IPNs. CV, convert ion-controlled IPN; SW, swelling ratio-controlled IPN. Source: Adachi and Kotaka [40]. Reproduced with permission of Nature Publishing Group

1.4.4 Kinetic Properties

In an ideal IPN, the two components are present as co-continuous, interlocking networks. This co-continuity can be achieved by the kinetic retention of the initial mixture of the monomers used to form network chains, with phase segregation inhibited by the network structure. Gel time, solidification time, and crystallization time are used to represent the kinetic formation of IPNs from its constituent polymeric networks [39]. Rheological methods are used to determine the gel point. The gel points at various temperatures can be used to calculate the activation energy of crosslinking. Kinetic results were used as a guideline to optimize the conditions of polymerization. Kinetic control and phase segregation are the chief controlling factors in the morphology of IPN.

1.4.5 Spectroscopic Techniques

The spectroscopic methods that are widely used in the structure elucidation are nuclear magnetic resonance (NMR), infrared (IR), and electron spin resonance (ESR) techniques. NMR spectroscopy is used to estimate the nanoscale interpenetration of polymers in IPN. The spin–lattice relaxation time measurements may also give some insight into the domain size [41]. Solid-state NMR has been proven a fruitful technique to investigate the microstructure and molecular dynamics of IPNs. Detailed information about morphology, miscibility, microstructure, and mobility can be directly accessed by NMR studies [39, 41]. The interaction between PMMA/NR (natural rubber) IPNs was analyzed using the C-13 relaxation measurements at room temperature by John *et al.* [28] (Figure 1.19).

The wide-line separation (WISE) NMR spectroscopy is used to measure the dynamics of solid polymers by correlating the proton line shape with

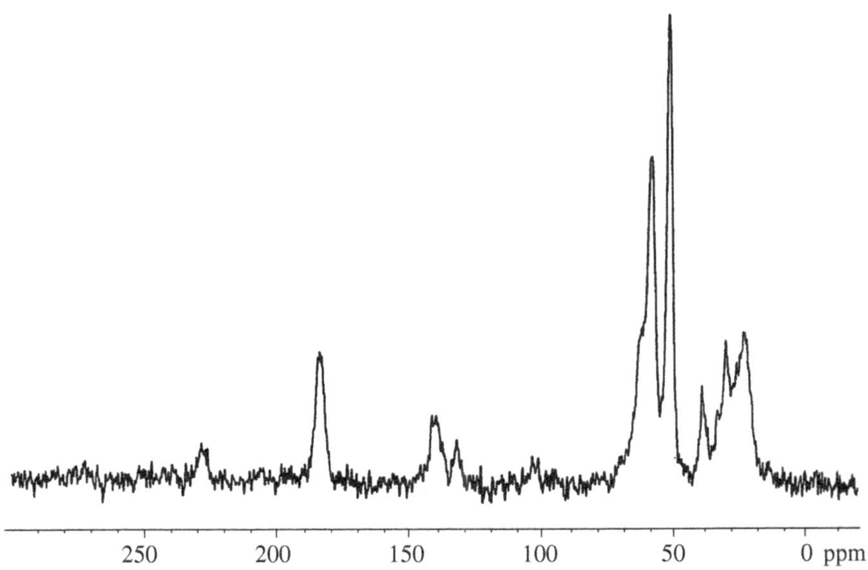

FIGURE 1.19 C-13 NMR spectrum of semi-IPN. Source: John *et al.* [28]. Reproduced with permission of Elsevier

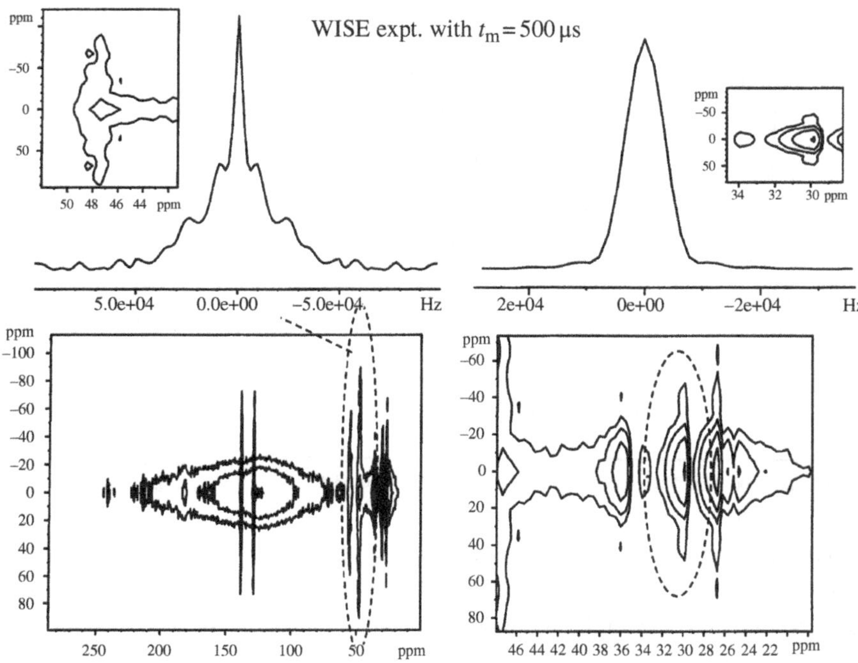

FIGURE 1.20 Contour plot of the WISE NMR spectrum of NR/PMMA semi-IPN. Source: John *et al.* [28]. Reproduced with permission of Elsevier

carbon chemical shift. This experiment yields information about dynamics by measuring the degree to which the proton line shapes are averaged by molecular motion. The WISE experiment does establish a correlation of the chemical structure and segmental mobility as reflected in the C-13 chemical shifts and the H-1 line shapes, respectively. Figure 1.20 shows contour plots of the WISE NMR spectrum of PI/PMMA 50/50 semi-IPN at 298 K [28].

Figure 1.20 shows contour plots of the WISE NMR spectrum of PI/PMMA 50/50 semi-IPN at 298 K. *Several well-resolved peaks are observed, including the PI proton peaks for* $-CH_2-$ *at 36 ppm, other* $-CH_2-$ *at 30 ppm, and —CH-group at 129 ppm. From the PMMA phase, protons of the* $-CH_2$ *group at 50 ppm,* $-OCH_3$ *at 54 ppm, and* $-CH_3$ *at 27 ppm are clearly resolved.* Peaks at 27 ppm contain overlapped peaks from methyl groups in PI and PMMA. The contour plot shows very different proton line widths for the PMMA and NR chains. This is the characteristic of a phase-separated multicomponent system consisting of rigid and mobile phases. *The WISE experiment shows that the protons in the ester group (54 ppm) of PMMA have the highest mobility and* $-CH_2-$ *protons at 50 ppm are the most rigid among protons in the PMMA phase.* 2D WISE experiment also showed the unusual high mobility in the

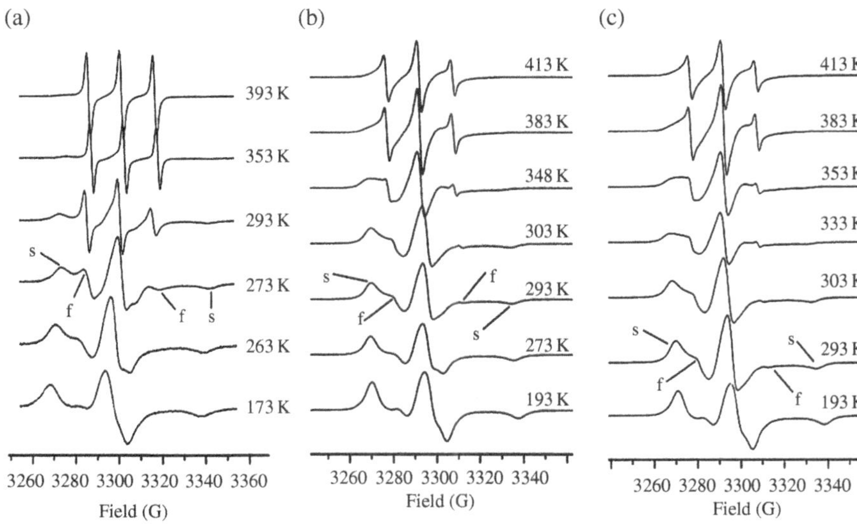

FIGURE 1.21 ESR spectra of probe measured as a function of temperature in (a) NR, (b) semi-IPN with 35 wt% of PMMA, and (c) semi-IPN with 50/50 composition. Source: John *et al.* [28]. Reproduced with permission of Elsevier

H atoms attached to the ester carbon of PMMA, proving the influence of highly flexible environment close to PMMA chains.

ESR spectroscopy is another technique used for the characterization of IPNs. ESR spectroscopy is widely used as a potential tool for the understanding of the structure and heterogeneities in IPN on a length scale that is less than 5 nm. ESR line shapes and spectral simulation data are used to arrive at some morphological conclusions along with the help of SEM and TEM. ESR spectra of the spin probes in semi-IPNs with different composition, crosslink density, and molecular weight were measured in the temperature range of 173–413 K at intervals of 10 K or less. ESR spectra of probe measured as a function of temperature for NR (poly isoprene) and for different semi-IPN of the NR-PMMA system [28] are shown in Figure 1.21. The free nitroxide radical 4-hydroxy-2,2,6,6-tetramethylpiperidine-1-oxyl was used as a spin probe for ESR measurements. In this study, the main conclusions from ESR investigations of PI-PMMA IPNs were as follows:

1. At lower temperatures, the ESR spectra approached the rigid limit spectrum and as the temperature increased, the spectral lines got narrowed and outer peaks shifted inward. In the range of 293–373 K, complex spectra were obtained, which is an indication of the presence of motionally distinct regions in semi-IPNs.

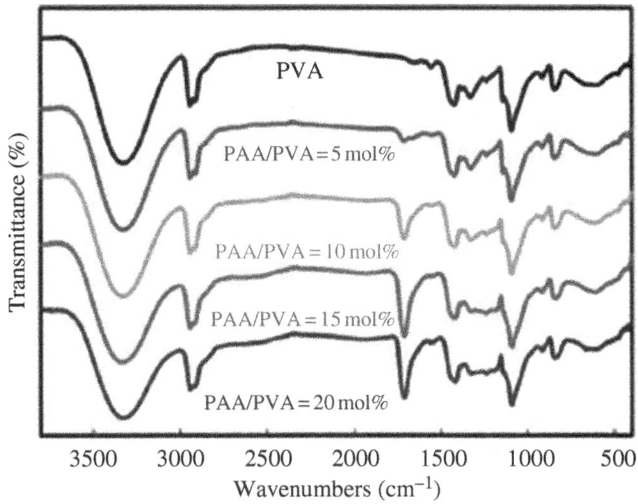

FIGURE 1.22 FT-IR spectrum of PVA-PAA IPNs. Source: Lu *et al.* [42]. Reproduced with permission of Elsevier

2. Semi-IPN synthesis resulted in a dramatic change in the motional behavior of both polymers, and this was due to the molecular level interpenetration between two polymer chains. The intimate mixing of the rigid PMMA chains with highly mobile NR networks strengthened the NR networks.

IR and FT-IR spectroscopic techniques have been widely used for identifying the components with their special groups in IPNs. This forms additional tools to confirm the completion of reaction and to diagnose the networks [42]. The FT-IR spectra of PVA/PAA IPN films with different PAA contents (0, 5, 10, 15, and 20 mol% of AA monomer per vinyl alcohol repeating unit) are shown in Figure 1.22.

The FT-IR spectrum of pure PVA hydrogel membrane shows the main characteristic bands as follows: 3300 cm^{-1} at the O—H stretching vibration, 2941 and 2910 cm^{-1} at the C—H stretching vibration, 1421 and 1328 cm^{-1} at the C—H deformation vibration, 1241 cm^{-1} at the C—O stretching vibration, 1090 cm^{-1} at the C—O stretching vibration, and 917 cm^{-1} at the O—H bend. With an increase of PAA content in IPN hydrogels, the C—O stretching band of PAA at 1711 cm^{-1} was boosted up and shifted gently toward the lower wave numbers, while the C—O stretching vibration of pure PVA at 1241 cm^{-1} was gradually enhanced and broadened. This indicates that new intermolecular hydrogen bond interaction in the IPNs was formed, which is in evidence that the hydrogen bond interactions of PVA hydrogel are replaced by hydrogen bond interactions between PVA and PAA [42]. Moreover, the peak of the O—H stretching vibration for the pure PVA membrane, at 3300 cm^{-1}, was gradually broadened and weakened, while the C—O stretching vibration at 1090 cm^{-1} was slightly strengthened. This may be caused by the etherification reaction between carboxylic acid groups in PAA and hydroxyl groups in PVA. The study with the FT-IR spectra confirmed the formation of the PVA/PAA IPN hydrogel.

1.4.6 Viscoelastic Measurements of Interpenetrating Polymer Networks

This is one of the major tools for characterization of IPNs. Dynamic mechanical analysis (DMA) is widely used for the measurement of viscoelastic behavior of IPNs. DMA is a technique used to characterize IPN's properties as a function of temperature, time, frequency, stress, or a combination of these parameters. It measures stiffness and damping (dissipation of energy) in a material under cyclic loading. In DMA experiments, the SINs show either two well-defined glass transitions [43]. This may be due to three possibilities via gelation of polymer 1, gelation of polymer 2, and phase separation of polymer 1 from polymer 2.

Reddy and Takahara reported on the DMA of thermosensitive transparent semi-IPNs for wound dressing and cell adhesion control applications [44]. In this work, semi-IPNs were composed of segmented polyurethane urea/poly (*N*-isopropylacrylamide) (SPUU/PNiPAAm). Figure 1.23 shows the storage modulus (E') and loss modulus (E'') versus temperature of SPUU and semi-IPNs. The damping peak is associated with the partial loosening of the polymer structure so that small chain segments can move, which occurs near T_g at low frequencies. The maximum in the loss modulus (E'') at low frequencies is, however, very close to T_g (here T_g is −19°C). The increase in T_g observed due to addition of PNiPAAm could be attributed to the more interaction of PNiPAAm chains with SPUU soft segments. It is also observed from Figure 1.23 that the damping peak height decreases with the addition of PNiPAAm (curves 2–5).

Figure 1.23 also shows dynamic storage modulus (E') as a function of temperature. A sharp decrease of E' in the case of parent polymer above the glassy region has been observed. However, in the case of semi-IPNs, a systematic decrease of rubbery plateau region or high rubbery plateau modulus (E') has been observed with an increase of PNiPAAm content, which implies that elastic behavior predominates rather than rubbery behavior with PNiPAAm. This may be due to the increased number of interactions between SPUU and PNiPAAm.

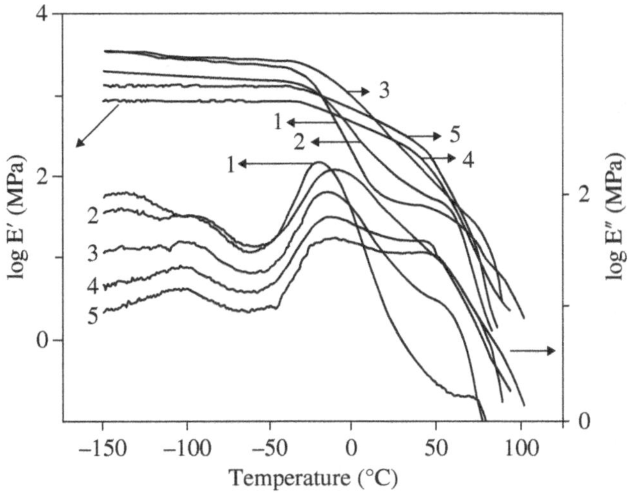

FIGURE 1.23 Dynamic mechanical analysis of SPUU and semi-IPNs: (1) NEAT SPPU, (2) SPPU with 20% PNiPAAm, (3) SPPU with 30% PNiPAAm, (4) SPPU with 40% PNiPAAm, and (5) SPPU with 50% PNiPAAm. Source: Reddy *et al.* [44]. Reproduced with permission of the American Chemical Society

1.5 APPLICATIONS OF INTERPENETRATING POLYMER NETWORKS

IPNs have variety of applications and they are in fact commercially successful form of polymer blends, probably owing to the crosslinked structure that provides better thermal stability, mechanical properties, chemical resistance, and so on [9]. IPNs are traditionally used as damping materials, impact-resistant materials, adhesives, and so on. The immense applications of IPNs are summarized in the following chart. Their unique properties of specificity, mechanical strength, swelling capacity, nutrient and oxygen permeability, durability in the body, and sensitivity can be identified [9] (Table 1.2).

1.6 FUTURE TRENDS

Recently, IPNs are combined with nanoparticles to produce high-performance nanocomposites. IPN-supported scaffold [45] is one of the growing area of polymer blends. Nanocomposite polymer IPN hydrogels are new-generation materials useful for a wide variety of applications [46]. They are used as responsive hydrogels [44, 46–48], medical implants, porous scaffolds [49] or

TABLE 1.2 Potential Applications of IPN [9, 10, 23]

Components	Application
Natural leather/rubber	Improved leather
Plastic/rubber	Noise damping, tough plastic
Plastic/plastic	Optically smooth surfaces
Rubber/rubber	Pressure-sensitive adhesives
Rubber/plastic	Impact modifier
Rubber/crystalline plastic	Thermoplastic, elastomer
Rubber/water-swellable phase	Arteriovenous shunt
Xanthan gum/PVA	IPN microsphere
Grafted Na alginate/Na alginate	IPN-based tablets
Poly acrylamide/PVA	IPN hydrogel capsule
Collagen/glycosaminoglycan	IPN-based sponges for wound dressing
PGA/collagen (3D scaffold)	Bioengineering tissue
Collagen/hydroxyapatite	Bone substitute
Collagen/chitosan	Cartilage scaffold
PMMA/poly(butyl acrylate)	Noise damping coating
PMMA/PMMA	Dental filling

catalyst supports, self-healing materials, thermosensitive materials, contact lenses [50], and so on. The new applications of IPN are controlled drug delivery systems [51], energy storage materials [48], corneal transplantation [52], and use in fuel cells [48]. IPNs have immense scope in tissue engineering and as an uteral stents [50]. Thus IPN holds many advancements and future applications that will be of immense help to mankind. The literature survey of latest works on IPN reveals that current thrust areas of IPN are hydrogel [47, 51, 53], polymer solar cell [22], controlled drug release [54, 55], and thermo-sensitive IPN for wound dressing [44]. IPN hydrogels are called hungry networks or intelligent polymers due to their potential technological application in medicine, industry, biology, and environmental cleanup [56].

One of the shining future aspects of IPN is in the manufacture of solar cell [57]. Here, incoming photon from the sun creates exciton (electron hole pair) that travels to interface where charge separation occurs, producing useful voltage. IPN has found immense use in various biomedical applications [58] and in biochemistry [59]. IPN electrolyte is another thrust area in the polymer research field [60].

The outlook for IPNs is exceedingly bright and is a hot research area. The data from ISI web of science indicate that China, United States, and India are the leaders in IPN research area, and currently about 500 papers per year are being published in this research field. Processing techniques for the synthesis of eco-friendly IPNs and clear differentiation of physical and chemical interpenetration using characterization techniques are the still remaining gaps in this field. Despite many advances, numerous challenges and opportunities remain for making an impact in the field of smart interpenetrating polymers.

ACKNOWLEDGEMENT

Financial support from Council of Scientific and Industrial Research is duly acknowledged.

REFERENCES

[1] Sperling, L.H. (1981) *Interpenetrating Polymer Networks and Related Materials*, Plenum Press, New York, pp. 1–30.

[2] Donatelli, A.A., Sperling, L.H. and Thomas, D.A. (1981) *Macromolecules*, **15**, 625.

[3] Klempner, D., Sperling, L.H. and Utracki, L.A. (1994) *Interpenetrating Polymer Networks*, American Chemical Society, Washington, DC, pp. 39–75.

[4] Aylsworth, J.W. (1914) Plastic Composition. US Patent 1,111,284A.

[5] Sperling, L.H. and Mishra, V. (1996) *Polymers for Advanced Technologies*, **7**, 197.

[6] Mishra, V. and Sperling, L.H. (1996) *The Polymeric Materials Encyclopedia: Synthesis, Properties, and Applications*, CRC Press, Boca Raton.

[7] Mathew, A.P. (2013) *Advances in Elastomers I*, Springer Verlag, Berlin/Heidelberg, pp. 282–300.

[8] Mathew, A.P. (1999) Interpenetrating polymer networks based on natural rubber and poly styrene. Ph.D. thesis. Mahatma Gandhi University.

[9] Shivashankar, M. and Mandal, B.K. (2012) *International Journal of Pharmacy and Pharmaceutical Sciences*, **4**, 1–7.

[10] Goujon, L.J., Khaldi, A., Maziz, A. *et al.* (2011) *Macromolecules*, **44**, 9683–9691.

[11] Samui, A.B., Suryavanshi, U.G., Patril, M. *et al.* (1998) *Journal of Applied Polymer Science*, **68**, 255–262.

[12] Mathew, A.P., Packrisamy, S., Kumaran, M.G. and Thomas, S. (1995) *Polymer*, **36**, 4935–4942.

[13] Ivankovic, M., Brnardic, I. and Gajovic, A. (2009) *Polymer*, **50**, 2544–2550.

[14] El-Aasser, M.S., Hu, R., Dimonie, V.L. and Sperling, L.H. (1999) *Journal of Colloids and Interfaces*, **153**, 241–253.

[15] Halls, J.J.M., Walsh, C.A., Greenham, N.C. *et al.* (1995) *Nature*, **376**, 498–500.

[16] Lee, Y.K., Sohn, I.S., Jeon, E.J. and Kim, S.C. (1991) *Polymer Journal*, **23** (5), 427–433.

[17] Nishi, S. and Kotaka, T. (1989) *Polymer*, **21** (5), 393–402.

[18] Chiu, H.T., Chiu, S.H., Jeng, R.E. and Chung, J.S. (2000) *Polymer Degradation and Stability*, **70**, 505–514.

[19] Gan, L.H., Gan, Y.Y. and Yin, W.S. (1999) *Polymer*, **40**, 4035–4039.

[20] Lu, Y., He, W., Cao, T., *et al.* (2014) *Scientific Reports 4*, 1–8, article number: 5792.

[21] Xie, H.Q. and Guo, J.S. (2002) *European Polymer Journal*, **38**, 2271–2277.

[22] Ma, W. and Heeger, A.J. (2005) *Advanced Functional Materials*, **15**, 1617–1622.

[23] Anandaraj, T., Mohan, P.S., Krishnan, S.M. and Raghavan, M. (2001) *Paintindia*, **4**, 43–54.

[24] Das, B. and Gangopadhyay, T. (1992) *European Polymer Journal*, **8**, 867.

[25] Thomas, S., Visakh, P.M., Chandra, A.K. and Mathew, A.P. (2013) *Advances in Elastomers II: Blends and IPNs*, Springer, Berlin.

[26] Ahmad, H., Nurunnab, M., Rahman, M.M. *et al.* (2014) *Colloids and Surfaces A: Physicochemical and Engineering Aspects*, **459**, 39–47.

[27] John, J., Suriyakala, R., Thomas, S. *et al.* (2010) *Journal of Materials Science*, **45**, 2892–2901.

[28] John, J., Klepac, D., Didovi, M. *et al.* (2010) *Polymer*, **51**, 2390–2402.

[29] Kim, J.M., Park, J.H., Lee, C.K. and Lee, S.Y. (2014) *Scientific Reports*, **4**, 4602.

[30] Vancaeyzeele, C., Fichet, O., Laskar, J. *et al.* (2005) *Polymer*, **46**, 6888–6896.

[31] Jones, D.S., Andrews, G.P., Caldwell, D.L. *et al.* (2012) *European Journal of Pharmaceutics and Biopharmaceutics*, **82**, 563–571.

[32] Lee, J.H. and Kim, S.C. (1984) *Polymer*, **16** (6), 453–459.

[33] Francke, I.F., HenriAubert, P., Alfonsi, S. *et al.* (2012) *Solar Energy Materials & Solar Cells*, **99**, 109–115.

[34] Turner, J.S. and Cheng, Y.L. (2003) *Macromolecules*, **36**, 1962–1966.

[35] Hou, S.S. and Kuo, P.L. (2001) *Polymer*, **42**, 9505–9511.

[36] Donatelli, A.A., Sperling, L.H. and Thomas, D.A. (1976) *Macromolecules*, **9**, 676.

[37] Yeo, J.K., Sperling, L.H. and Thomas, D.A. (1982) *Polymer Engineering & Science*, **22**, 190.

[38] Thomas, S., Chandra, A.P. and Visakh, P.M. (2012) *Advances in Elastomers I: Blends and Interpenetrating Networks*, Springer, Berlin, pp. 1–70.

[39] Pfeifer, C.S., Shelton, Z.R., Szczpanski, C.R., Barros, M.D., Wilson, N.D. and Stansbury, J.W. (2014) *Journal of Polymer Science, Part A: Polymer Chemistry*, **52**, 1796–1806.

[40] Adachi, H. and Kotaka, T. (1982) *Polymer*, **14** (5), 379–390.

[41] Parizel, N., Meyer, G. and Weill, G. (1995) *Polymer*, **36** (12), 2323–2330.

[42] Lu, Y., Wang, D., Li, T. *et al.* (2009) *Journal of Biomaterials*, **30**, 4143–4151.

[43] Sperling, L.H. (2003) *Encyclopedia of Polymer Science and Technology*, vol. **10**, John Wiley & Sons, New York, pp. 272–311.

[44] Reddy, T.T., Kano, A., Maruyama, A., Hadano, M. and Takahara, A. (2008) *Journal of Biomacromolecules*, **9**, 1313–1321.

[45] Botero, A.J., Blanco, M., Li, Y.Y., McGunnies, G. and Goodard, W.A. (2010) *Journal of Computational and Theoretical Nanoscience*, **7**, 1238–1256.

[46] Therian, H. and Kumacheva, E. (2013) *Journal of the American Chemical Society*, **135**, 4834–4839.

[47] Thankam, F.G., Muthu, J., Sankar, V. and Gopal, R.K. (2013) *Journal of Colloids and Surfaces B: Biointerfaces*, **107**, 137–145.

[48] Noel, V. and Randriamahazaka, H.N. (2012) *Journal of Electrochemistry Communications*, **19**, 32–35.

[49] Aduba, D.C., Jr, Hammer, J.A., Yuan, Q. *et al.* (2013) *Journal of Acta Biomaterialia*, **9**, 6576–6584.

[50] Venkatesan, N., Shroff, S., Jayachandran, K. and Doble, M. (2010) *Journal of Endourology*, **4**, 191–198.

[51] Huang, Y., Liu, M., Chen, J., Gao, C., Gong, Q. and Huang, Y. (2012) *European Polymer Journal*, **48**, 1734–1744.

[52] Zhang, Q., Fang, Z., Cao, Y. *et al.* (2012) *ACS Macro Letters*, **1**, 876–881.

[53] Murthy, P.S.K., Murali Mohan, Y., Varaprasada, K. *et al.* (2008) *Journal of Colloid and Interface Science*, **318**, 217–224.

[54] Rani, M. and Negi, Y.S. (2010) *African Journal of Pharmacy and Pharmacology*, **4** (2), 35–54.

[55] Anirudhan, T.S. and Parvathy, J. (2014) *International Journal of Biological Macromolecules*, **67**, 238–245.

[56] Shidhaye, S., Surve, C., Dhone, A. and Budhkar, T. (2013) *International Journal of Research and Reviews in Pharmacy and Applied Science*, **2**, 637–650.

[57] Jiang, B., Pu, H., Pana, H., Changa, Z. and Jina, M. (2014) *Journal of Electrochimica Acta*, **132**, 457–464.

[58] Zhao, N. and Liu, S. (2011) *European Polymer Journal*, **47**, 1654–1663.

[59] Liu, X.D., Kubo, T., Diao, H.Y., Benjamas, J., Yonemichi, T. and Nishi, N. (2009) *Journal of Analytical Biochemistry*, **393**, 67–72.

[60] Bar, N. and Basak, P. (2014) *The Journal of Physical Chemistry C*, **118**, 10640–10650.

2

MISCIBILITY, MORPHOLOGY, AND PHASE BEHAVIOR OF IPNs

XUEMEI WU, GAOHONG HE, XIAOMING YAN, XIANGCUN LI, WU XIAO AND XIAOBIN JIANG

State Key Laboratory of Fine Chemicals, Research and Development Center of Membrane Science and Technology, Dalian University of Technology, Dalian, China

2.1 INTRODUCTION

Interpenetrating polymer networks (IPNs) are special polymer blends with the characteristics of intimate combination of two polymers both in network form. In the cases that one linear polymer is interpenetrated with another crosslinked polymer, the obtained materials are semi-interpenetrating polymer networks (sIPNs). Both IPNs and sIPNs have been widely used as impact-resistant polymers [1–3], reinforced elastomers [4], sound and vibration damping materials [2, 3], and coatings [5–8] and are being developed for their new applications as fuel cell ion exchange membranes [9–15] and biochemical and biomedical materials [16–18]. The advantageous properties of IPNs, such as enhanced mechanical strength, broad frequency range of damping, high conductivity, and stimuli-responsive behavior, originate from excellent compatibility that results in fine disperse microphase-separated morphologies. It is well established that IPNs usually have pronounced structural heterogeneity. That is, except for homogeneous IPNs (homo-IPNs), substantially all of the IPNs undergo phase

Micro- and Nano-structured Interpenetrating Polymer Networks: From Design to Applications, First Edition. Edited by Sabu Thomas, Daniel Grande, Uroš Cvelbar, K.V.S.N. Raju, Ramanuj Narayan, Selvin P. Thomas and Akhina H.

separation. This is important in the development of good properties. However, the type of phase separation, in terms of phase size, continuous phase, and the frequent bonding between the different kinds of chains, differs widely.

IPNs cannot be considered as systems with a molecular level of mixing of the two networks. The degree of mixing in IPNs is controlled by the balance between both thermodynamic immiscibility and kinetic "forced compatibility" of IPNs. That is, the phase separation would have been largely prevented when substantial crosslinking of both components occurs before significant phase-separated diffusion can occur. Therefore, depending on the component miscibility, as well as relative kinetics of polymerization, crosslinking, and phase separation, the obtained IPNs exhibit various morphologies. IPNs with dispersed phase domains ranging from a few micrometers (incompatible) to a few tens of nanometers (intermediate) to those with no resolvable domains structure (complete mixing) have been reported [19]. By comparison, polymer alloys and blends with diameter less than 100 nm are difficult to prepare.

This chapter focuses on the fundamentals and experimental advancements of the thermodynamic miscibility and kinetics of IPN formation. As a result, phase diagrams and different morphologies exhibited will be discussed. In the literature, the research groups of Sperling, Lipatov, Utracki, and Frisch have done pioneer works and reviewed on the relevant topics [19–27].

2.2 MISCIBILITY OF INTERPENETRATING POLYMER NETWORKS

Miscibility of IPNs implies the complete chain entanglement or mixing of two polymers at the monomer level (i.e., complete chain entanglement) [19]. It influences greatly the degree of phase separation, and then the morphology and properties of IPNs. Most multicomponent polymers are thermodynamically immiscible, which have macrophase or microphase separations with various morphologies. In principle, IPNs should be considered as a special class of polymer alloys that is characterized by fine dispersion of phases. Except for homo-IPNs, substantially all of the IPNs undergo immiscibility and phase separation.

2.2.1 Thermodynamic Immiscibility of Interpenetrating Polymer Networks

The present state of the thermodynamics of polymer mixing was reviewed [22, 25–28]. Based on the general thermodynamic principles, the thermodynamic miscibility of two-component linear polymers or networks is governed by the Gibbs equation, as shown in Equation 2.1. The thermodynamic criterion of miscibility is the negative value of the Gibbs free energy.

$$\Delta G_m = \Delta H_m - T\Delta S_m \tag{2.1}$$

where ΔG_m is the Gibbs free energy of mixing, ΔH_m is the enthalpy of mixing, ΔS_m is the entropy of mixing, and T is the absolute temperature. From Equation 2.1, ΔG_m can be expressed as ΔH_m minus the product of ΔS_m times T.

For Van der Waals polymer mixtures with no strong hydrogen bonding, donor–acceptor, or acid–base interaction, the enthalpy of mixing is expressed according to the van Laar–Hildebrand–Scatchard equation (Eq. 2.2) in the theory of regular solutions [25]:

$$\Delta H_m = V\phi_1\phi_2(\delta_1 - \delta_2)^2 \tag{2.2}$$

where V is the total volume of the system and ϕ_1, ϕ_2 and δ_1, δ_2 are the volume fractions and Hildebrand solubility parameters of polymers 1 and 2, respectively, in the mixing system, which reflect their chemical structure.

The entropy of mixing, ΔS_m, normally represents the configurational entropy of mixing and is expressed according to the Flory–Huggins equation [19, 25]:

$$\Delta S_m = -k\left(N_1 \ln\phi_1 - N_2 \ln\phi_2\right) \tag{2.3}$$

where k is Boltzmann's constant and N is the number of molecules of species 1 and 2, respectively. The volume selected for Equations 2.2 and 2.3 should be identical.

The thermodynamic criterion of miscibility is the negative value of the Gibbs free energy. Based on Equation 2.1, the value of the Gibbs free energy is determined by the sign and magnitude of the enthalpy and entropy of mixing times temperature. Calculations from Equation 2.2 have shown that enthalpy of mixing, ΔH_m, is never negative. And the entropy of mixing ΔS_m is always positive because there is always an increase in entropy upon mixing. However, the value of ΔS_m in the mixing of polymers is negligibly small for polymer blends due to the high molecular weight of polymers and subsequent contiguity restrictions [19, 25]. Therefore, a very small positive enthalpy of mixing is sufficient to generate a positive Gibbs free energy and results in immiscibility. From the aforementioned equilibrium thermodynamic analysis, most of the mixing of polymers are immiscible [29].

Several points should be made regarding the preceding discussion:

1. Since the entropy of mixing, ΔS_m, for polymer blends is negligibly small [26], a rough guide to the magnitude of the Gibbs free energy can be provided by the square of the difference of the solubility parameters, δ, of the polymers in the mixture. The smaller this value is, the better would be the miscibility of the two polymers and the tendency to thermodynamic mixing. The overall solubility parameter can be calculated either by the sum of the dispersive, δ_d, the polar, δ_p, and the hydrogen bonding, δ_h, or it can roughly be computed from group contributions as listed in Table 2.1 [30, 31].
2. Although Equation 2.1 predicts that all polymer blends are immiscible, there are at least 400 miscible polymer blends [25, 32] existed. Evidently the assumptions on which Gibbs equation is based are not adequate for these systems. A number of modifications have been proposed, focusing on the binary interaction parameter and specific interaction effects. References [25, 32–35] are available for more detailed discussion.
3. Compared with linear polymer blend systems, the crosslinked systems are less miscible due to the reduction of the entropic contribution to the free energy of mixing [25]. It is implicitly assumed that crosslinking does not change molecular conformation of the two polymeric components. This assumption is valid for most systems.

As mentioned earlier, the thermodynamic criterion of miscibility is the negative value of the Gibbs free energy, ΔG_m (or of the enthalpy, ΔH_m). The free energy of mixing of the constituent network needs to be estimated to study the immiscibility of the polymer mixing systems. Several methods were employed for this purpose, such as vapor sorption [22, 36] and inverse gas chromatography [37]. The enthalpy of two networks mixing was first measured [38, 39] using inverse gas chromatography. As a result, the two constituent networks are thermodynamically immiscible, in spite of forming the monolithic system.

TABLE 2.1 Experimental and Calculated Values of Solubility Parameter for Some Polymers

Polymer	δ exp. $(MJ/m^3)^{1/2}$ From	To	δ calc. (H.+v. K.) $(MJ/m^3)^{1/2}$
Polyethylene	15.8	17.1	16.0
Polypropylene	16.8	18.8	17.0
Polyisobutylene	16.0	16.6	16.4
Polystyrene	17.4	19.0	19.1

TABLE 2.1 (*Continued*)

Polymer	δ exp. $(MJ/m^3)^{1/2}$		δ calc. (H. + v. K.) $(MJ/m^3)^{1/2}$
	From	To	
Poly(vinyl chloride)	19.2	22.1	19.7
Poly(vinyl bromide)	19.4	—	20.3
Poly(vinylidene chloride)	20.3	25.0	20.6
Poly(tetrafluoroethylene)	12.7	—	11.7
Poly(chlorotrifluoroethylene)	14.7	16.2	15.7
Poly(vinyl alcohol)	25.8	29.1	—
Poly(vinyl acetate)	19.1	22.6	19.6
Poly(vinyl propionate)	18.0	—	18.8
Poly(methyl acrylate)	19.9	21.3	19.9
Poly(ethyl acrylate)	18.8	19.2	19.2
Poly(propyl acrylate)	18.5	—	18.7
Poly(butyl acrylate)	18.0	18.6	18.3
Poly(isobutyl acrylate)	17.8	22.5	18.7
Poly(2,2,3,3,4,4,4-heptafluorobutyl acrylate)	13.7	—	15.8
Poly(methyl methacrylate)	18.6	26.2	19.0
Poly(ethyl methacrylate)	18.2	18.7	18.6
Poly(butyl methacrylate)	17.8	18.4	17.9
Poly(isobutyl methacrylate)	16.8	21.5	18.3
Poly(tert-butyl methacrylate)	17.0	—	18.0
Poly(benzyl methacrylate)	20.1	20.5	19.3
Poly(ethoxyethyl methacrylate)	18.4	20.3	18.6
Polyacrylonitrile	25.6	31.5	25.7
Polymethacrylonitrile	21.9	—	22.8
Poly(α-cyanomethyl acrylate)	28.7	29.7	23.8
Polybutadiene	16.6	17.6	17.5
Polyisoprene	16.2	20.5	17.4
Polychloroprene	16.8	18.9	19.2
Poly(oxymethylene)	20.9	22.5	20.5
Poly(tetramethylene oxide)	17.0	17.5	17.6
Poly(propylene oxide)	15.4	20.3	18.9
Polyepichlorohydrin	19.2	—	20.1
Poly(ethylene sulfide)	18.4	19.2	18.9
Poly(styrene sulfide)	19.0	—	19.6
Poly(ethylene terephthalate)	19.9	21.9	20.5
Poly(8-aminocaprylic acid)	26.0	—	25.7
Poly(hexamethylene adipamide)	27.8	—	28.0

Source: Krevelen [30]. Reproduced with permission of Elsevier.

IPNs synthesized to date exhibit varying degrees of phase separation depending principally on the miscibility of the polymers [40]. With highly incompatible polymers, the thermodynamic phase separation nature is so powerful that it occurs substantially before the kinetic force compatibility can prevent it. In these cases, only small gains in phase mixing can be obtained. In cases where the polymers are more thermodynamically miscible, phase separation can be almost completely avoided, since the permanent entanglements (catenation) can effectively prevent phase separation [19]. Therefore, complete miscibility is not necessary to achieve complete phase mixing. Intermediate situation of miscibility results in complex phase behavior and different domain size and shape [41].

2.2.2 Kinetically "Forced Compatibility" of Interpenetrating Polymer Networks

Equilibrium thermodynamics indicates that IPNs cannot be considered as systems with a molecular level of mixing of the two networks. Distinct from almost all polymer blend materials, IPNs possess kinetically "forced compatibility." That is, thermodynamic immiscibility would be kinetically limited. Structure and morphology of IPNs depend on chemical factors, rather than the ways of processing, heat treatment like traditional polymer blend materials [19, 22, 42, 43]. There are two processes in the formation of IPNs, chemical reaction of polymerization/crosslinking and physical reaction of microphase separation. They proceed simultaneously and interconnected each other. The initial system for the synthesis of IPNs represents a homogeneous mixture of components, which are able to form two independent networks in the course of chemical reactions proceeding according to different mechanisms. With the increase of the conversion, thermodynamic incompatibility of the propagating fragments of the constituent networks arises in the system. When the free energy of mixing becomes positive, the onset of phase separation takes place. The thermodynamic immiscibility of the IPN components and the driving force for phase separation increase with the reaction proceeding. From the onset of phase separation, the chemical polymerization/crosslinking and the physical microphase separation proceed simultaneously in the formation of the IPNs. While the phase separation process occurs under nonequilibrium conditions, the composition of the phases and their ratio changes with time. This process determines the characteristic structures which develop and transform at different stages of the reaction and phase separation. The kinetics allows the realization of the path and predetermines the real structure far from equilibrium.

The kinetics of IPN formation is a governing factor in the development of the system morphology, the degree of interpenetration of the polymer networks, and their composition and property. IPNs with various degrees of phase separation, from fully immiscible to apparently compatible, have been reported. Several reviews on this subject have been published in the literature [19, 22, 44–47]. However, future kinetic studies of IPN formation require still more extensive and detailed investigations. With the potential modifications in the state-of-art techniques and the application of the theoretical kinetic models, a higher degree of understanding of IPN formation can be made. The effects of the kinetic factors, such as types of catalyst, concentration of components and crosslinker, temperature, and so forth, on the morphologies of IPNs will be discussed later in Section 2.3.

2.3 PHASE DIAGRAM

Equilibrium thermodynamics describes the overall immiscibility of IPNs. The positive value of Gibbs free energy of mixing that appears in the course of IPN formation will lead to the microphase separation of the system. The best way to characterize the binary system is via a phase diagram [22]. It gives information of how ΔG_m varies with independent variables, such as temperature, pressure, composition, crosslinking density, and so on. Corresponding information from the phase diagram will aid to tailor specific morphologies and properties of IPNs [21]. The phase diagram implies thermodynamic equilibrium and should be independent of exact method of synthesis [20]. However, the thermodynamic equilibrium phase diagram for IPNs is practically difficult because phase separation proceeds simultaneously with IPN formation, so that the composition and the molecular weight of network fragments change constantly, thus creating thermodynamic immiscibility [22]. In spite of this, many efforts have been made to construct phase diagrams to give information about the phase separation in IPNs.

2.3.1 Types of Phase Diagrams

There are several types of phase diagrams for IPNs, such as temperature–composition diagram, monomer–polymer phase diagrams, and even polymer–polymer phase diagrams, as shown in Figure 2.1 [20, 44]. Many research works in the 1990s focus on phase diagrams, and there are several reviews for Refs. [20, 22, 27].

2.3.2 Temperature–Composition Phase Diagram

The first temperature–composition classical phase diagram for the sIPN net-polyurethane–inter-poly(vinyl chloride) (PU–PVC) is shown in Figure 2.2 [27, 48–50]. For this system, a lower critical solution temperature was found

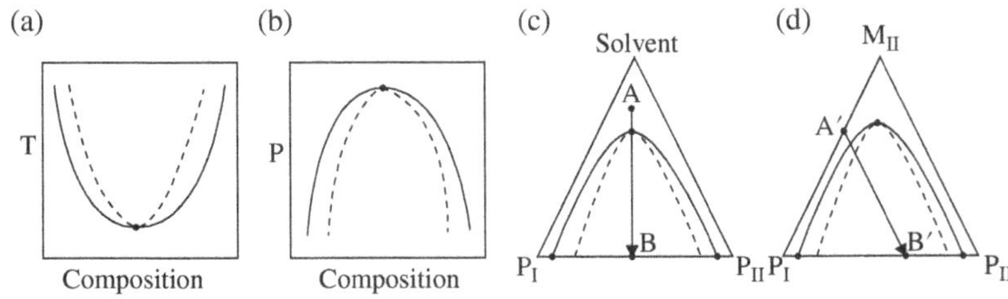

FIGURE 2.1 Phase diagrams for multicomponent polymer materials: (a) lower critical solution temperature; (b) upper critical solution pressure; (c) ternary phase diagram including a solvent undergoing evaporation from A to B; (d) monomer II polymerizing from A′ to B′ [20, 44]

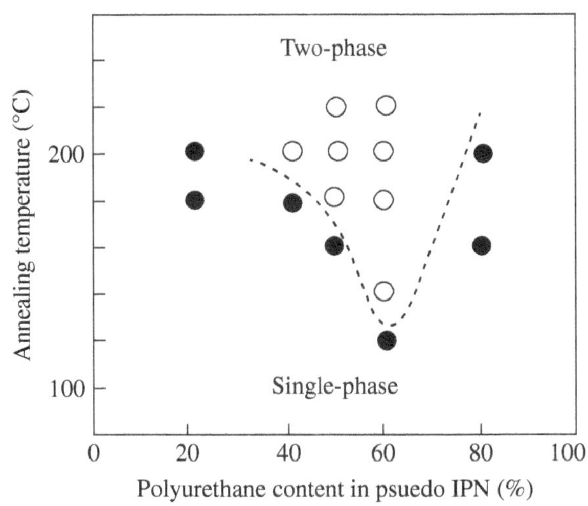

FIGURE 2.2 A phase diagram of a sIPN composed of net-polyurethane–inter-poly(vinyl chloride), showing a lower critical solution temperature [27, 48–50]

at approximately 120°C. Below this temperature the system is one-phase and above this temperature the system is two-phase. Due to the very small entropy of mixing for very large molecules, polymer blends and IPNs become less soluble with the increase in temperature.

A typical time–temperature cure cycle for the castor oil materials is shown in Figure 2.3 [21, 51]. Four generalized factors were determined to play roles in the development of morphology in the PET/triglyceride oil sIPNs: bond interchange, phase separation, crystallization of the PET, and network formation. Bond interchange reaction proceeded between castor oil and PET and increased the miscibility of the PET/triglyceride oil sIPNs. By means of controlling the reaction time and temperature cycle of polymerization and crosslinking, microstructure of the material could be attained.

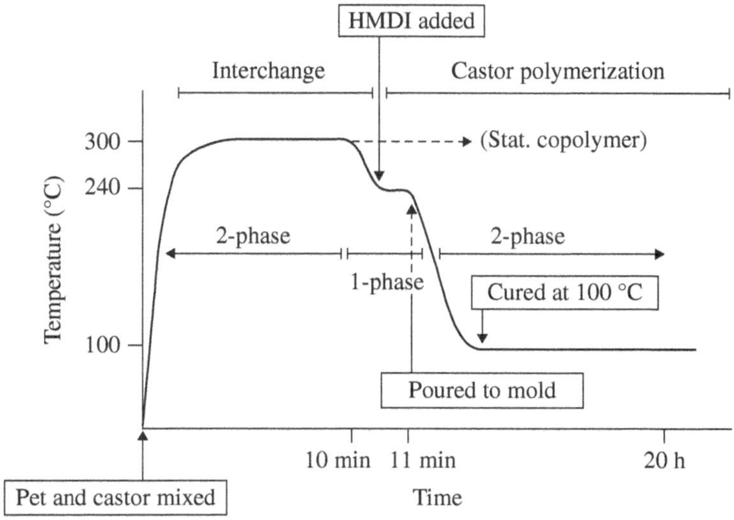

FIGURE 2.3 Bond interchange and phase behavior of PET-castor oil-based IPNs [21, 51]

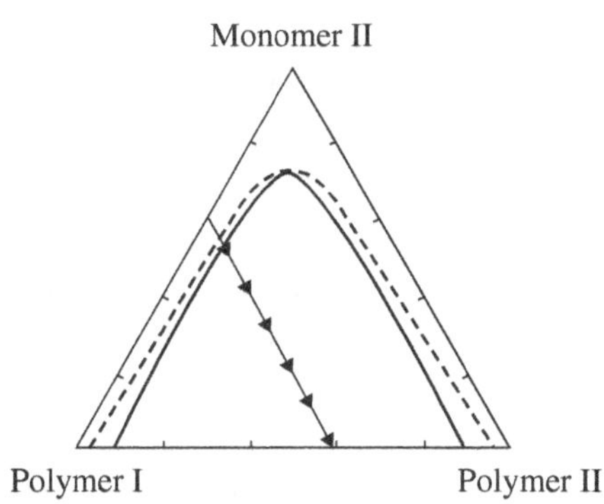

FIGURE 2.4 The path of polymerization of monomer II in the presence of polymer I, illustrating the intercept with the binodal and then the spinodal, all at constant temperature [21, 44, 52, 53]

2.3.3 Monomer–Polymer Phase Diagram

Sperling *et al.* illustrated partly complete phase diagram for a sequential IPNs. As shown in Figure 2.4, polymerization of styrene plus divinyl benzene crosslinker (monomer II) is in the presence of crosslinked polybutadiene (polymer I) [21, 44, 52, 53]. The authors used transmission electron micrograph (TEM) and small angle neutron scattering (SANS) techniques to detect the phase morphology and place numerical values in Figure 2.4. Below about 3% conversion of polystyrene, the low neutron scattering intensity and small

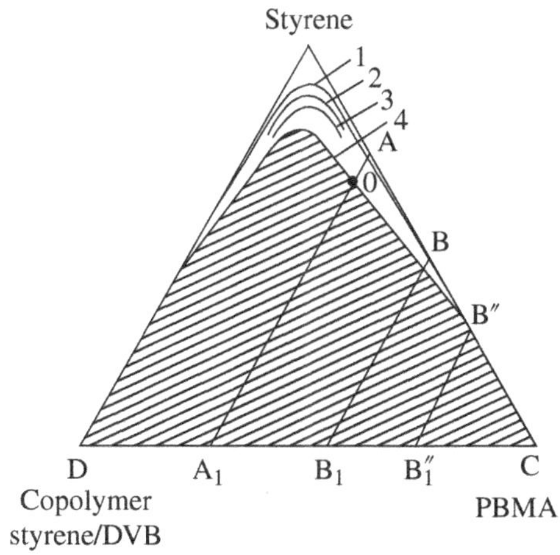

FIGURE 2.5 A phase diagram of sIPN composed of styrene-DVB-PBMA at (1) 333 K, (2) 343 K, (3) 353 K, and (4) 363 K [22, 54]

angular dependence suggest that phase separation has not yet taken place, and the system is presumed to be one-phased. Spherical domains are seen at about 7% conversion of the polystyrene, indicative of nucleation and growth morphology. Therefore, nucleation and growth begins in the range of approximately 3–6% conversion, as the binodal curve (dashed line) in Figure 2.4. Spinodal decomposition begins at approximately 20–25% polystyrene according to SANS experiments, as the spinodal curve (solid line) in Figure 2.4. Polymerization is completed as the line joining polymer I and polymer II is reached, producing a sample of 80% polystyrene. These experiments have been reviewed in Refs. [21, 44, 52, 53].

Lipatov *et al.* illustrated a monomer–polymer phase diagram for the curing sIPN based on poly(butyl methacrylate) (PBMA) and styrene–divinylbenzene (DVB) copolymer, as shown in Figure 2.5 [22, 54]. In this system, polystyrene (PS)/PBMA shows limited compatibility at low concentrations and low molecular mass of components and full incompatibility at higher concentrations and molecular weights (CD line), which results in the much larger regions of two-phase sIPN (hatched areas) as compared with the regions of one-phase state. Consideration of the polymerizing system passes from the initial mixture point A to the final IPN point A_1. After the border of the two-phase system (binodal curve, point O) is reached, phase separation can occur, and the thermodynamic incompatibility of this polymerizing system appears at a very low conversion degree. Analysis of the region of IPN compositions situated inside the triangle $B''B_1''C$ shows that polymerization of styrene mixtures with PBMA is accompanied by phase separation almost from the very

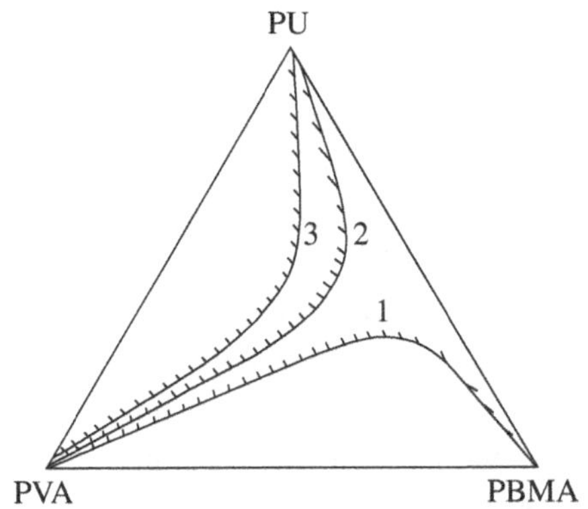

FIGURE 2.6 A phase diagram of the ternary system PVA-PBMA-PU at (1) 384 K, (2) 400 K, and (3) 476 K [22, 55]

beginning. From the phase diagram it is seen that an increase in the reaction temperature does not change the shape of the diagram. The area of the one-phase state slightly increases with the increasing temperature, which is typical for systems with an upper critical solution temperature.

Lipatov *et al.* [22, 55] also studies the influence of a third-polymer component on the phase separation of the crosslinked polyurethane (PU) and linear poly(vinyl acetate) (PVA) sIPNs. The ternary phase diagram is shown in Figure 2.6. In this case, the third-component PBMA was introduced into the curing system and led to a region of component compatibility (curve 1). With increasing temperature, the compatible region broadens (curves 2 and 3). The authors explain the enhanced compatibility by the Flory–Huggins theory. PBMA is compatible with both PVA and PU with the binary interaction parameter less than 0. By introducing PBMA, the total interaction parameter may become negative and the region of compatibility is created. With increasing temperature, the values of interaction parameter become more negative. Since PBMA is introduced into the curing system, it may have influenced the kinetics of the reaction. It is an example of the "forced compatibility" effect of IPNs.

It is easy to draw the phase diagram of sequential IPNs (usually sIPNs as shown previously) within a triangle. However, for the simultaneous IPNs (SINs), in which two networks polymerize or crosslink at the same time, the triangle phase diagram is difficult to express all the components involved. Moreover, during the synthesis of SINs, there are three significant events: gelation of polymer I, gelation of polymer II, and phase separation of polymer I from polymer II. The time order of these events is expected to produce a specific morphology and a concomitant specific set of physical and mechanical properties. Phase diagram will help to understand the different time order.

A form of tetrahedron metastable phase diagram was developed for the PU-poly(methyl methacrylate) (PMMA) SINs by the Sperling group [21, 22, 56, 57], as shown in Figure 2.7. Four vertices represent four pure components: two polymers, PMMA and PU, and two monomers, methyl methacrylate (MMA) and "U." Four triangular faces of the tetrahedron represent ternary systems. For example, the phase separation curve for the ternary system MMA-PMMA-"U" is represented by the front left triangle (points C-D-A-E). The phase separation curve for the MMA-PMMA-PU system is indicated by points J-B-K-L (the rear triangle). The entire tetrahedron volume is divided into two regions: single-phase region and separated-phase region, represented by points C-D-A-E-L-K-B-J and characterized by the "sail-like" shape. A–B curve is the intersection of the PMMA gelation plane with that of the phase-separated sail-like surface and is a critical curve representing simultaneous PMMA gelation and phase separation of the PU from the PMMA. Reactions moving to the left of A–B (as shown in Figure 2.7a for the miscible region) will have the PMMA gel before phase separation, while reactions to the right of A–B will phase separate before gelation. Figure 2.7b shows the gelation planes of both the PU (plane G_2-MMA-PMMA) and the PMMA (plane G_1-"U"-PU). Especially, the intersection of the two planes describes the line of simultaneous gelation of both polymers. Reactions passing first to one side of this line will have the corresponding polymer gel first.

It is noticed that Figure 2.7 works for both simultaneous and sequential synthesis of IPN systems. All SINs and sequential IPNs must begin somewhere along the edge line MMA-"U" and end along the edge line PMMA-PU, whereas the polymerization lines between these two end regions might not be straight. Based on the order in which the reacting system meets the three eventful surfaces inside the tetrahedron, IPNs with different possible morphology could be prepared.

2.3.4 Phase Continuity Diagram

Phase continuity and phase inversion are important morphological structures for polymer blends, solution graft copolymers, and IPNs, which qualitatively compares with physical properties of IPNs. For example, a 50:50 wt.% glassy-rubbery system with dual-phase continuity can raise the modulus half that of the glassy phase. Values are close to the Takayanagi upper-bound model [20, 22, 58–66].

According to Sperling *et al.*, dual-phase continuity is defined as the continuity of both polymer I and polymer II domains throughout the macroscopic sample [20]. In terms of spinodal decomposition concepts, the most usual

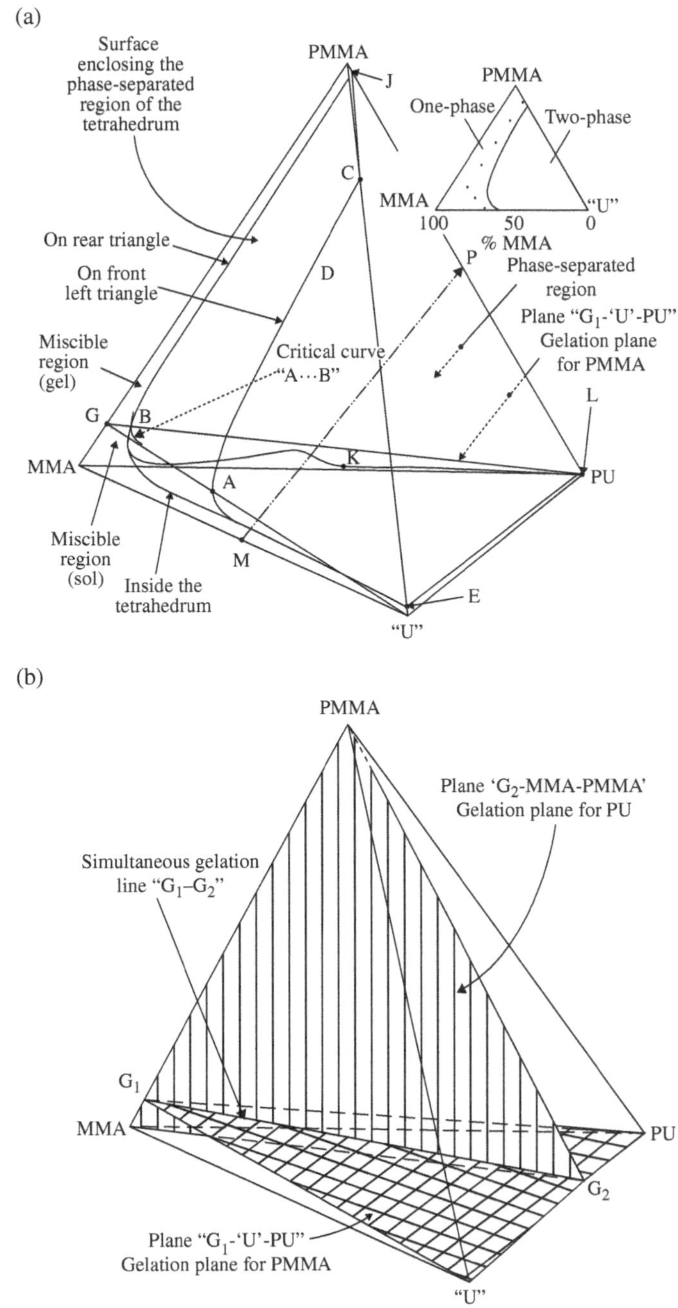

FIGURE 2.7 Two expressions of the tetrahedron metastable phase diagram for the PU-poly(methyl methacrylate) (PMMA) SINs. Tetrahedron for the simultaneous IPN from PU and PMMA shows the gelation plane of the PMMA and the curvilinear plane of phase separation. (a) Gelation plane of the PMMA and (b) curvilinear plane of phase separation. Inset: experimental data for the phase-separating part of the front left triangle, showing the appearance of turbidity during MMA polymerization as an indication of phase separation [21, 22, 56, 57]

physical situation is expected to be a system of interconnected cylinders of polymer II stretching through a matrix of polymer I. Phase inversion is defined as a transposition of the continuous and discontinuous phases. They developed a semiempirical model for predicting phase continuity and phase inversion in a low shear rates system, based on the volume fraction, ϕ, and viscosity, η, of the two phases, as shown in Equation 2.4, where the subscripts 1 and 2 represent phases 1 and 2, respectively.

$$\begin{aligned} &\frac{\phi_1}{\phi_2}\times\frac{\eta_2}{\eta_1}>1 \quad \text{phase 1 continuous}\\ &\frac{\phi_1}{\phi_2}\times\frac{\eta_2}{\eta_1}=1 \quad \text{dual phase continuity or phase inversion}\\ &\frac{\phi_1}{\phi_2}\times\frac{\eta_2}{\eta_1}<1 \quad \text{phase 2 continuous} \end{aligned} \tag{2.4}$$

Experimental data from various phase-separated systems confirm this relationship, as shown in Figure 2.8; either the viscosity ratio or the volume fraction ratio is near 1, dual-phase continuity is favorable. Since the viscosity and the volume fractions may change with time of curing, the dual-phase

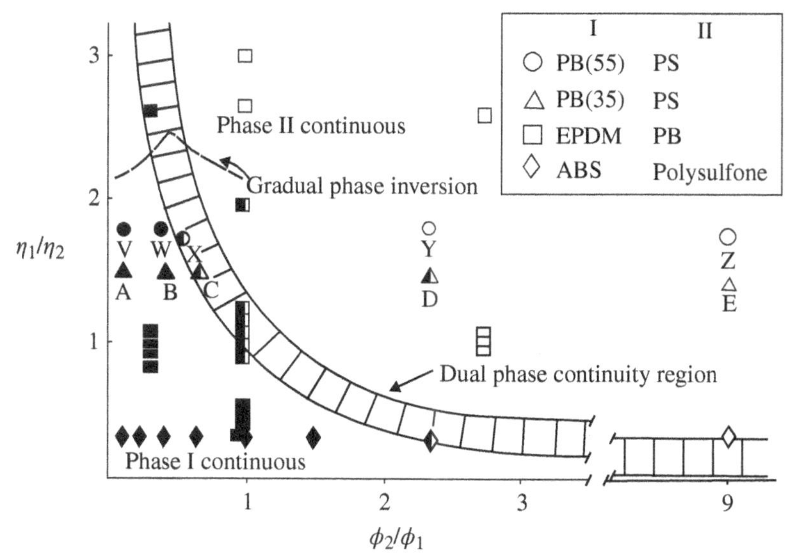

FIGURE 2.8 Phase continuity diagram for blends and IPNs undergoing phase inversion during polymerization or under shear [10, 20, 58]

continuity morphology will be dependent on the reaction conditions [22]. For a number of industrial materials, especially the thermoplastic IPNs, one of the components is being crosslinked. Dual-phase continuity achieves if the reaction is stopped as Equation 2.4 equals to 1.

2.4 MORPHOLOGY OF INTERPENETRATING POLYMER NETWORKS

There are many applications of IPNs, such as impact-resistant polymers, reinforced elastomers, sound and vibration damping materials, rubber blends, electrical insulators, diffusion-selective compositions, coatings, and ion exchange resins [25]. The advantageous properties of IPNs originate from their pronounced structural heterogeneity, but excellent compatibilization that obtained by co-reactions. Except for homo-IPNs, substantially all of the IPNs undergo phase separation, while the type of phase separation, in terms of phase size, continuous phase, and the frequent bonding between the different kinds of chains, differs widely. Depending on the component miscibility, as well as relative kinetics of polymerization, crosslinking, and phase separation, IPNs with different morphologies, for example, co-continuity and one phase dispersed in another phase are prepared. Dispersed phase domains ranging from a few micrometers (incompatible), to a few tens of nanometers (intermediate) [19, 41], to those with no resoluble domain structure (complete mixing) have been reported [19, 44–46]. The obtained IPNs exhibit divers and fine dispersed morphologies that are important in the development of good properties. This is also the advantages of IPNs over polymer alloy and blends, in which domains less than 100 nm are difficult to prepare.

2.4.1 Phase Separation Mechanism

Different phase separation mechanism leads to different morphologies in the IPN materials. According to Sperling, nucleation and growth (NG) mechanism often results in spherical domains dispersing in the continuous phase, while spinodal decomposition (SD) mechanism often results in interconnected cylinders' phase separation structure.

From the equilibrium thermodynamics aspect, IPNs are usually phase-separated systems because of the positive enthalpy of mixing [25]. Actually, even if the value of Gibbs free energy is negative, the IPN systems are also likely to be immiscible. Thermodynamic stability of the multicomponent systems is determined by the concentration dependence of the Gibbs free energy. As

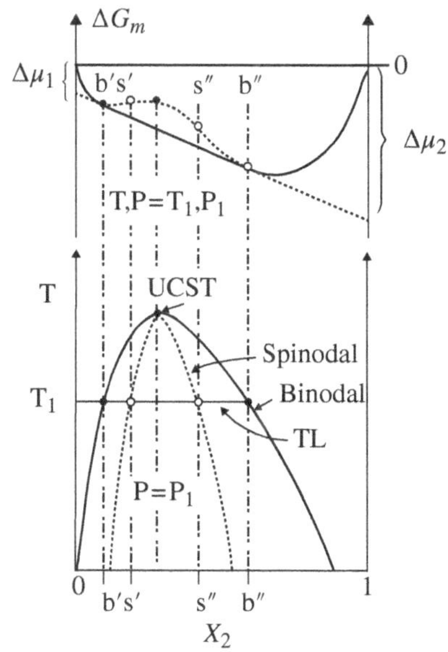

FIGURE 2.9 Construction of phase diagram for two-component systems

shown in Figure 2.9, for two-component systems, the upper part of Figure 2.9 is the Gibbs free energy of mixing as a function of composition at constant temperature and pressure. The lower part is a plot of the phase separation temperature as a function of composition at constant pressure. Collections of point b and point s give the binodal curve (solid line) and spinodal curve (dotted line), respectively. Either binodal and spinodal curves meet at the critical point, exhibited lower critical solution temperature (LCST) or upper critical solution temperature (UCST). Therefore, three regions are present in the phase diagram of two-component systems [22, 54, 67–75]:

1. Single-phase, miscible region on the outside of the binodal curve, where $\Delta G_m < 0$, and $\partial^2 \Delta G_m / \partial \phi^2 > 0$. The system is thermodynamically miscible.
2. Metastable phase region, between the binodal and spinodal curves (see the region in between s′ and s″ in the upper part in Figure 2.9) $\left(\partial^2 \Delta G_m / \partial \phi^2\right) > 0$, which indicates that decomposition would lead to the increase in ΔG_m. Therefore, the system in this region is thermodynamically immiscible, but phase separation will not take place spontaneously except for large concentration fluctuation (e.g., impurity, vibration, and supersaturation).
3. Phase-separated (or spinodal) region enclosed by the spinodal (see the regions in between b′, s′ and s″, b″, respectively, in the upper part in

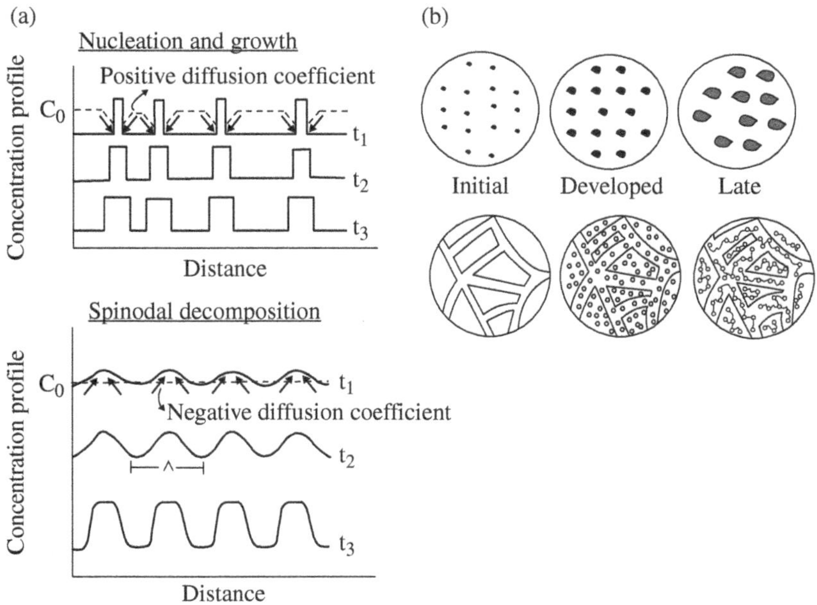

FIGURE 2.10 Illustration of the kinetic mechanisms of phase domain formation. (a) Kinetic mechanisms and (b) development of phase domain formation [20, 22, 52, 75]

Figure 2.9) $(\partial^2 \Delta G_m / \partial \phi^2) < 0$, which indicates that decomposition would lead to the decrease in ΔG_m. The system is thermodynamically immiscible and spontaneously decomposed for even small fluctuation.

Two major mechanisms of phase separation are developed: spinodal decomposition and nucleation and growth, according to quite different, time-dependent concentration fluctuations [22]. The resulting morphology of IPNs is quite different. As shown in Figure 2.10a, for nucleation and growth mechanism, nuclei will form in the system if there are large concentration fluctuations. During nucleation, the diffusion flows are diminishing the composition fluctuations. The decomposed nuclei will rich in the dissolved component and surround by the dispersion medium. Therefore, nucleation and growth mechanism tends to produce spheres of the second phase within the matrix of the first phase. These spheres grow by increasing their diameter [20]. For the spinodal decomposition mechanism, the system is unstable to small fluctuation in composition and separate into two phases simultaneously in the whole volume of the system without forming the nuclei. During spinodal decomposition, the diffusion flows are increasing the composition fluctuations because the Gibbs free energy of mixing can be reduced in this way. The polymerization component is the spontaneous formation and continuous growth in the instable existing phase, resulting in a high degree of interpenetration of evolving phases. Therefore, spinodal decomposition mechanism tends to produce interconnected cylinders of the second phase within the matrix of the first phase. These cylinders grow by increasing their wave amplitude. Figure 2.10b shows the schematics of nucleation and growth and spinodal decomposition at various reaction stages [22, 76]. Several works have been done to line the morphology of IPNs to the phase separation mechanism in the phase diagram, as we have shown in Figure 2.4 [20, 22, 52, 75].

2.4.2 Typical Morphologies of Interpenetrating Polymer Networks

From the preceding discussion, it is evident that the morphology of IPNs should be determined by several factors, such as the thermodynamic miscibility of two networks, the kinetic conditions of the IPN formation, and the mechanism of phase separation [22]. These result in three typical morphologies of IPN structure: one-phase IPNs (or homo-IPNs), embedded particles in the matrix, and dual-phase continuity structures. Microscopic structures such as spatial inhomogeneity are not only attractive objects from the scientific standpoints, but also important factors that should be designed to control properties of IPN materials.

In order to identify the morphology of IPNs, several methods can be used. For example, one single broad glass transition temperature, T_g, detected by differential scanning calorimetry (DSC) or dynamic mechanical analysis (DMA), may indicate the one-phase morphology. With appropriate staining of one component, such as mercury stains for PPO, osmium tetroxide for PB, lead acetate for sulfonic acid group, chlorpalladium for hydroxide ion group, and so on, transmission electron microscopy (TEM) can be viewed as a direct method to visualize the size of the phase-separated domains (~10 nm). However, it could not give the exact shape of the domains sometimes because of the cross-sectioning effects of the very thin (within 100 nm) slice sample [20]. Several techniques such as light scattering, small-angle neutron scattering (SANS), and small-angle X-ray scattering (SAXS) allow us to better understand the microscopic structures of IPNs.

2.4.2.1 One-Phase Morphology One-phase morphology is achieved in two cases. For thermodynamically miscible IPNs, phase separation can be almost completely avoided, which means the full interpenetrating of different polymer chains. In this case, the IPNs are also called homo-IPNs. In another case, one-phase morphology is observed due to the absence of any structure for phase separation with the present technique, which means the size of phase-separated domain is extremely small (say, <10 nm) [20, 22, 76]. One-phase IPNs can have superior properties (such as ultimate mechanical strength) compared to the pure crosslinked networks or any blend of the corresponding linear polymers.

Many IPNs with one-phase morphology are reported in the literature [19, 20, 26, 27, 44–46, 54]. Some examples are poly(2,6-dimethyl-1,4-phenylene oxide)-polystyrene (PPO-PS) and polycarbonate-polyurethane-polybutadiene (PCU-PB). Some of them have already been widely used, such as poly(methyl methacrylate) (PMMA)-based homo-IPN, which is a kind of false teeth material commercially known as Bioform IPN 200 Posteriors.

Since T_g is a measurement of mobility of polymer in the level of chain segments, there should be one single T_g experimentally revealed for the one-phase IPNs. Wu *et al.* [77–80] prepared two kinds of one-phase sIPNs proton exchange membrane materials: sulfonated poly(phthalazinone ether sulfone ketone)-crosslinked polyacrylic acid (SPPESK/CrPAA) and poly(ether ether ketone)-crosslinked poly(styrene sulfonic acid) (SPEEK-CrPSSA), through the *in situ* free radical polymerization and crosslinking of acrylic acid or styrene sulfonic acid monomers in the presence of linear SPPESK or SPEEK. The values of glass transition temperature T_g of the sIPN membranes are determined by DSC at a heating rate of 10°C/min. From Figure 2.11, it can be seen that all SPPESK/PAA sIPN membranes exhibit one single T_g for the investigated compositions. One single T_g combined with the homogeneous surface morphology scanning electron microscopy (SEM) images (Figure 2.12) are evidence for the compatibility of the two components of the sIPN. The authors attribute the compatibility to the chain entanglements in the formation of the sIPNs. Higher experimental T_g values as compared with the calculated values from Fox formula are also investigated in the SPPESK/CrPAA and SPEEK-CrPSSA sIPNs, which indicates the extra interactions due to the closely chain entanglements in the IPNs.

Frisch group studies the phase morphology of series of simultaneous IPNs (SINs) [26, 81–86]. They conclude that one-phase IPNs can be made from components whose linear polymers are wholly immiscible. They developed the criterion to correlate with the phase morphology of fully or partially miscible IPNs, roughly based on the differences in solubility parameter and T_gs. Square of the difference of the solubility parameters of the polymers in the mixture can roughly guide to the magnitude of the excess Gibbs free energy. The less of the difference means a smaller thermodynamically demixing tendency in the IPNs, thus expects to produce either complete or microphase separation. The difference of the T_gs is the magnitude of the difference of chain mobility of the IPN polymers. Therefore, polymer chains with well matched T_gs, particularly if both T_gs are glassy, should show a smaller dynamically demixing tendency.

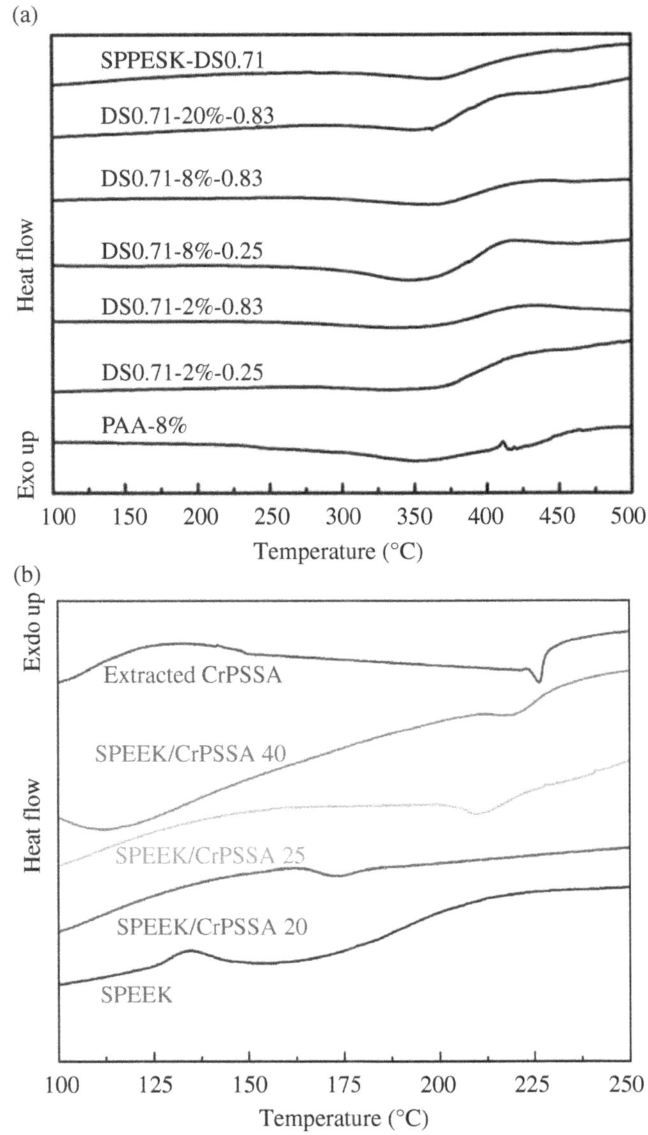

FIGURE 2.11 One single T_g investigated in the SPPESK-CrPAA (Source: Wu *et al.* [77]. Reproduced with permission of Elsevier) and SPEEK-CrPSSA sIPNs (Source: Wu *et al.* [78]. Reproduced with permission of Elsevier)

Two series of glassy and rubbery simultaneous IPNs were systematically investigated of their solubility parameters and T_g. As shown in Tables 2.2 and 2.3, in the glassy series, PPO, a glassy material at room temperature, was crosslinked by methyl bromination of PPO and condensation of the resultant bromide of PPO

SPPESK-DS0.71 DS0.71-8%-0.83 DS0.71-8%-0.25

FIGURE 2.12 Surface morphology of SPPESK and SPPESK/PAA sIPN membranes. Source: Wu *et al.* [77]. Reproduced with permission of Elsevier

TABLE 2.2 Phase Morphology of the Glassy and Rubbery Series of IPNs and That of the Corresponding Blend of Linear Polymers

IPN	Composition Range (wt%)	Number of Distinct T_g s	y_c (wt%)	Number of Phases	
				In IPN	In Blends[a]
PPO-PS	Whole	1	75 (PPO)	1	1
PPO-PMMA	Whole	1	80 (PPO)	1	2
PPO-PUA	Whole	1	80 (PPO)	1	2
PPO-PB	Whole	1	90 (PPO)	1	2
PPO-PDMS	Whole	2	90 (PPO)	2	2
PCU-PMMA	Whole	1	30 (PCU)	1	2
PCU-PB	Whole	1	50 (PCU)	1	2
PCU-PS	100–60 PCU	1	30 (PCU)	1	2
50 PCU or less		2		2	2
PCU-PVP	100–60 PCU	1	70 (PCU)	1	2
50 PCU or less		2		2	2

Source: Frisch and Zhou [26]. Reproduced with permission of American Chemical Society.
[a] The composition of the linear blends is the same as the IPN.

with a diamine. The simultaneous formation of the other network was made by free radical crosslinking of the other component, such as PMMA, PUA, and PB and did not interfere with the condensation crosslinking of PPO. In the rubbery (PCU) series, the PCU network was prepared by crosslinking poly(1,6-hexanediol carbonate) with the biuret triisocyanate derived from hexamethylene diisocyanate via an addition reaction. The other free radical crosslinked network was made from PMMA, PB, PVP, and PS and did not interfere with the simultaneous addition crosslinking of PCU. From Tables 2.2 and 2.3, in most cases the linear polymer blends are immiscible, while the corresponding SINs are one-phase

TABLE 2.3 T_g and Solubility Parameter of the Linear Polymer Chains Found in the IPNs

	Solubility Parameters $((cal/cm^3)^{1/2})$				
Polymer	δ	δ_d	δ_p	δ_h	T_g (°C)
PPO		8.6	3.0	2.0	211
PVP		9.6	5.1		142
PMMA	9.1–12.8	9.2	5.0	4.2	105
PS	8.5–9.3	8.6	3.0	2.0	100
PUA	9.1[a]				
PCU	8.8[a]				−40
PB	8.1–8.6	8.8	2.5	1.2	−95
PDMS	7.3–7.6				−127

Source: Frisch and Zhou [26]. Reproduced with permission of American Chemical Society.
[a] δ computed from group contributions.

either in the whole composition range (e.g., PPO-PMMA, PPO-polyurethane-polyacrylate (PUA), PPO-PB, PCU-PMMA, and PCU-PB) or over a portion of the composition range (e.g., PCU-natural rubber, PCU-PS, and PCU-poly (4-vinylpyridine) (PVP)). The small mismatch in solubility parameters in the PPO-PMMA system allows full miscibility in the IPNs, although prevents miscibility of the linear blends. With a large mismatch in solubility parameters and T_gs, phase separation is found in PPO-PDMS system for both the blends and the IPNs. The relatively small difference in T_gs is in accordance with the one-phase morphology of the PCU-PB IPNs, even though the difference in solubility parameters prevents any miscibility of the linear blends of the somewhat polar PCU and nonpolar PB. For polar PCU and PMMA, the solubility parameter mismatch is sufficiently small that the IPNs are fully miscible even though there is a significant mismatch in the T_gs. The dynamics of IPN formation plays an important role for the enhanced miscibility in IPNs. Dynamic factors lead to interpenetration of the system between each other, and thus far behind the equilibrium thermodynamic state.

2.4.2.2 One Phase Dispersed in Another The properties of multicomponent polymeric systems obtained from thermodynamically incompatible polymers strongly depend on their morphology, including the degree of phase separation and the size, shape, distribution, and type of arrangement of the microregions in the phase-separated material.

Sequential Interpenetrating Polymer Networks Morphology of IPNs changes with the reaction proceeding. For sequential IPNs, Sperling describes four distinct stages of morphology development [44, 87]. At first, monomer II may be soluble

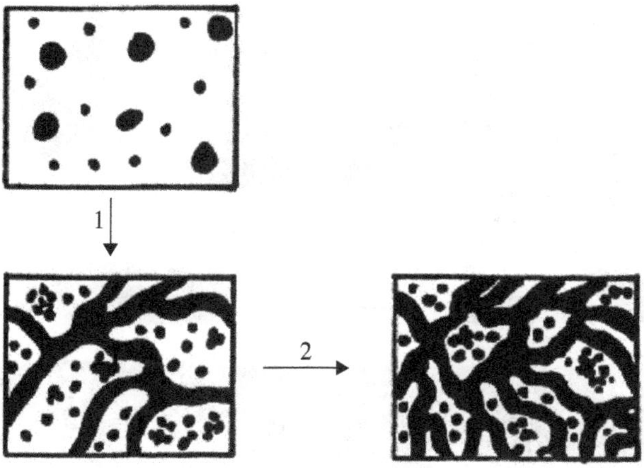

FIGURE 2.13 Model of the morphological development in sequential IPNs. Fernandez [88]. Reproduced with permission of the American Chemical Society

in polymer (or network) I. During this stage, polymer II also remains soluble, therefore, the polymerizing system may be optically clear. Stage 2 is characterized by the suddenly clouding up, indicating phase separation. This phase separation may be nucleation and growth mechanism and results in spheres morphology [88]. With the increase of polymerization, interconnected cylinders develop and increase in number during the stages 3 of polymerization, indicative of spinodal decomposition. In stage 4, the morphology may become less distinct because the high viscosity of the system very significantly reduces diffusion toward the equilibrium state. The progress of morphology development is modeled in Figure 2.13 [88].

Sperling also summarized generalizations for sequential IPNs based on the results of many investigations, as follows:

1. Polymer I tends to be continuous in space for all compositions. Midrange and high concentrations of polymer II tend to have dual-phase continuity that is most likely interconnected cylinders for polymer II.
2. Phase separation probably proceeds through a brief period of nucleation and growth, followed by a region of spinodal decomposition.
3. Domain sizes vary from about 100nm for highly immiscible systems to about 10nm for microheterogeneous systems. The domain size decreases with the increase of crosslinking density. For many compositions, the effect is about 10 times as large for crosslinkings in network I as for network II.

The amount of crosslinking introduced during network formation has great effect on the level of phase separation. With the increase of crosslinking

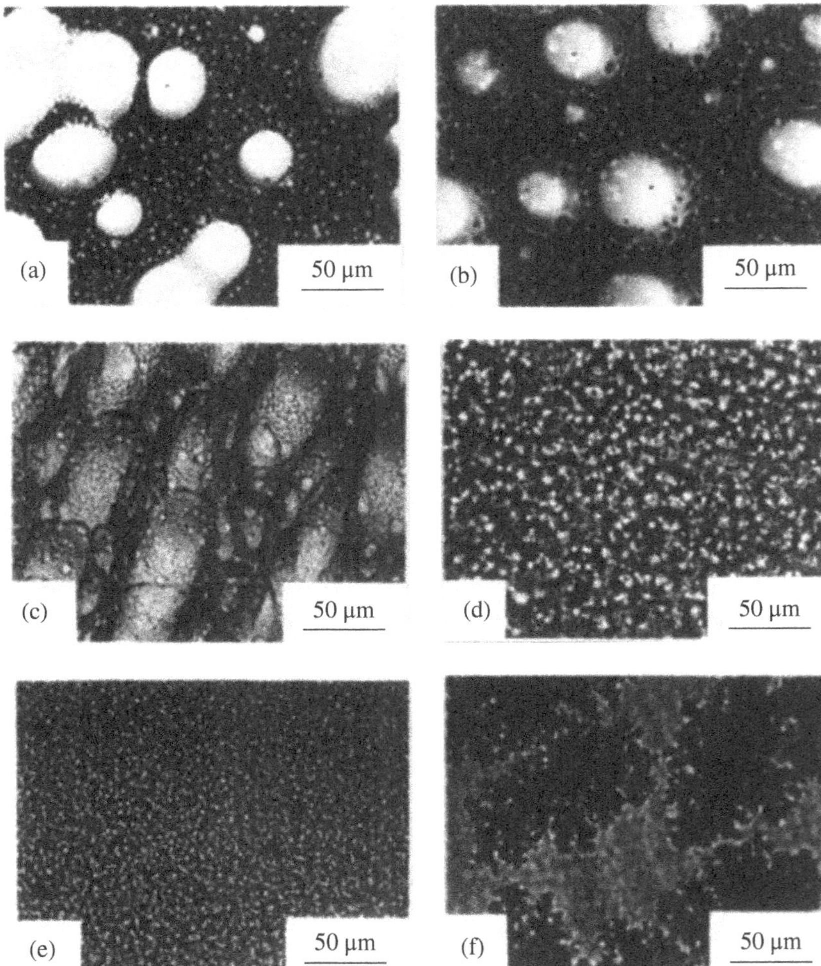

FIGURE 2.14 Optical micrographs of the sIPNs and related materials: (a) sIPN (20% P3OT)/0% DVB; (b) sIPN (20% P3OT)/1% DVB; (c) sIPN (20% P3OT)/2% DVB; (d) sIPN (20% P3OT)/5% DVB; (e) sIPN (20% P3OT)/10% DVB; (f) sIPN (20% P3OT)/16% DVB. Source: Wang and Rubner [89]. Reproduced with permission of American Chemical Society

degree, the original immiscible polymer blend system, for example, PMA/PMMA, can become homogeneous IPN by forced compatibility of the both networks. sIPNs based on linear poly(3-octylthiophene) and crosslinked PS (DVB as crosslinker) were investigated of their morphology by varying DVB content [89]. As shown in Figure 2.14, in the optical microscopy images, domains of the polystyrene phase is in the bright-field. Distinctive spherical

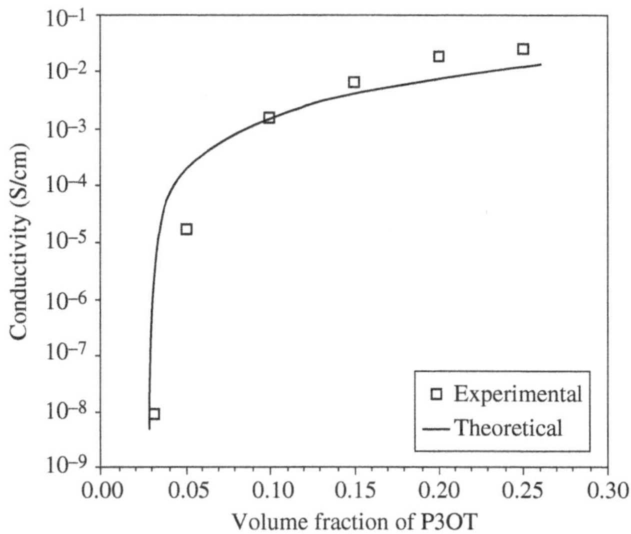

FIGURE 2.15 Electrical conduction threshold of the sIPN/10% DVB samples. Source: Wang and Rubner [89]. Reproduced with permission of American Chemical Society

phases can be clearly observed in the sample of simple blend and sIPN of low DVB content (Figure 2.14a and b). With the increase of crosslinker DVB content, the large domains of PS start to break up and become better dispersed throughout the sIPN. The sIPN achieves highest level of phase dispersion as the amount of DVB increases to 10% (Figure 2.14e), while at extremely high DVB content (23% in Figure 2.14f), the crosslink density of the polystyrene network has become so high that syneresis occurs and a large portion of the P3OT is rejected from the crosslinked matrix.

Since the sIPN is a semiconductor, conductivity was measured as a function of P3OT volume fraction, as shown in Figure 2.15. The obtained data were fitted according to the percolation theory to get the percolation threshold. The authors got a very low value (2.8%) of percolation threshold, which suggests a very fine dispersion of the P3OT domains in the sIPN.

Donatelli *et al.* [20, 25, 90–93] derived relations to estimate the domain size in sequential IPNs, on the basis of the equilibrium thermodynamics, as shown in Equation 2.5. The domain diameter of polymer II, D_2, is related to the interfacial tension, γ, the absolute temperature, T, the gas constant, R, the concentration of effective network chains ν_1 and ν_2, and the volume fractions v_1 and v_2 for networks I and II, respectively. Equation 2.5 is later generalized by Yeo *et al.* (Eq. 2.6) [20, 25, 93–96] to describe the drop formation in any IPN system. Both equations give nearly the same numerical results for many

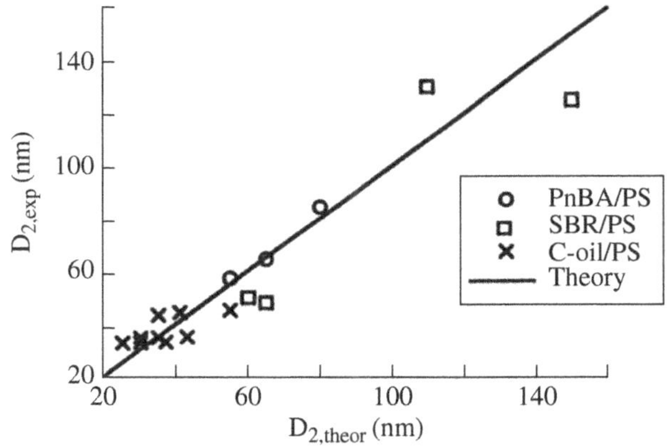

FIGURE 2.16 Measured versus calculated diameters of dispersed drops in IPNs; poly(cross-(*n*-butyl acrylate))-inter-poly(cross-styrene) (PnBA/PS) [25, 94–96], styrene butadiene rubber–polystyrene (SBR/PS), and castor oil–polystyrene (C-oil/PS) [25, 97]

systems of interest and both agree substantially with experimental results, as shown in Figure 2.16 [20, 25, 94–97]. Although both equations assume spheres of polymer II embedded in a continuous matrix of polymer I, they provide the approximate corresponding dimension of the cylinders [20].

$$D_2 = \frac{2\gamma\upsilon_2}{RTv_1\upsilon_1\left[1/v_1^{2/3} - \frac{1}{2}\right]} \tag{2.5}$$

$$D_2 = \frac{4\gamma}{RT\left(Av_1 + Bv_2\right)} \tag{2.6}$$

where $A = \frac{1}{2}(1/v_2)\left(3\upsilon_1^{1/3} - 3\upsilon_1^{4/3} - \upsilon_1 \ln\upsilon_1\right)$ and $B = \frac{1}{2}\left(\ln\upsilon_2 - 3\upsilon_2^{2/3} + 3\right)$

Formation sequence is a determine factor which influences the morphology of IPNs [19, 98–103]. IPNs based on cellulose acetate butyrate (CAB) and α,ω-divinyl-poly(dimethyl siloxane) (PDMS) have been synthesized [104, 105]. Two sIPNs, linear CAB in crosslinked PDMS and linear PDMS in crosslinked CAB, and PDMS/CAB IPNs with two crosslinking networks were prepared. Through DMA analysis, T_g was detected as a detector of the phase separation. Both sIPNs exhibit phase-separated structure as it has been illustrated earlier. The continuous phase in the sIPNs is the crosslinked phase, which is formed first and that the non-crosslinked polymer is the dispersed phase [106, 107], while in the PDMS/CAB IPNs, phase-separated morphology is controlled by a complex interplay between the polymerization/crosslinking reaction rates and the occurrence of the phase separation and varies as functions of catalyst content (dibutyltindilaurate (DBTDL) for CBA crosslinking) and system weight composition. As shown in Figure 2.17, the formation of the two crosslinking networks must be well matched in order to obtain a transparent material showing only one mechanical relaxation by DMA analysis. For different compositions, macrophase separation can be avoided by controlling the formation rate of silicone and CAB networks, for example, the crosslinking of the CAB network must be sufficiently advanced and at least 40% conversion of the NCO functions before the PDMS gel point (plateau in Figure 2.17).

Phase separation is also greatly influenced by IPN weight composition, as shown in Figure 2.18. PDMS/CAB (75/25) IPN displays two broad relaxations at maximum 50 and 5°C, respectively, which indicates interpenetrated polymer domains with various composition as a rich PDMS phase containing CAB and a rich CAB one containing PDMS, respectively, while only one mechanical relaxation is detected for PDMS/CAB (50–50) and (25–75) IPNs, located at 49 and 70°C, respectively. In this case, phase separation is considerably limited during the IPN synthesis. These IPNs exhibit some characteristics of homo-IPNs [107].

Simultaneous Interpenetrating Polymer Networks In SINs, the overall morphology is determined by the degree of mixing, which is controlled by the balance between the thermodynamics and kinetics. With the polymerization of the SIN components, molecular weights of both crosslinked networks increase and their miscibility decreases. Covalent bonds make diffusion and subsequent phase separation cannot occur after crosslinking. If substantial crosslinking of both components occurs before significant diffusion can occur, phase separation would have been largely prevented and a high degree of mixing (close to one-phase morphology) would obtain. In other words, thermodynamic equilibrium would be kinetically limited [19].

Many research works focus on kinetics of the IPN formation and its influence on the obtained morphologies. The effects of catalyst concentration, reaction temperature, components ratio, and the amount of crosslinkers on the formation of IPNs were investigated [25, 27, 108–119].

Klempner, Frisch, and coworkers [19, 44, 119] have thoroughly investigated polyurethane/polyacrylic ester (PU/PAE) SIN systems. They show that increasing the catalyst concentration decreases the gelation time as expected (Table 2.4). They also observe the increased inward shifts in T_g (Table 2.5),

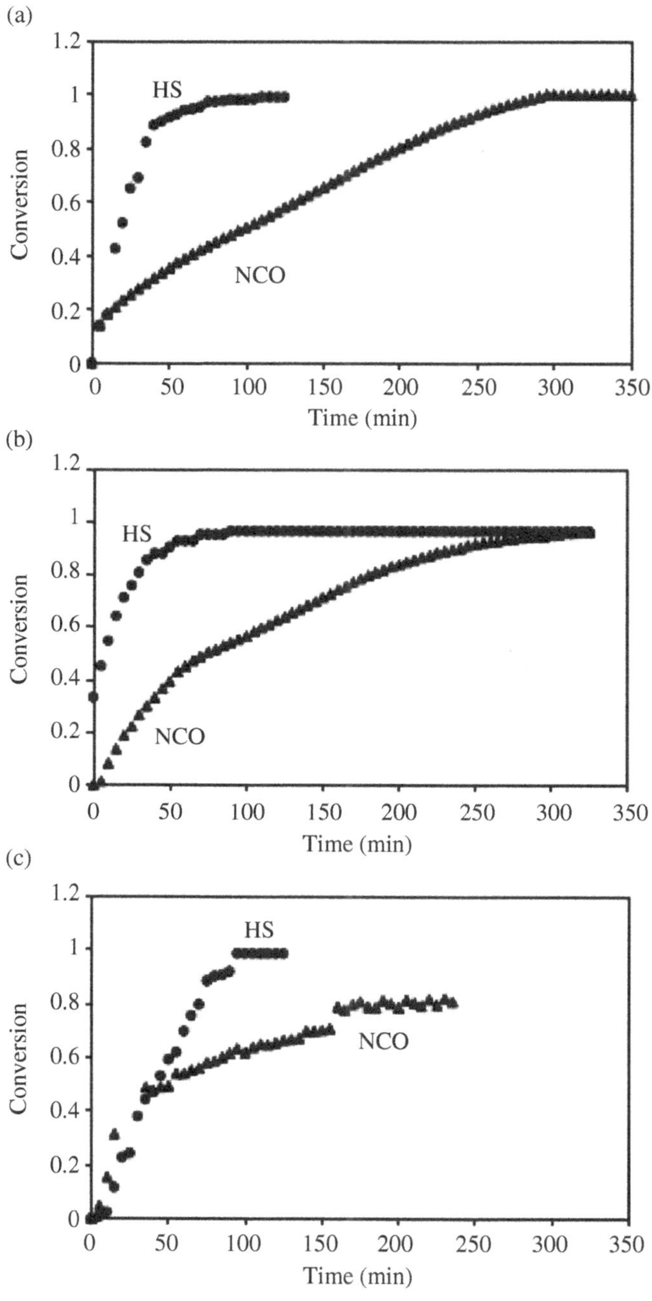

FIGURE 2.17 Thiol and NCO group conversions during the synthesis of (a) PDMS/CAB (75–25) IPN with [DBTDL]/[OH] = 0.075; (b) PDMS/CAB (50–50) IPN with [DBTDL]/[OH] = 0.050; and (c) PDMS/CAB (25–75) IPN with [DBTDL]/[OH] = 0.025, [NCO]/[OH] = 0.60, [AIBN]/[CaC] = 0.05, [HS]/[CaC] = 1.1, T = 50°C. Source: Fichet *et al.* [104]. Reproduced with permission of Elsevier

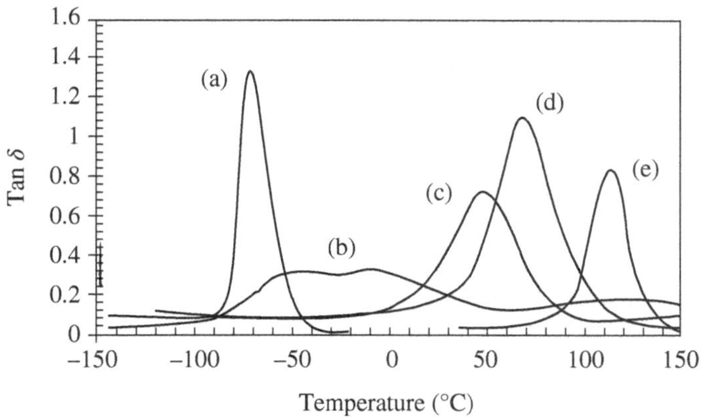

FIGURE 2.18 Tan δ versus temperature of PDMS/CAB IPNs with different weight compositions (b) 75/25; (c) 50/50 and (d) 25/75; PDMS (a) and CAB (e) single network curves are reported for comparison. Source: Fichet *et al.* [104]. Reproduced with permission of Elsevier

TABLE 2.4 Effect of Catalyst Concentration on the Gel Times of Crosslinked Polymers [19, 119]

Component	Catalyst (%)	Gel Time (min)[a]
Component I		
AC-1	BPO, 0.5	4.25
AC-2	BPO, 1.0	4.00
AC-3	BPO, 2.0	3.2
Component II		
PU-1	T-12, 0.02	12.3
PU-2	T-12, 0.1	7.25
PU-3	T-12, 0.5	3.15

[a] At 110°C.

which indicates reduction in the extent of phase separation (i.e., increased mixing) and increased simultaneity of the independent polymerization during IPN formation. This is due to the fact that the increased reaction rates lower the time available for diffusion and subsequent phase separation. When fully crosslinked, further phase separation cannot take place due to topological constraints imposed by the crosslinking of the component polymers (Table 2.6).

Segregation degree is proposed by the Lipatov group [22, 23, 42, 74, 120] to indicate the fraction of the system mass that is phase separated: $\alpha=1$ for fully separated systems and $\alpha=0$ for miscible systems. For most IPNs, α is in the range of 0.3–0.6, which means that a large fraction of the system mass is

TABLE 2.5 Effect of Catalyst Concentration on the Glass Transition Temperature of IPNs [19, 119]

		Glass Transition Temperature T_g (K)			
IPN	Composition of IPN	AC-40/PU-60		AC-60/PU-40	
1.1	AC-1/PU-1	212.2	318.5	212	319
1.2	AC-1/PU-2	211	318	210	318
1.3	AC-1/PU-3	215	314	214	314
2.1	AC-2/PU-1	212	317	211	318
2.2	AC-2/PU-2	212.5	316	214	315.5
2.3	AC-2/PU-3	216	313	215	316
3.1	AC-3/PU-1	211	314	211.5	316
3.2	AC-3/PU-2	214	308	215	307
3.3	AC-3/PU-3	219	293	220	298

TABLE 2.6 Effect of Temperature and Composition in sIPN Poly(Styrene-*co*-Divinylbenzene)/PBMA on the Interdiffusion Coefficients (D) and Microheterogeneity Region Sizes (μm) [19, 87]

	$-D \times 10^{13}$ (cm^2/s)		λm (μm)	
Temp. (K)	Stage 1	Stage 2	Stage 1	Stage 2
10 wt% PBMA				
333	13	50	1.2	1.1
343	22.5	56	1.2	1.0
363	61.5	113	1.0	1.4
20 wt% PBMA				
333	35	46	1.6	1.4
343	42	54	1.5	1.1

preserved in the unseparated state and distributed between two phases. Segregation degree strongly depends on the kinetic conditions and can be calculated by T_g. As shown in Figure 2.19, in a two-phase IPN system, dynamic mechanical losses would become closer as compared with the completely phase-separated system (case a) and the fully miscible system (case e). A semiempirical equation (Equation 2.7) is proposed to calculate the extent of phase separation:

$$\alpha = \frac{h_1 + h_2 - \left(l_1 h_1 + l_2 h_2 + l_m h_m\right) / L}{\left(h_1^0 + h_2^0\right)} \tag{2.7}$$

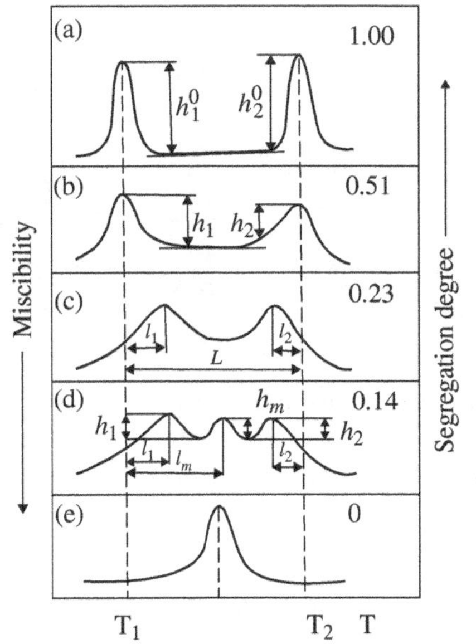

FIGURE 2.19 Temperature dependence of mechanical loss at various degrees of component segregation. (a–c) No mixed phase peak, (d) partially miscible, and (e) fully miscible [23, 120, 121]

where l_1 and l_2 represent the temperature shift of the lower and upper glass transitions, respectively; h_1 and h_2 represent the intensities of the transitions above the plateau region between the transitions; L is the initial temperature difference between the peaks; the superscript zero refers to the pure components; and the subscript m refers to the temperature and intensity of an intermediate mixed phase peak, if any.

For the simultaneous IPNs, Vipatov [22, 122, 123] reviewed phase separation and its relationship with the kinetics of network formation. The following conclusions provide a basis for the later investigations of IPN morphology. The domain dimensions are a function of the curing rates and of the time of reaching gel point (e.g., the onset of phase separation time). The simultaneous formation of two incompatible networks from the homogeneous mixture of initial components frustrates phase separation and promotes the seizing of small amounts of one component by large regions of another. The crosslinking density greatly influences the domain size. The ratio of the length to the width of domains also increases with the amount of crosslinking agent. The smallest particles appear if the phase separation occurs near or after the gel point (e.g., simultaneous curing).

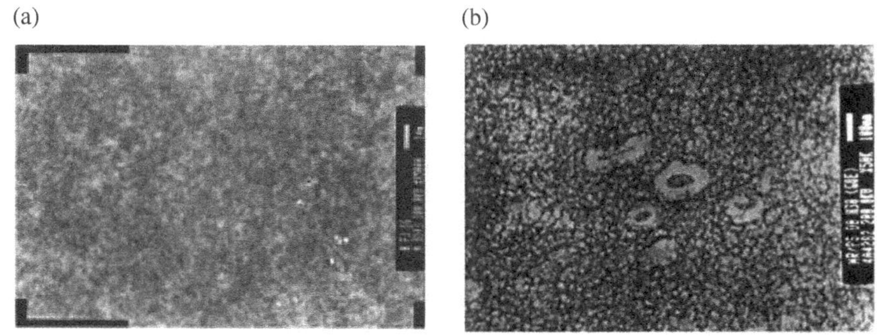

FIGURE 2.20 TEM images of dual-phase-separated morphology in IPNs. (a) Epoxy/UAR 50/50 SIN. Source: Hua and Hu [128]. Reproduced with permission of Elsevier. (b) NR/PS IPN. Source: Mathew *et al.* [129]. Reproduced with permission of Elsevier

2.4.2.3 Dual-Phase Continuity Dual-phase continuity is a special and important morphology of IPNs. It renders the IPNs with nanoscale mixing of the two components, resulting in excellent properties, such as mechanical and conductive properties, sound and vibration damping properties, and so on. As it has been discussed in the phase diagram section, dual-phase continuity is defined as the continuity of both polymer I and polymer II domains throughout the macroscopic sample [20]. The semiempirical model developed by Sperling related dual-phase continuity with the product of the volume fraction ratio and the viscosity ratio of the two phases. Here in this section, some applications with dual-phase continuity are reviewed.

IPNs technique is useful to prepare reinforced polymer materials. For different types of IPNs, including simultaneous IPNs and sequential IPNs, dual-phase continuous morphology can be observed providing suitable conditions, such as composition, rate of network formation, and so on [124–127]. Figure 2.20 shows simultaneous and sequential IPN examples with dual-phase continuous morphology. Figure 2.20a is simultaneous IPNs composed of epoxy resin and urethane acrylate resin (UAR). Although two T_gs for different compositions indicate the phase-separated morphology, the miscibility between the two networks for weight ratio 50/50 SIN is really much better than those SINs with other compositions. The dual-phase continuous morphology shows the best reinforce effect. That is, on one hand, the epoxy SIN network could be toughened obviously by UAR network. The impact strength increases appreciably and the tensile strength is very close to that of the pure epoxy network. On the other hand, the UAR network is greatly reinforced by the epoxy network. Figure 2.20b is the

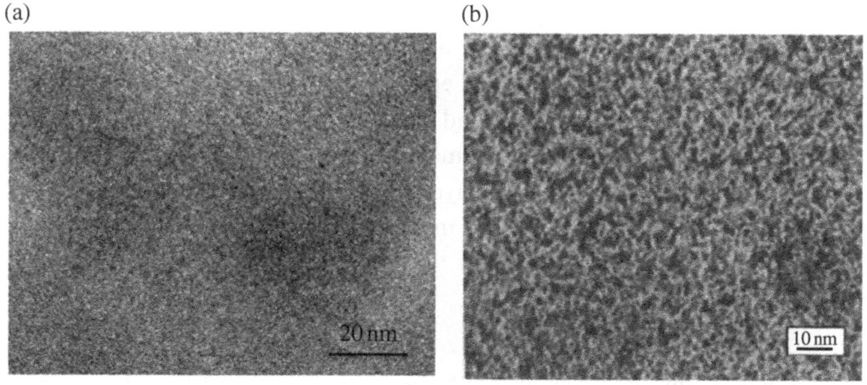

FIGURE 2.21 TEM images of dual-phase-separated morphology in (a) BPPO/SPPO sIPN (Source: Fang *et al.* [137]. Reproduced with permission of Elsevier) and (b) Nafion membranes (Source: Smitha *et al.* [138]. Reproduced with permission of Elsevier)

sequential IPNs prepared by first crosslinking of natural rubber (NR), and then polymerization and crosslinking of polystyrene (PS). The IPNs with the composition of 1% dicumyl peroxide (DCP) initiator, 30% NR, and 4% DVB crosslinker exhibited nano-structured morphology (10–15 nm), where the two phases are intimately mixed with phase separation. The obtained IPN material exhibited highest tensile strength (16.66 ± 0.49 MPa), hardness (97 ± 0.8), and tear strength (110.1 ± 1.21 N/mm).

Ionic exchange membrane is the key component in the newly electrochemical devices such as fuel cell, redox flow battery, electrochemical hydrogen pump and reactor, electrodialysis, and electrolysis. Nafion™ is the most popularly commercial cation exchange membrane, which has a highly hydrophobic polytetrafluoroethylene (PTFE) backbone and highly hydrophilic grafted sulfonic acid side chains. Nafion exhibits a dual-phase continuous morphology, in which the hydrophilic ionic clusters form interconnected pathway in the hydrophobic polymer backbone. Although the shape of the ionic cluster is still debatable (spherical, rod-like, laminar) [130, 131], the dual-phase continuous morphology renders Nafion with excellent proton conductivity, as well as high mechanical stability. Later, nonfluorinated ionic exchange membranes are developed to overcome the withdraws of Nafion, such as high crossover and low T_g. IPNs are viewed as promising materials because of their ability to force compatibility and to form dual-phase continuous morphology similar to Nafion. Several nice reviews about IPN's proton and hydroxide ion exchange membranes for fuel cell applications are published [132–136].

Many IPNs reported have the dual-phase continuous morphology. As shown in Figure 2.21 [137, 138], well connected hydrophilic channels are found in the BPPO (poly(2,6-dimethyl-1,4-phenylene oxide)/SPPO (sulfonated PPO) sIPN membranes, which fabricated by thermally crosslinking aminated BPPO with different epoxide crosslinkers in the presence of SPPO. The crosslinking structure and hydrophobicity are found to impact the membrane morphology strongly. Smaller and more hydrophobic crosslinks form narrow and well-connected hydrophilic channels, which can support facile proton transport and suppress methanol crossover.

ACKNOWLEDGMENTS

The authors thank the Program for National Science Fund for Distinguished Young Scholars of China (Grant no. 21125628), the National Science Foundation of China (Grant nos. 20976027 and 21176044), and the Fundamental Research Funds for the Central Universities of China (DUT13ZD204) for financial supports of this work.

REFERENCES

[1] Greenhill, D.A., Hourston, D.J. and Waters, J.A. (1990) Chapter 22 Synthesis of Composite Latex Particles and Their Role in Sound and Vibration Damping, pp397–414. in *Sound and Vibration Damping with Polymers* (eds R.D. Corsaro and L.H. Sperling), American Chemical Society, Washington, DC.

[2] Huelck, V., Thomas, D.A. and Sperling, L.H. (1972) *Macromolecules*, **5**, 340–348.

[3] Sperling, L.H., Chiu, T.W., Gramlich, R.G. and Thomas, D.A. (1974) *Journal of Paint Technology*, **46**, 47.

[4] Sperling, L.H. (1981) *Interpenetrating Polymer Networks and Related Materials*, Plenum, New York.

[5] Simic, S., Dunjic, B., Tasic, S. *et al.* (2008) *Progress in Organic Coatings*, **63**, 43–48.

[6] Marinovic, S., Popovic, I., Dunjic, B. *et al.* (2010) *Progress in Organic Coatings*, **68**, 293–298.

[7] Asthana, K.K., Aggarwal, L.K. and Lakhani, R. (1999) *Cement and Concrete Research*, **29**, 1541–1548.

[8] Cleary, D.B. and Ramirez, J.A. (1989) *ACI Materials Journal*, **86** (2), 167–174.

[9] Wu, X., Chen, W., Yan, X. *et al.* (2014) *Journal of Materials Chemistry A*, **2**, 12222–12231.

[10] Gu, S., Cai, R., Luo, T. *et al.* (2009) *Angewandte Chemie International Edition*, **48**, 6499–502.

[11] Kim, D.S., Labouriau, A., Guiver, M.D. and Kim, Y.S. (2011) *Chemistry of Materials*, **23**, 3795–3797.

[12] Wang, J., He, G., Wu, X. *et al.* (2014) *Journal of Membrane Science*, **459**, 86–95.

[13] Wu, X., Wang, X., He, G. and Benziger, J. (2011) *Journal of Polymer Science Part B: Polymer Physics*, **49**, 1437–1445.

[14] Yan, X., He, G., Wu, X. and Benziger, J. (2013) *Journal of Membrane Science*, **429**, 13–22.

[15] Gu, S., Skovgard, J. and Yan, Y. (2012) *ChemSusChem*, **5**, 843–848.

[16] Miyata, T. (2002) *Gel and Interpenetrating Polymer Networks*, CRC Press LLC, Florida.

[17] Dong, Z., Kennedy, S.J. and Wu, Y. (2011) *Journal of Power Sources*, **196**, 4886–4904.

[18] Farris, S., Schaich, K.M., Liu, L. *et al.* (2009) *Trends in Food Science & Technology*, **20**, 316–332.

[19] Suthar, B., Xiao, H.X., Klempner, D. and Frisch, K.C. (1996) *Polymers for Advanced Technologies*, **7**, 221–233.

[20] Sperling, L.H. (1994) in *Interpenetrating Polymer Networks*, Advances in Chemistry (eds L.H. Sperling, D. Klempner and L.A. Utracki), American Chemical Society, Washington, DC, pp. 3–38.

[21] Sperling, L.H. and Mishra, V. (1996) *Polymers for Advanced Technologies*, **7**, 197–208.

[22] Lipatov, Y.S. and Alekseeva, T.T. (2007) *Advances in Polymer Science*, **208**, 1–227.

[23] Lipatov, Y.S. (1994) in *Interpenetrating Polymer Networks*, Advances in Chemistry (eds L.H. Sperling, D. Klempner and L.A. Utracki), American Chemical Society, Washington, DC, pp. 125–139.

[24] Lipatov, Y.S. (2002) *Progress in Polymer Science*, **27**, 1721–1801.

[25] Utracki, L.A. (1994) in *Interpenetrating Polymer Networks*, Advances in Chemistry (eds L.H. Sperling, D. Klempner and L.A. Utracki), American Chemical Society, Washington, DC, pp. 77–123.

[26] Frisch, H.L. and Zhou, P. (1994) in *Interpenetrating Polymer Networks*, Advances in Chemistry (eds L.H. Sperling, D. Klempner and L.A. Utracki), American Chemical Society, Washington, DC, pp. 269–283.

[27] Sperling, L.H. (2011), ACS Symposium Series, American Chemical Society, Washington, DC, pp. 69–82.

[28] Nesterov, A.E. and Lipatov, Y.S. (1997) *Thermodynamics of Polymer Blends*, ChemTec, Toronto.

[29] Flory, P.J. (1953) *Principles of Polymer Chemistry*, Cornell University Press, New York.

[30] Krevelen, D.W.V. (2008) *Properties of Polymers*, Elsevier, Oxford.

[31] Utracki, L.A. (1989) *Polymer Alloys and Blends*, Hanser, Munich.

[32] Patterson, D.D. and Robard, A. (1978) *Macromolecules*, **11**, 690.

[33] Huggins, M.L. (1976) *The Journal of Physical Chemistry*, **80**, 1317.

[34] Ratzseh, M.T. and Kehlen, H. (1989) *Progress in Polymer Science*, **14**, 1.

[35] Walker, J.S. and Vause, C.A. In Eighth Symposium Thermodynamic Properties, J.V. Sengers, ed. American Society of Mechanical Engineers, New York, 1982; Vol. **1**, p. 411.

[36] Lipatov, Y.S., Karabanova, L.V., Gorbach, L.A. *et al.* (1992) *Polymer International*, **28**, 99.

[37] Mukhopadhyay, P. and Schreiber, H.P. (1994) in *Aspects of polymer surface characterization by inverse gas chromatography*, Journal of Polymer Science Part B Polymer Physics, **32**(9): pp. 1653–1656.

[38] Lipatov, Y.S., Nesterov, A.E. and Sergeeva, L.M. (1974) *Doklady Akademii Nauk SSSR*, **220**, 637.

[39] Lipatov, Y.S., Sergeeva, L.M., Karabanova, L.V. *et al.* (1976) *Vysokomol Soedin A*, **18**, 1025 in Russian.

[40] Klempner, D., Wang, C.L., Ashtiani, M. and Frisch, K.C. (1986) *Journal of Applied Polymer Science*, **32**, 4197.

[41] Sperling, L.H. (1986) in *Multi-Component Polymer Materials*, ACS 211 (eds D.R. Paul and L.H. Sperling), American Chemical Society, Washington, DC.

[42] Lipatov, Y.S., Alekseeva, T.T. and Shilov, V.V. (1991) *Polymer Networks and Blends*, **1**, 129.

[43] Lipatov, Y.S., Babich, V.F., Karabanova, L.V. *et al.* (1976) *Representation of Academy of Science Ukrainian SSR*, **B1**, 39.

[44] Sperling, L.H. (1992) Chain Structure and Configuration, in *Introduction to Physical Polymer Science*, 2nd edn (eds L.H. Sperling, S.L. Cooper and A. Richard), Wiley-Interscience, New York, pp. 29–70.

[45] Sperling, L.H., Heck, C.S. and An, J.H. (1989) in *Multiphase Polymers: Blends and lonomers*, ACS Symposium 395 (eds L.A. Utracki and R.A. Weiss), American Chemical Society, Washington, DC.

[46] Fay, J.J., Murphy, C.J., Thomas, D.A. and Sperling, L.H. (1990) in *Sound and Vibration of Polymers* (eds R.D. Corsaro and L.H. Sperling), American Chemical Society, Washington, DC.

[47] Lipatov, Y.S. and Alekseeva, T.T. (1992) *Progress in Chemistry*, **61**, 2187 in Russian.

[48] Sophia, D., Klempner, D., Suthar, B. and Frisch, K.C. (1994) in *Interpenetrating Polymer Networks*, Advances in Chemistry Series 239 (eds K. Klempner, L.H. Sperling and L.A. Utracki), American Chemical Society, Washington, DC.

[49] Millar, J.R. (1960) *Journal of the Chemical Society*, **3**, 1311.

[50] Park, J.W. and Kim, S.S. (1996) *Polymers for Advanced Technologies*, **7**, 209.

[51] Barrett, L.W., Sperling, L.H., Gilmer, J.W. and Mylonakis, S.G. (1994) in *Interpenetrating Polymer Networks*, ACS Books (eds D. Klempner, L.H. Sperling and L.A. Utracki), American Chemical Society, Washington, DC.

[52] An, J.H. and Sperling, L.H. (1988) in *Cross-Linked Polymers: Chemistry, Properties, and Applications*, ACS Symposium 367 (eds R.A. Dickie, S.S. Labana and R.S. Bauer), American Chemical Society, Washington, DC.

[53] Sperling, L.H., An, J.H., Chang, M. and Thomas, D.A. (1988) in *International Symposium on Polymers for Advanced Technologies* (ed M. Lewin), VCH, New York.

[54] Lipatov, Y.S., Grigoryeva, O.P., Kovernik, G. *et al.* (1985) *Makromolekulare Chemie*, **186**, 1401.

[55] Lipatov, Y.S., Nesterov, A.E., Kuzmina, G.P. and Ignatova, T.D. (1993) *Polymer Networks and Blends*, **3**, 89.

[56] Mishra, V., Prez, F.D., Gosen, E. *et al.* (1995) *Journal of Applied Polymer Science*, **58**, 331.

[57] Sperling, L.H. and Mischra, V. (1996) *Polymers for Advanced Technologies*, **76**, 197.

[58] Jordhamo, G.M., Manson, J.A. and Sperling, L.H. (1986) *Polymer Engineering and Science*, **26**, 517.

[59] Paul, D.R. and Barlow, J.W. (1980) *Journal of Macromolecular Science: Chemistry*, **C18**, 109.

[60] Metelkin, V.I. and Blecht, V.S. (1984) *Kolloidnyi Zhurnal*, **46**, 476.

[61] Utracki, L.A. (1991) *Journal of Rheology*, **35**, 1615.

[62] Termonia, Y. (1991) *Macromolecules*, **24**, 1392.

[63] Sperling, L.H. and Widmaier, J.M. (1983) *Polymer Engineering and Science*, **23**, 694.

[64] Widmaier, J.M. and Sperling, L.H. (1982) *Journal of Applied Polymer Science*, **27**, 3513.

[65] Widmaier, J.M. and Sperling, L.H. (1982) *Macromolecules*, **15**, 625.

[66] Widmaier, J.M., Yeo, J.K. and Sperling, L.H. (1982) *Colloid and Polymer Science*, **260**, 678.

[67] Frisch, H.L. and Wasserman, E. (1961) *Journal of the American Chemical Society*, **83**, 3789.

[68] Lipatov, Y.S. and Shilov, V.V. (1984) *Uspekhi Khimii*, **53**, 1197.

[69] Lipatov, Y.S., Grigoryeva, O.P., Shilov, V.V. and Sergeeva, L.M. (1984) *Doklady Akademii Nauk SSSR*, **278**, 409.

[70] Lipatov, Y.S., Rosovitsky, V.F., Maslak, Y.V. and Zakharenko, S.A. (1984) *Doklady Akademii Nauk SSSR*, **278**, 1408.

[71] Lipatov, Y.S., Grigoryeva, O.P., Sergeeva, L.M. and Shilov, V.V. (1986) *Vysokomol Soedin A*, **28**, 385.

[72] Lipatov, Y.S. (1983) *Mechanics of Composite Materials*, **5**, 771.

[73] Lipatov, Y.S., Shilov, V.V., Gomza, Y.P. and Kovernik, G.P. (1984) *Makromolekulare Chemie*, **185**, 947.

[74] Lipatov, Y.S. (1989) in *Advances in Interpenetrating Polymer Networks* (eds D. Klempner and K. Frisch), Technomic, Lancaster, PA, p. 261.

[75] Alekseeva, T.T., Lipatova, T.E. and Lipatov, Y.S. (1995) *Journal of Polymer Science: Part A*, **37**, 1194.

[76] Kim, S.C., Klempner, D., Frisch, K.C. *et al.* (1976) *Macromolecules*, **9**, 258.

[77] Wu, X., He, G., Gu, S. *et al.* (2007) *Journal of Membrane Science*, **295**, 80–87.

[78] Wu, X., He, G., Li, X. *et al.* (2014) *Journal of Power Sources*, **246**, 482–490.

[79] Wu, X., He, G., Gu, S. *et al.* (2010) *Chemical Engineering Journal*, **156**, 578–581.

[80] Wu, X., He, G., Gao, L. *et al.* (2006) *Chinese Chemical Letters*, **17**, 965–968.

[81] Frisch, H.L., Zhou, P., Frisch, K.C. *et al.* (1991) *Journal of Polymer Science, Polymer Chemistry Edition*, **29**, 1031–1038.

[82] Frisch, H.L. and Zhou, P. (1992) *Journal of Polymer Science, Polymer Chemistry Edition*, **30**, 2794.

[83] Zhou, P. and Frisch, H.L. (1992) *Journal of Polymer Science, Polymer Chemistry Edition*, **30**, 835.

[84] Zhou, P. and Frisch, H.L. (1992) *Journal of Polymer Science, Polymer Chemistry Edition*, **30**, 887.

[85] Frisch, H.L. (1990) *Materials Research Society Symposium Proceedings*, **171**, 231.

[86] de Barros, G.G., Huang, M.W. and Frisch, H.L. (1992) *Journal of Applied Polymer Science*, **44**, 255.

[87] Sanchez, M.S., Ferrer, G.G., Cabanilles, C.T. *et al.* (2001) *Polymer*, **42**, 10071–10075.

[88] Fernandez, A.M. (1984) Ph.D. thesis. Lehigh University.

[89] Wang, Y. and Rubner, M.F. (1992) *Macromolecules*, **25**, 3284–3290.

[90] Donatelli, A.A., Sperling, L.H. and Thomas, D.A. (1977) *Journal of Applied Polymer Science*, **21**, 1189.

[91] Yeo, J.K., Sperling, L.H. and Thomas, D.A. (1983) *Polymer*, **24**, 307.

[92] Michel, J., Hargest, S.C. and Sperling, L.H. (1981) *Journal of Applied Polymer Science*, **26**, 743.

[93] Yeo, J.K., Sperling, L.H. and Thomas, D.A. (1981) *Polymer Engineering and Science*, **21**, 696.

[94] Yeo, J.K., Sperling, L.H. and Thomas, D.A. (1982) *Polymer Engineering and Science*, **22**, 190.

[95] Yeo, J.K., Sperling, L.H. and Thomas, D.A. (1981) *Journal of Applied Polymer Science*, **26**, 3283.

[96] Yenwo, G.M., Sperling, L.H., Pulido, J. *et al.* (1977) *Polymer Engineering and Science*, **17**, 251.

[97] Vancaeyzeele, C., Fichet, O., Laskar, J. *et al.* (2006) *Polymer*, **47**, 2046–2060.

[98] Nowers, J.R. and Narasimhan, B. (2006) *Polymer*, **47**, 1108–1118.

[99] Andrzejewska, E. (2001) *Progress in Polymer Science*, **26**, 605–665.

[100] Dusek, K. and Duskova-Smrckova, M. (2000) *Progress in Polymer Science*, **25**, 1215–1260.

[101] Widmaier, J.M. and Drillieres, S. (1996) *Journal of Applied Polymer Science*, **63**, 951–958.

[102] Zhou, P. and Frisch, H.L. (1994) *Macromolecules*, **27**, 1788–1794.

[103] Vidal, F., Fichet, O., Laskar, J. and Teyssie, D. (2006) *Polymer*, **47**, 3747–3753.

[104] Fichet, O., Vidal, F., Laskar, J. and Teyssie, D. (2005) *Polymer*, **46**, 37–47.

[105] Klempner, D., Sperling, L.H. and Utracki, L.A. (1994) *Interpenetrating Polymer Networks*, Advances in Chemistry 239, American Chemical Society, Washington, DC.

[106] Sperling, L.H. and Mishra, V. (1997) in *IPNs around the World: Science and Engineering* (eds S.C. Kim and L.H. Sperling), John Wiley & Sons, Inc., New York, pp. 1–25.

[107] Touhsaent, R.E., Thomas, D.A. and Sperling, L.H. (1974) *Journal of Polymer Science*, **46C**, 175.

[108] Stroeks, A. and Nies, E. (1990) *Macromolecules*, **23**, 4092.

[109] John, J., Suriyakala, R., Thomas, S. *et al.* (2010) *Journal of Materials Science*, **45**, 2892–2901.

[110] Zygadło-Monikowska, E., Florjanczyk, Z., Wielgus-Barry, E. and Pasniewski, J. (2006) *Journal of Power Sources*, **159**, 385–391.

[111] Lungu, A., Florea, N.M. and Iovu, H. (2012) *Polymer*, **53**, 300–307.

[112] Jia, Q., Zheng, M., Chen, H. and Shen, R. (2006) *Materials Letters*, **60**, 1306–1309.

[113] Hashimoto, T. (1987) in *Current Topics in Polymer Science* (eds R. Ottenbrite, L.A. Utracki and S. Inoue), Carl Hanser Verlag, Munich.

[114] Fu, R., Hong, L. and Lee, J.Y. (2008) *Fuel Cells*, **8**, 52–61.

[115] Hashimoto, T. (1988) *Phase Transitions*, **12**, 47.

[116] Larbi, F.B.C., Halary, J.L. and Monnerie, L. (1991) *Macromolecules*, **24**, 867.

[117] Wen, X., Cao, X., Yin, Z. *et al.* (2009) *Carbohydrate Polymers*, **78**, 193–198.

[118] Voigt-Martin, I.G., Leister, K.H., Rosenau, R. and Koningsveld, R. (1986) *Journal of Polymer Science Polymer Physics Edition*, **24**, 723.

[119] Klempner, D., Yoon, H.K., Frisch, K.C. and Frisch, H.L. (1980) in *Polymer Alloys II* (eds D. Klempner and K.C. Frisch), Plenum Press, New York, p. 185.

[120] Rosovitsky, V.F. and Lipatov, Y.S. (1985) *Doklady Akademii Nauk SSSR*, **283**, 910.

[121] Lipatov, Y.S. (1985) *Pure and Applied Chemistry*, **57**, 1691.

[122] Sperling, L.H. (1977) *Macromolecular Reviews*, **12**, 141.

[123] Touhsaent, R.E., Thomas, D.A. and Sperling, L.H. (1974) *Journal of Polymer Science Part C*, **46**, 175.
[124] Plesse, C., Vidal, F., Randriamahazaka, H. *et al.* (2005) *Polymer*, **46**, 7771–7778.
[125] Vidal, F., Plesse, C., Teyssie, D. and Chevrot, C. (2004) *Synthetic Metals*, **142**, 287.
[126] Randriamahazaka, H., Plesse, C., Vidal, F., *et al.* (2004) Proceedings of the SPIE Conference on Smart Structures and Materials, Vol. **5385**, p. 294.
[127] Paul, D. (1985) *Multicomponent Polymer Materials*, Advances in Chemistry, American Chemical Society, Washington, DC.
[128] Hua, F.J. and Hu, C.P. (2000) *European Polymer Journal*, **36**, 27–33.
[129] Mathew, A.P., Packirisamy, S., Radusch, H.J. and Thomas, S. (2001) *European Polymer Journal*, **37**, 1921–1934.
[130] Kong, X. and Schmidt-Rohr, K. (2011) *Polymer*, **52**, 1971–1974.
[131] Rohr, K.S. and Chen, Q. (2008) *Nature Materials*, **7**, 75–83.
[132] Couture, G., Alaaeddine, A., Boschet, F. and Ameduri, B. (2011) *Progress in Polymer Science*, **36**, 1521–1557.
[133] Chikh, L., Delhorbe, V. and Fichet, O. (2011) *Journal of Membrane Science*, **368**, 1–17.
[134] Merle, G., Wessling, M. and Nijmeijer, K. (2011) *Journal of Membrane Science*, **377**, 1–35.
[135] Rikukawa, M. and Sanui, K. (2000) *Progress in Polymer Science*, **25**, 1463–1502.
[136] Peckham, T.J. and Holdcroft, S. (2010) *Advanced Materials*, **22**, 4667–4690.
[137] Fang, C., Ni Toh, X., Yao, Q. *et al.* (2013) *Journal of Power Sources*, **226**, 289–298.
[138] Smitha, B., Sridhar, S. and Khan, A.A. (2005) *Journal of Membrane Science*, **259**, 10–26.

3

SYNTHETIC RUBBER-BASED IPNs

QIHUA WANG, SHOUBING CHEN AND TINGMEI WANG
State Key Laboratory of Solid Lubrication, Lanzhou Institute of Chemical Physics, Chinese Academy of Sciences, Lanzhou, China

3.1 INTRODUCTION

Synthetic rubber is any type of artificial elastomer that can undergo much more elastic deformation under stress than most materials and still return to its previous size without permanent deformation. Among the synthetic polymers in use today, synthetic rubber is unique in that it was developed not as an interesting new material but to fill a dire need of the modern world. The reason for this unique position of synthetic rubber is its unique property: the only substance that exhibits long-range elasticity [1]. Therefore, it can fill a special need in modern technology. It is widely used in industry, agriculture, defense, transportation, and daily life, serving as a substitute for natural rubber in many cases, especially when improved material properties are required.

Synthetic rubber can be made from the polymerization of a variety of monomers. The common synthetic rubbers are listed in Table 3.1. Synthetic rubber is usually divided into two categories: general-purpose rubbers and specialty rubbers. General-purpose rubbers mainly include styrene-butadiene rubber (SBR), polybutadiene rubber (BR), polyisoprene rubber (IR), ethylene

Micro- and Nano-structured Interpenetrating Polymer Networks: From Design to Applications, First Edition. Edited by Sabu Thomas, Daniel Grande, Uroš Cvelbar, K.V.S.N. Raju, Ramanuj Narayan, Selvin P. Thomas and Akhina H.

TABLE 3.1 List of Part Common Synthetic Rubbers

ISO 1629 Code[a]	Technical Name	Common Names
BR	Polybutadiene	Buna CB
CR	Polychloroprene	Chloroprene, neoprene
EP	Ethylene propylene	
EPDM	Ethylene propylene diene monomer	EPDM
FKM	Fluorinated hydrocarbon	Fluororubber
IR	Polyisoprene	(Synthetic) natural rubber
IIR	Isobutylene isoprene butyl	Butyl rubber
NBR	Acrylonitrile butadiene	NBR, nitrile rubber, Perbunan, Buna-N
PU	Polyurethane	PU, polyurethane
SBR	Styrene butadiene	SBR, Buna S, GRS, Buna VSL, Buna SE
SI	Polysiloxane	Silicone rubber
VMQ	Vinyl methyl silicone	Silicone rubber

[a]ISO 1629:1995—Rubber Test Standard.

propylene diene monomer (EPDM) rubber, chloroprene rubber (CR), and so on. Specialty rubbers mainly include nitrile rubber (NBR), butyl rubber (IIR), silicone rubber, fluororubber (FKM), PU rubber, and so on.

Interpenetrating polymer networks (IPNs) are polymer alloys consisting of two polymers in a network form, held together by permanent entanglements with only occasional covalent bonds between the chains of the two different types of polymers [2]. In the IPNs, at least one of them was polymerized and/or crosslinked in the immediate presence of the other. In most cases, phase separation occurs, resulting in a two-phase polymer system. The IPNs can be prepared by sequential polymerization, simultaneous polymerization, or a latex blending technique [3, 4].

The IPNs show excellent thermal stability and mechanical properties because of a synergistic effect induced by the forced compatibility of individual components. The IPN technology can conquer the disadvantages of sole polymer to obtain integrated high-performance polymer composites, and the properties of the IPNs can be designed ranging from a filler-reinforced elastomeric material, to a rubber-reinforced plastic material via used original polymers with different properties. Therefore, attention has been given to IPNs in recent years.

The crosslinking density of the IPN system strongly influences the phase morphology of the IPNs, and thus the physical and mechanical properties of a given IPN can be modified by varying the crosslinking density [3]. The crosslinking density is directly related to the composition of the IPNs. For the IPNs, of particular interest are systems based on brittle plastics and elastomers, which may give rise to tough IPNs. There is much information regarding the influence of composition and method of synthesis on mechanical and damping properties [5, 6].

Synthetic rubbers (including general-purpose rubbers and specialty rubbers) are artificial elastomers that have many types and uses: they can not only serve as substitute of natural rubber but also fill a special need in modern technology. Therefore, synthetic rubbers are often selected as rubber phases and the synthetic rubber-based IPNs are interested by many researchers and have had many important advances in recent years.

3.2 SYNTHETIC RUBBER-BASED IPNs

3.2.1 The Synthesis Methods of Synthetic Rubber-Based IPNs

As previously mentioned, the basic synthesis methods for the IPNs are sequential polymerization, simultaneous polymerization (shown in Figure 3.1) [4], or a latex blending technique:

1. *Sequential IPN (SIPN)*: Polymer A is swollen in polymer B, as well as its crosslinking agent, and initiator, polymerizing B *in situ.*
2. *Simultaneous IPN (SIN)*: Monomers A and B, and their respective crosslinking agents and initiators, are polymerized simultaneously by way of noninterfering modes.
3. *Latex IPN*: The polymers are made in the form of latexes. Thus, each particle constitutes a micro-IPN. Frequently, a core-shell structure develops. And after film formation, further crosslinking between the particles may take place.

For synthetic rubber-based IPNs, sequential polymerization method is the most frequently used. The forming of crosslinking rubbers needs relatively high temperature, a certain pressure, and special apparatus (curometer). Thus, the basic synthesis process of synthetic rubber-based IPNs is as follows [7–9]: synthesis of synthetic rubber network, swelling of the synthetic rubber network in monomer solution of polymer B, polymer B polymerization, and then the synthetic rubber-polymer B IPN formation.

3.2.2 General-Purpose Rubber-Based IPNs

3.2.2.1 SBR-Based IPNs SBR is the largest general-purpose synthetic rubber varieties that consist of styrene and butadiene. It has good abrasion resistance, good aging stability when protected by additives, and is widely used in car tires. Polystyrene (PS), a relatively low-cost thermoplastic material,

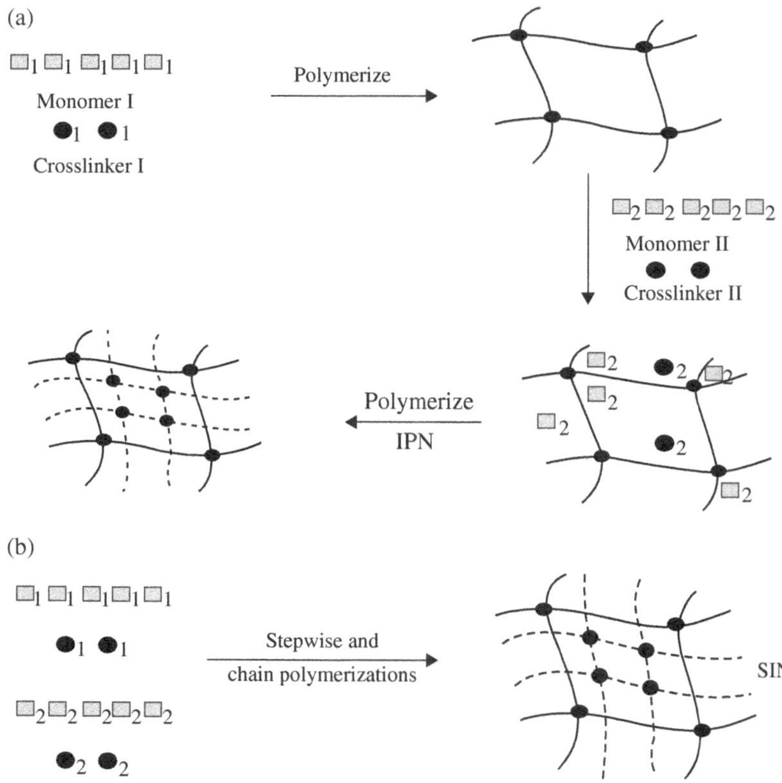

FIGURE 3.1 The basic synthesis methods for IPNs: (a) Sequential IPN and (b) simultaneous interpenetrating network. Source: Sperling and Mishra [4]. Reproduced with permission of John Wiley & Sons

having some excellent properties, has a prominent limitation in its brittleness. Therefore, combination with synthetic rubbers can improve the impact property of PS and thus complement the properties of each other.

Donatelli and coworkers [10, 11] prepared IPNs and semi-IPNs based on SBR as polymer A and PS as polymer B. The IPNs and semi-IPNs were synthesized by thermal polymerization techniques. The rubber phase for the IPN was prepared by dissolving the SBR in benzene, adding the appropriate amount of dicumyl peroxide (DiCup) for crosslinking, and then evaporating the solvent. The SBR was then cured in a compression molding operation. Then the obtained crosslinked SBR was swollen in the monomer of styrene before the styrene was polymerized thermally. The studies showed that the polymer synthesized first tends to control morphology and forms the more continuous phase when the polymerizing system is not stirred. The size of the cellular structure is a function of the degree of crosslinking of the polymer A component, with an increase in crosslinking, producing

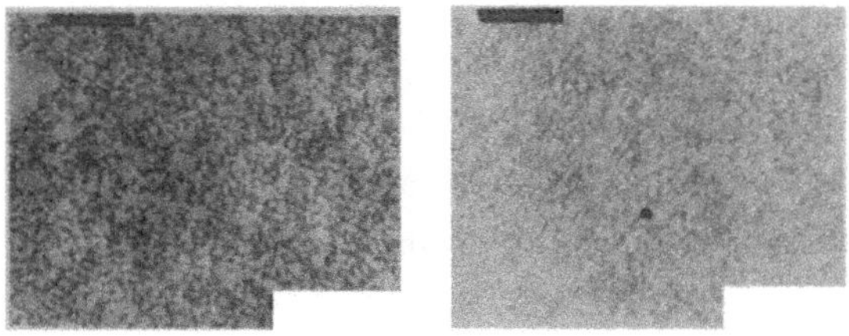

IPN, 0.1% Coss-linking IPN, 0.2% Coss-linking

FIGURE 3.2 Phase morphologies of SBR/PS IPN (20% SBR and 80% PS: the elastomer is stained dark with osmium tetroxide). Source: Donatelli, Sperling, and Thomas [10]. Reproduced with permission of American Chemical Society

(a) (b)

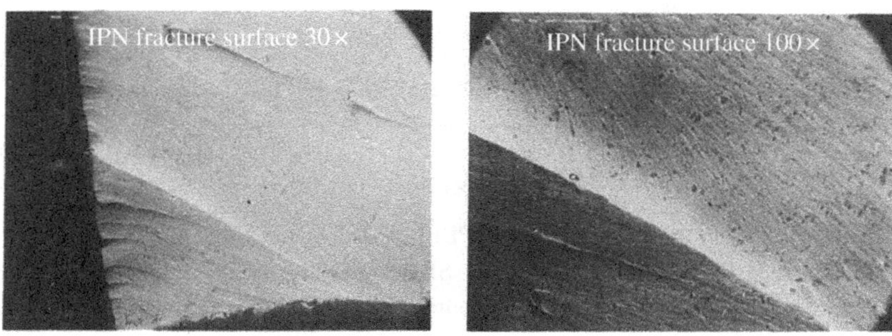

FIGURE 3.3 IPN (70% PS and 30% SBR) fracture surface (stained with OsO_4): (a) ×9.9 and (b) ×132. Source: Burford *et al.* [5]. Reproduced with permission of Springer

a finer structure (Figure 3.2). Dynamic mechanical behavior as a function of temperature shows two distinct transitions, which indicated that the SBR/PS system is incompatible. Yield strength and modulus decrease with increasing SBR content but increase with the degree of crosslinking of the rubber phase. Impact strength tends to increase with SBR content. The results presented the relation between crosslinking, composition, chemical compatibility, morphology, and their properties that directed the research about SBR/PS IPNs.

Burford and coworkers [5] described their work to determine the fracture toughness, K_c, of the IPNs based on PS and SBR. At high plastic levels, the material behaves in a similar manner to conventional rubber-toughened plastics, and the impact strength is superior to that of the unmodified plastic (Figure 3.3). Therefore, the determination of K_c is helpful to understand the mechanics of fracture of the elastomeric network.

Besides the previously discussed SBR/PS IPN systems, the synthesis and characterization of sequential IPN based on SBR-polyalkyl methacrylates [7] were performed by Patri *et al.* The IPNs based on SBR and polyalkyl (methyl,

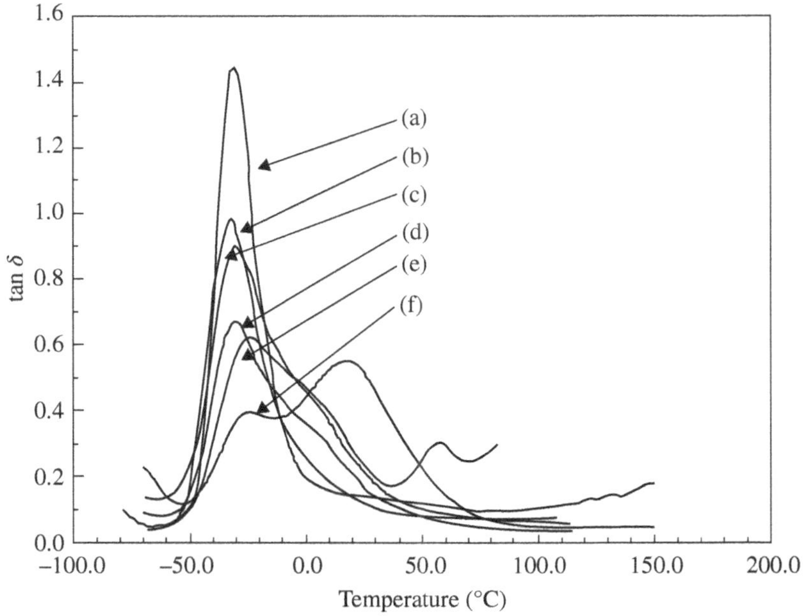

FIGURE 3.4 Damping plot of SBR/PBuMA IPNs: (a) SBR, (b) SBR/PBuMA (87/13), (c) SBR/PBuMA (80/20), (d) SBR/PBuMA (73/27), (e) SBR/PBuMA (72/28), and (f) SBR/PBuMA (57/42). Source: Patri *et al.* [7]. Reproduced with permission of John Wiley & Sons

ethyl, and butyl) methacrylates were prepared by using tetraethylene glycol dimethacrylate as crosslinker. The IPNs based on PMMA have shown higher tensile strength compared with others. Dynamic mechanical analysis (DMA) showed that the IPNs have superior damping properties than SBR (Figure 3.4). Because of IPN formation, tan δ peak shifted inward between SBR and acrylates when IPN was forming. Moreover, the peaks were broadened because of micro heterogeneous phase separation. At higher concentration of methacrylate, splitting in tan δ peak was noticed because of phase separation.

3.2.2.2 BR-Based IPNs BR is a synthetic rubber that is a polymer formed from the polymerization process of the monomer 1,3-butadiene. Cured BR imparts excellent abrasion resistance (good tread wear), and low rolling resistance (good fuel economy) due to its low glass transition temperature (T_g). It is the second largest volume synthetic rubber produced, next to SBR. Seventy percent of total production is used in tire compounds with another 20% being used for modifying plastics [12].

Ghosh and Ray [8] used BR to prepare BR/PS semi- and full-PNs by sequential polymerization method to modify PS, respectively. Dicumyl peroxide

(DCP) was used to form the initial PBR network. The semi-IPNs were formed by allowing the initially formed BR network to swell in styrene monomer and then completing polymerization of the monomer through the swollen network. For full-IPN formation, low doses of divinylbenzene (DVB) were used in the monomer styrene and copolymerization of the styrene-DVB system in the swollen BR network subsequently. Effects of variation of DCP dose, DVB dose, and PBR/PS blend ratio on the physico-mechanical properties of the blends were examined. Analysis of the micrographs (Figure 3.5) shows that the effect of variation in blend ratio points to a critical ranges of BR/PS blend ratio around 50 : 50 to 40 : 60. Around this blend ratio, the phase morphological pattern undergoes a notable change akin to phase inversion. For full IPNs when PS has high content (>50%), the PS phase distributed as continuous long ridge-like or fibrillar structures. For BR>50%, the elastomer phase is apparently the continuous phase. The full-IPNs exhibited higher tensile strength and modulus, while the semi-IPNs exhibited greater toughness.

Xue *et al.* [13] report on the synthesis and characterization of a novel simultaneous pseudo-IPN (semi-IPN) based on crosslinked CR and BR. They mixed proportional amounts of linear CR (potassium iodide pretreated in benzene and dimethyl sulfoxide-mixed solvent) and BR (which was dissolved in dry benzene) to give the desired compositions followed by adding the diamine crosslinking agent for CR. Subsequently, the mixtures were sufficiently stirred and then were carefully poured into aluminum dishes and placed in an oven to obtain solidified films. All IPNs studied (scanning electron microscopy or SEM and differential scanning calorimetry or DSC) revealed two distinct glass transitions and a melting point that corresponded to the linear PB component which suggest microphase separation in the IPNs. Pure crosslinked CR exhibits a uniform morphology (Figure 3.6a). IPN-1 showed a complex morphology with a small number of micrometer-sized disperses spherical domains and some large irregular "holes" (Figure 3.6b). IPN-2 and IPN-3 showed interesting wormlike disperse fibrous domains. The average thickness of the dispersed fibrous domains decreases as the weight fraction of PB component increases (Figure 3.6c and d). For ultimate mechanical properties of the IPNs, the tensile strengths and elongations at break are decreased as the content of BR increased. The pseudo-IPN structure could be an effective way to combine two rubbers.

3.2.2.3 CR-Based IPNs Polychloroprene (poly(2-chloro-1,3-butadiene)), CR, is an elastomer widely used in the electrical and automobile industries. CR is not characterized by one outstanding property, but its balance of properties is unique among the synthetic elastomers. It has outstanding physical

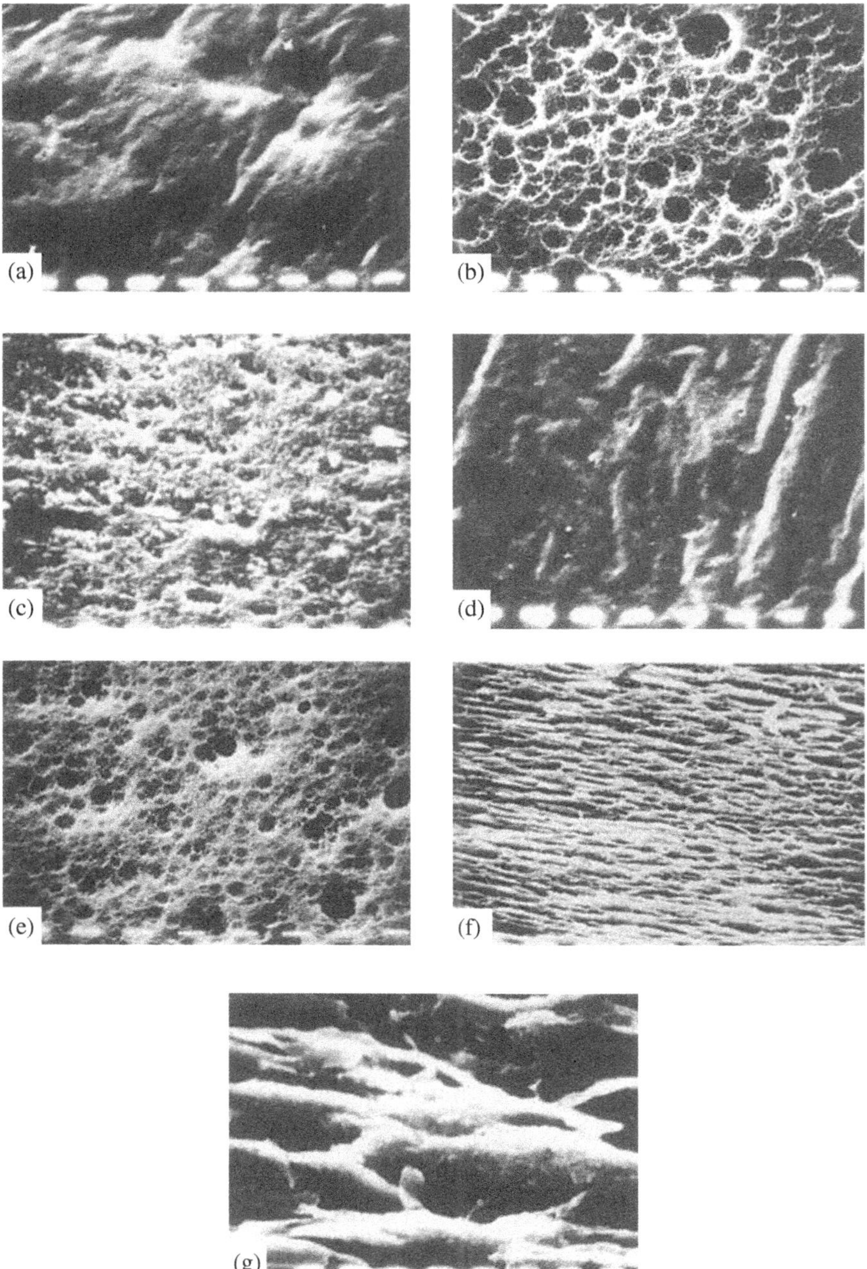

FIGURE 3.5 Scanning electron micrographs of BR/PS FIPNs and semi-IPNs (800×). Semi-IPNs, DCP 0.05 phr, DVB 0 wt % (a) 70 : 30, (b) 50 : 50, and (c) 20 : 80. Full-IPNs, DCP 0.05 phr, DVB 5 wt %, (d) 70 : 30, (e) 50 : 50, and (f) 20 : 80. Full-IPNs, DCP 0.05 phr, DVB 5 wt %, (6400×) (g) 20 : 80. Source: Ghosh and Ray [8]. Reproduced with permission of Springer

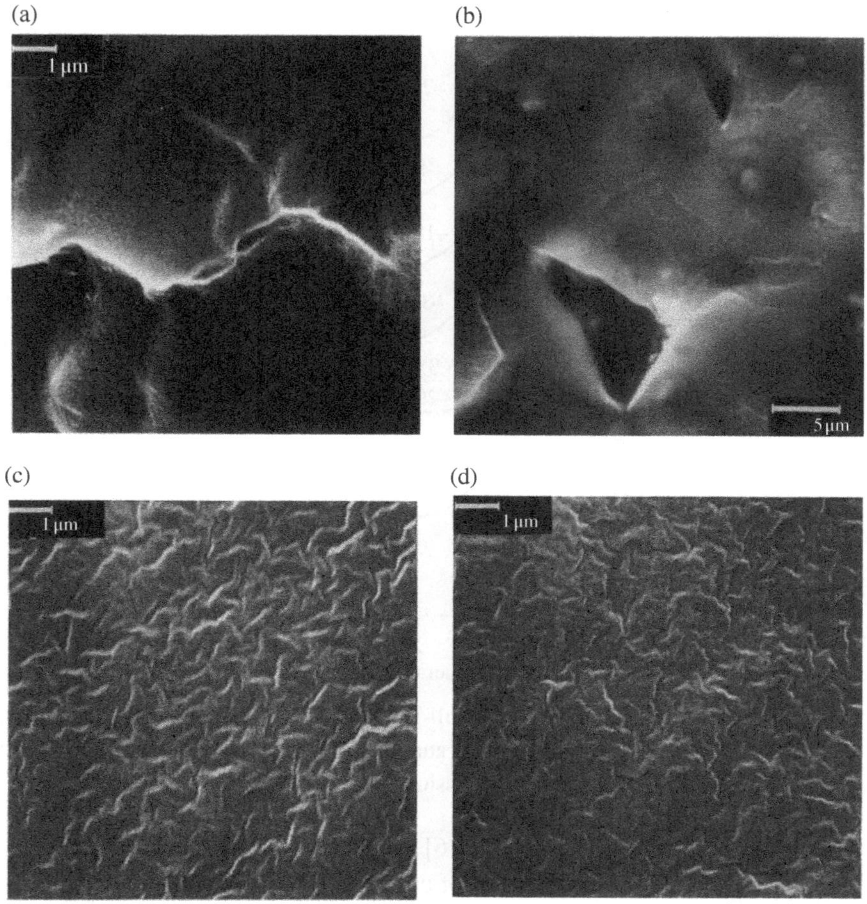

FIGURE 3.6 SEM of CR and pseudo-IPN: (a) CR, (b) IPN-1(CR/BR = 75 : 25), (c) IPN-2 (CR/BR = 50 : 50), and (d) IPN-3 (CR/BR = 25 : 75). Source: Xue, Chen, and Frisch [13]. Reproduced with permission of John Wiley & Sons

toughness, a wider operating temperature range, and excellent resistance to ozone, sun, and general weather conditions [14].

Besides the previously discussed CR/BR semi-IPN [13], Xue and Frisch [15] also prepared CR/poly(carbonate-urethane) (PCU) simultaneous IPN (SIN). In this study, the crosslinking of poly(chloroprene) (CR) use ethylene glycol dimethacrylate (EGDMA) and benzoyl peroxide (BPO) under mild temperature conditions. The effect of composition on the phase morphology of full-IPNs of CR/PCU has been studied by DSC (Figure 3.7) and SEM. The microphase separation of the component networks in the IPNs occurred in samples whose weight percentage of the CR component was 50% and higher.

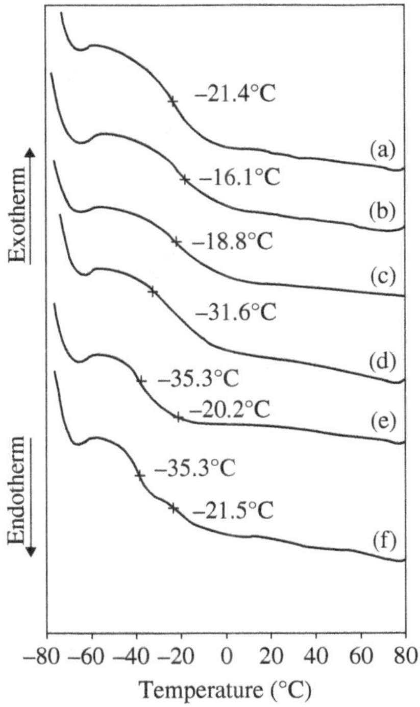

FIGURE 3.7 DSC thermograms of full-IPNs of crosslinked CR/PCU: (a) 15/85, (b) 25/75, (c) 30/70, (d) 40/60, (e) 50/50, and (f) 75/25 wt% composition. Source: Xue and Frisch [15]. Reproduced with permission of John Wiley & Sons

Rajalingam and Radhakrishnan [16] reported on another polychloroprene-based semi-IPN with the formation of castor oil (CO)-based PUs as adhesive. Polychloroprene was the first synthetic elastomer to be used as an adhesive for leather, rubber, and so on. Similarly, thermosetting and thermoplastic PUs are also used as adhesives for these materials. Results showed that the order of peel and shear strength is leather–leather > leather–rubber > rubber–rubber.

Latex IPNs (LIPNs) have been prepared using a crosslinked polychloroprene latex as the seed emulsion, followed by the *in situ* polymerization of styrene, typically with a 10% DVB crosslinker by Burford and Vo [17]. They found that the type of initiator contributed significantly in the morphology development of the LIPNs. For the water-soluble initiator ($K_2S_2O_8$), domain structures predominated for 70/30(Neoprene/XPS), whereas PS-rich shells and domain structures have been found in 60/40(Neoprene/XPS) and lower ratios of Neoprene/XPS. For the oil-soluble initiator (AIBN), larger domain sizes have been found in all cases, with thinner and less continuous shells for the 50/50 ratio. The reason is the different hydrophobic nature of the employed initiators.

Damping materials are an important application for synthetic rubber-based materials. Ping and Wang [18] report the dynamic mechanical and thermal decomposition behavior of CR/polybutyl methacrylate (PBMA) SIPNs. By varying the swelling time in the solution of butyl methacrylate and BPO, CR/PBMA SIPNs with different composition (20, 26, and 44% PBMA) were obtained. The IPNs show broader temperature distribution in loss factor. For IPNs with 44% PBMA, tan δ above 0.3 from −49 to 104°C was observed, which indicated an excellent damping property. The thermal decomposition is influenced by the components. Both the decomposition rate and initial decomposition temperature decreased.

3.2.3 Specialty Rubber-Based IPNs

3.2.3.1 NBR-Based IPNs NBR is a synthetic rubber copolymer of acrylonitrile (ACN) and butadiene synthesized by emulsion polymerization technology. Its physical and chemical properties vary depending on the polymer's composition of nitrile; this form of synthetic rubber is generally resistant to oil, fuel, and other chemicals (the more nitrile within the polymer, the higher the resistance to oils but lower the flexibility of the material) but always had problems with ultraviolet light and ozone attack [19]. Therefore, NBR is always used in binding with other polymers to avoid its disadvantages and to modify other polymers with its advantages. NBR-based IPNs are studied by many researchers.

NBR/Polymethacrylate IPNs Polyalkyl methacrylates such as poly(*n*-butyl methacrylate) (PBuMA), poly(ethyl methacrylate) (PEMA), and PMMA are well known for their good mechanical properties, but these have some drawbacks such as poor low-temperature resistance to solvents [20]. They are brittle in nature and they are incompatibly immiscible with synthetic rubbers thermodynamically. Therefore, rubber/polyalkyl methacrylate IPNs are used to conquer their drawbacks.

As early as 1983, Adachi and Kotaka [21] synthesized two types of IPNs consisting of incompatible NBR and PMMA: one was semi-IPNs in which only NBR was crosslinked, and the other was full-IPNs in which both NBR and PMMA were crosslinked. For a full-IPN, the NBR film was allowed to swell in an MMA monomer solution containing 0.1 mol% EGDMA and 1 wt% AIBN as a crosslinking agent and initiator, respectively. After swollen, the film was then heated at 70°C for 180 min to complete the PMMA network formation. Then the IPNs were dried in vacuo at 60°C for 4 days. For the semi-IPN preparation, the procedure was the same as the full-IPN preparation, except that the

MMA monomer mixture contained only 1 wt% AIBN and no EGDMA. Heterogeneous and fairly homogeneous structures containing a continuous NBR phase for semi- and full-IPNs have existed, respectively. The damping properties are especially studied. For the full-IPNs, as the PMMA content increases, the tan δ peak of the NBR component shrinks but its position does not change, while another peak nearly 400 K (PMMA) gradually becomes larger and shifts to high temperature. When the content of PMMA is 52%, the value of tan δ is almost constant over the temperature range 250–400 K, which shows good damping properties. To explore damping properties of NBR-based IPNs, Patri *et al.* [22] prepared sequential IPNs based on NBR and various types of polyalkyl methacrylates such as PBuMA, PEMA, and PMMA. And the compositions of the IPNs are varied by varying the reaction parameters such as swelling time and concentration of crosslinker. The results showed that the dynamic modulus and loss tangent of the IPNs decrease with increase in bulkiness of the ester group of the acrylates (Table 3.2). In addition, the tensile properties of the IPNs also show similar trend.

Similar to the earlier discussion, Manoj and coworkers [23–25] studied a series of IPNs based on carboxylated nitrile rubber (XNBR) and poly(alky methacrylates) such as PMMA, PEMA, and PBuMA. The compositions of the IPNs were varied by changing the swelling time of the rubber in the methacrylate monomer. The IPNs showed broad tan δ values as shown in Figures 3.8 and 3.9, which indicates the probable use as vibration and acoustic

TABLE 3.2 Damping Characteristics of NBR/Polyalkyl Methacrylate IPNs

Composition	%TEGDM[a]	tan δ_{max}	1/2 Peak Width of tan δ Curve (°C)
NBR	—	1.4	15
NBR/PBuMA			
85/15	2	0.61	26.5
80/20	4	0.63	30.0
80/20	6	0.45	30.0
65/35	8	0.43	43.0
NBR/PEMA			
50/50	2	0.46	58.0
50/50	4	0.38	76.0
NBR/PMMA			
70/30	4	0.45	32.0
65/35	6	0.36	57.0
65/35	8	0.34	75.0

Source: Patri *et al.* [22]. Reproduced with permission of John Wiley & Sons

[a]Tetraethylene glycol dimethacrylate (TEGDM) was used as crosslinker for acrylates during IPN synthesis.

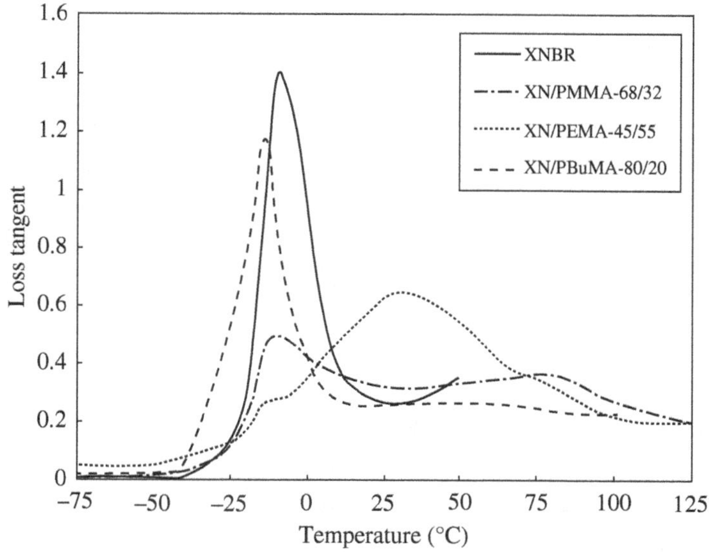

FIGURE 3.8 Plot of loss tangent versus temperature of XNBR and the IPNs. Source: Manoj *et al.* [23]. Reproduced with permission of John Wiley & Sons

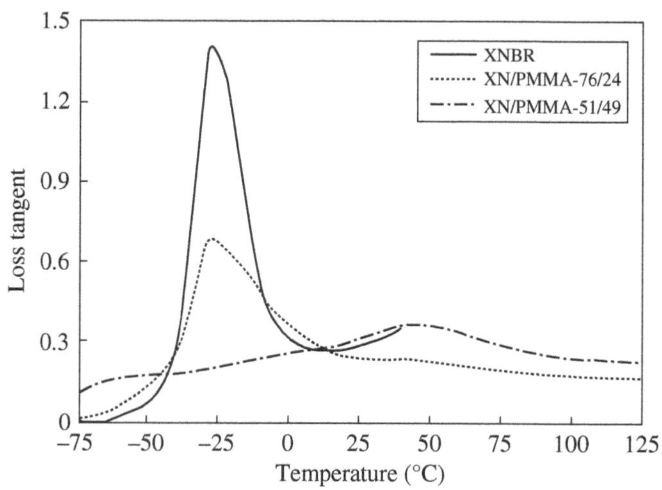

FIGURE 3.9 Plot of loss tangent versus temperature of XNBR and the IPNs. Source: Manoj *et al.* [24]. Reproduced with permission of John Wiley & Sons

dampers. For the superior mechanical properties, especially damping properties, NBR/polyalkyl methacrylate IPNs are extensively studied—some details, for example, morphology (Figure 3.10) of the IPNs—are given by Samui *et al.* [26, 27]. SIPNs of 83/17 and 62/38 compositions (NBR/MMA) show uniform distribution for both phases. For damping properties of polymethacrylate/NBR IPNs [26–28], similar results were obtained.

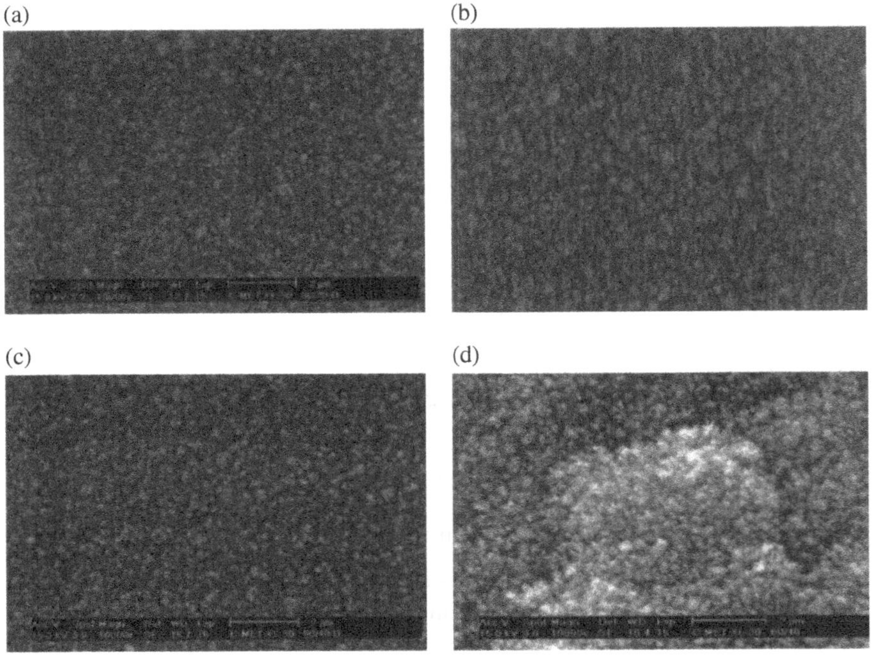

FIGURE 3.10 SEM micrographs of NBR-PMMA IPNs synthesized by single-step (SIPN) and multistep modes: (a) 83/17 SIPN, (b) 82/18 MIPN, (c) 62/38 SIPN, and (d) 62/38 MIPN. Source: Samui *et al.* [26]. Reproduced with permission of John Wiley & Sons

NBR/Fluoroelastomer IPNs Besides the formation of the IPNs with polyalkyl methacrylates, the NBR/fluoroelastomer (FKM) IPNs were prepared by molten blending at a high temperature [29]. Combination of the two highly immiscible polymers (FKM and NBR) by IPN technology might enhance the low-temperature flexibility of FKM and obtain the best compromise in compound process ability and cost. The respective chemical crosslinking of two components occurred in the IPNs. The morphology study shows that dark NBR domains (stained with O_sO_4) exist in the bright continuous phase of FKM when the component ratios are 90/10, 80/20, and 70/30 (w/w) FKM/NBR samples as shown in Figure 3.11. In the case of 70/30 FKM/NBR composition, there exist two transitions at –19 and –23°C, and the value of T_g between –18.5 and –24.5°C (Figure 3.12) shows that the microphase separation of the IPNs between the FKM and NBR phase has existed. This can be the directions for damping material design. The IPNs show better mechanical properties when the two components are in good intermixing. As the NBR content increases, the NBR phase becomes more and more agglomerate and the microphase separation becomes serious. Then the mechanical properties

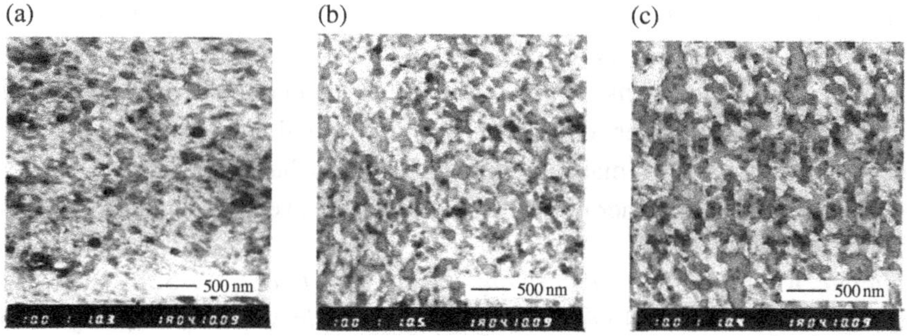

FIGURE 3.11 TEM of different component ratios of FKM/NBR IPNs: (a) 90/10, (b) 80/20, and (c) 70/30 FKM/NBR at magnification of 10000. Source: Chen *et al.* [29]. Reproduced with permission of Springer

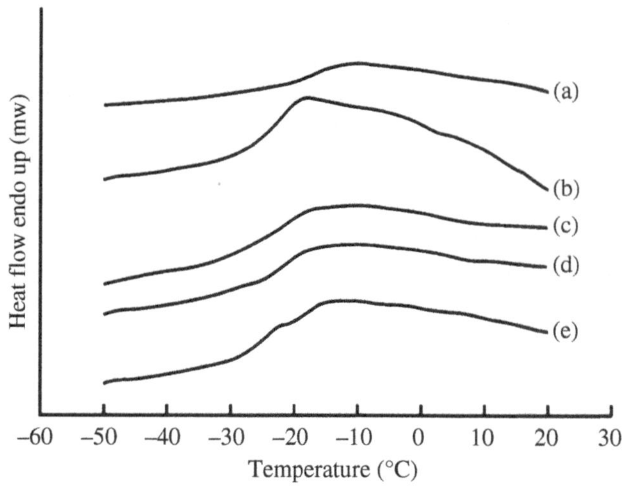

FIGURE 3.12 DSC curves of FKM/NBR IPNs: (a) 100/0, (b) 0/100, (c) 90/10, (d) 80/20, and (e) 70/30 FKM/NBR. Source: Chen *et al.* [29]. Reproduced with permission of Springer

decrease. For all the IPNs, the mechanical properties nearly consistent with the law: under a certain ratios of the component, the mechanical properties attain best and then decrease with the change of microphase separation.

3.2.3.2 Silicone Rubber-Based IPNs Silicone rubber is an elastomer composed of silicone-itself polymer-containing silicon together with carbon, hydrogen, and oxygen that differ from other polymers in that their backbones consist of Si–O–Si units. It has extensively been researched in recent years for their unusual properties such as high thermal and oxidative stability, chemical resistance, excellent dielectric, low glass transition temperature, low surface energy, high permeability and biocompatibility [30]. It is resistant to temperatures from –55 to +300°C, while still maintaining its useful properties. However, their mechanical strengths are rather poor because of nonpolar siloxane bond in the main chain. In order to improve the mechanical strength or to use their advantages effectively, many modifying methods have been found such as interpenetrating with other polymers that are extensively researched.

Silicone Rubber/Poly Methacrylate IPNs He *et al.* [31, 32] synthesized polydimethylsiloxane (PDMS)/PMMA IPNs by *in situ* sequential synthetic method with the following steps. A mixture of α,ω-dihydroxy-PDMS, cross-linker (using tetraethyl orthosilicate (TEOS) to form full-IPNs or using trimethoxysilylpropyl methacrylate (TMSPM) to form graft IPNs) stannous octoate, MMA, 1,1,1-trimethylolpropane trimethacrylate (TRIM), and AIBN was poured into a mold. The ratio K = [SiOR]/[–SiOH], where R = CH_3 or C_2H_5, which represents the number of alkoxy functions per hydroxyl function, was varied from 1 to 4. The PDMS network was formed first at room temperature (Scheme 3.1). After gelation of the medium, the temperature was raised to 60°C to initiate the copolymerization of the methacrylic monomers. The resulting IPN was cured overnight at 75°C and post-cured under vacuum at 120°C for 3 h. The studies showed that when PDMS networks are prepared in the presence of monomers that are able to undergo radical polymerization,

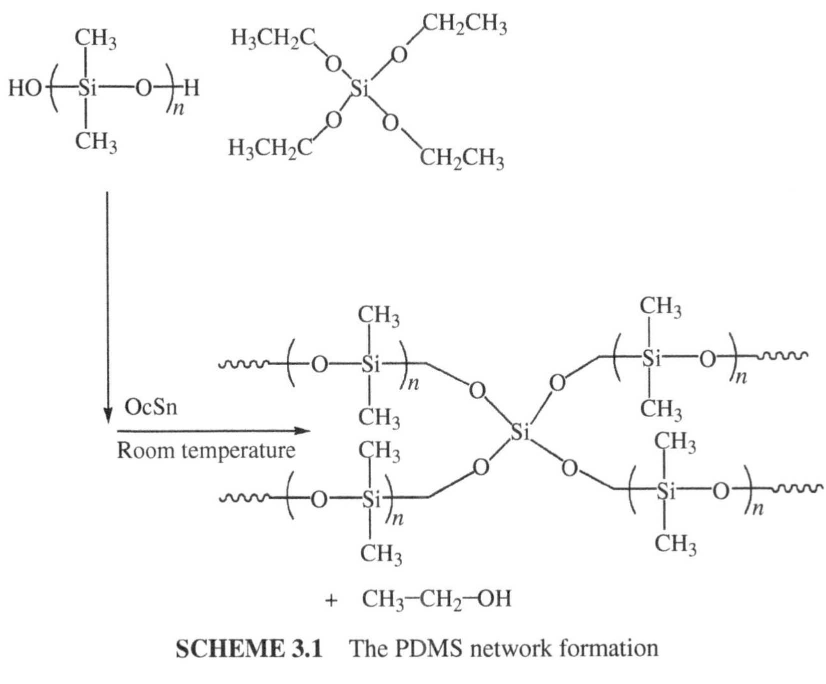

SCHEME 3.1 The PDMS network formation

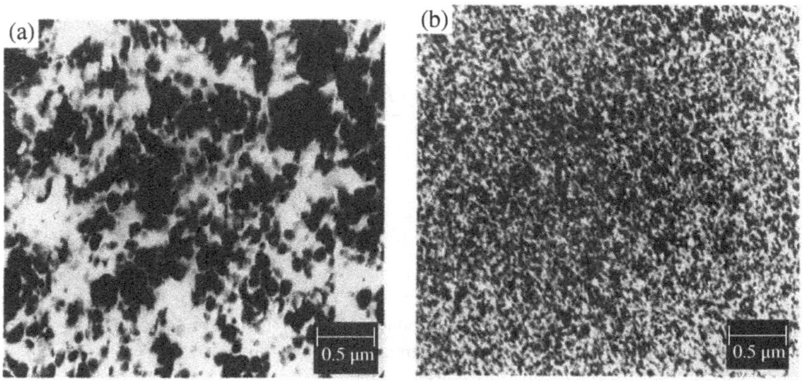

FIGURE 3.13 Transmission electron micrographs (ΔE = 165 eV) of 30/70 PDMS/PMMA IPNs: (a) $M_{n\ PDMS}$ = 4200 g mol^{-1} and (b) $M_{n\ PDMS}$ = 860 g mol^{-1}. Source: He *et al.* [32]. Reproduced with permission of Elsevier

deactivation of the catalyst can be avoided either by eliminating oxygen from the reaction medium or by adding a radical inhibitor. Therefore, it becomes possible to synthesize IPNs based on PDMS and vinyl polymers *in situ*. The morphology (Figure 3.13) study showed that the first-formed network (PDMS) constitutes the continuous phase. Increasing the level of crosslinking by decreasing the molar mass of the PDMS precursor leads to more dispersion and reduction of the size of the methacrylic domains. And the formation of PDMS/PMMA IPNs improves the poor mechanical properties of the silicone network. For example, when the PMMA content is 30%, the tensile modulus increases from 2 to 47 MPa, and the tensile strength increases from 0.15 to 3.23 MPa. In addition, the surface properties are PDMS-like, and the gas permeability of these materials remains high.

Latter, Hou and Kuo [33] used linear PMMA as the guest polymer and silicone as the host polymer to prepare a novel PMMA/silicone semi-IPN. The silicone network formed by the self-condensation reaction among the pendant-reactive methoxysilane groups on the polymethylphenylsiloxane (DC3074). PMMA and DC3074 were miscible at any blended ration, but phase separation occurred when the silicone network formed. DSC was used to investigate the T_g of the IPNs (Table 3.3). An inward-shift T_g for the pure component in the blends and a particular broad T_g were observed in the 50/50 blend. The SEM morphology confirms the results shown in Figure 3.14. The crosslinked silicone was transparent and shown to be a uniform morphology (Figure 3.14e). The PMMA/silicone (75/25) and (50/50) blend both exhibited a heterogeneous morphology consisting of dispersed spherical particles that comprise PMMA. In the 50/50 blend, the boundary of the two domains

TABLE 3.3 Glass Transition Temperatures of the PMMA/Silicone IPNs Measured from DSC

PMMA/Silicone	T_{g1} (K)[a]	T_{g2} (K)[a]
100/0	382.1	—
75/25	372.9	329.2
50/50	368.8	327.8
25/75	377.6	326.9
0/100	—	321.9

Source: Hou and Kuo [33]. Reproduced with permission of Elsevier
[a]T_{g1} is the glass transition temperature for the PMMA-rich domain and T_{g2} for silicone-rich domain.

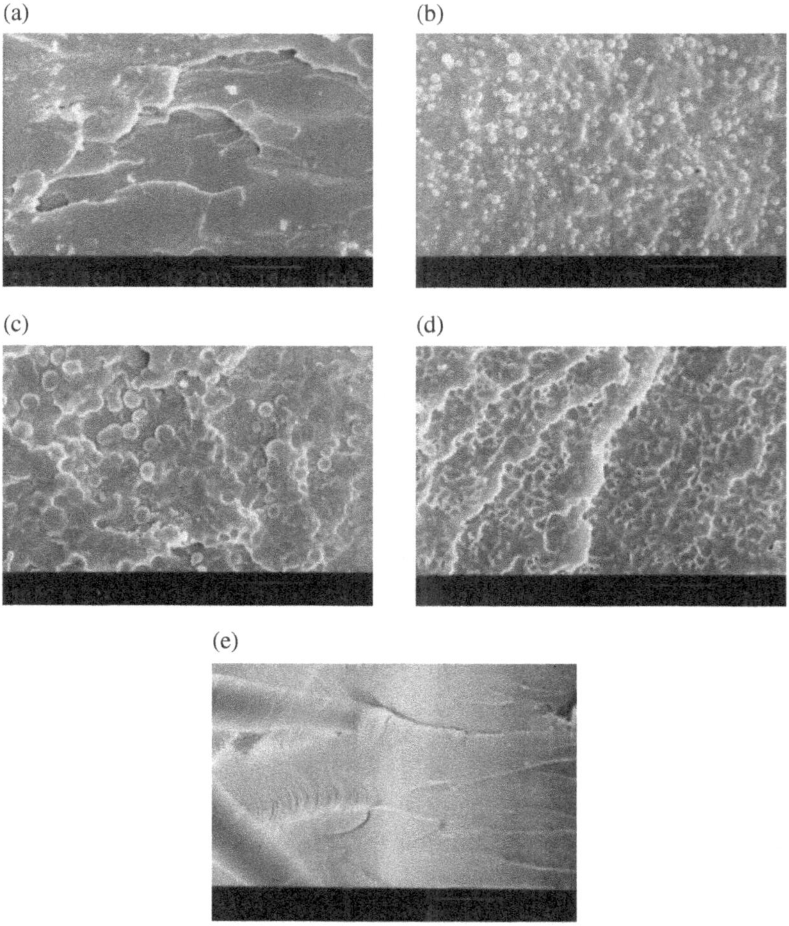

FIGURE 3.14 SEM of micrographs of the fracture surfaces of PMMA/silicone semi-IPN. (a) Pure MMA, (b) 75/25, (c) 50/50, (d) 25/75, and (e) pure crosslinked silicone. Source: Hou and Kuo [33]. Reproduced with permission of Elsevier

is less distinguishable, which indicted better compatibility of the PMMA and crosslinked silicone network than that of the other ratios.

The polymethacrylates can be modified by special organic functional groups before forming IPNs with the silicone rubber. Darras *et al.* [34] synthesized IPNs based on PDMS and poly(fluorinated acrylate) (polyAcRf6) shown in Scheme 3.2. The IPNs with different relative weight proportions of the fluorinated versus silicone partners were characterized by DMTA. The IPN mechanical relaxations are very broad, which indicated good damping properties of the formed IPNs. Vuillequez *et al.* [35] reported on an IPN combining a PU methacrylate (PUMA) network and a silicone network. The silica's presence increases the thermal stability and surface hardness and T_g decrease with increasing silica content. The facility modification of methacrylate monomer is conducive to the design and improvement of these IPNs.

As state previously, besides the mechanical and thermal properties, damping properties are important properties of synthetic rubber-based IPNs. The damping properties of silicone rubber/polyacrylate IPNs are studied by many researchers [36–38]. Abbasi *et al.* reported the preparation of the IPNs from poly(2-hydroxyethyl methacrylate) (PHEMA) and PDMS by sequential method [36]. The HEMA monomer was polymerized with simultaneous crosslink formation in the presence of the swollen silicone rubber (Scheme 3.3). The study discovered the optimum conditions for the preparation of silicone rubber/PHEMA IPNs that contain about 33% PHEMA. These compositions were reported as an optimum composition in which the obtained biomaterial has a good biocompatibility as well as elastomeric properties. Then, the damping factor (tan δ) and tensile properties for the PDMS/PHEMA IPNs and their variations with the structural composition were evaluated [37]. The results for the IPNs were compared with corresponding

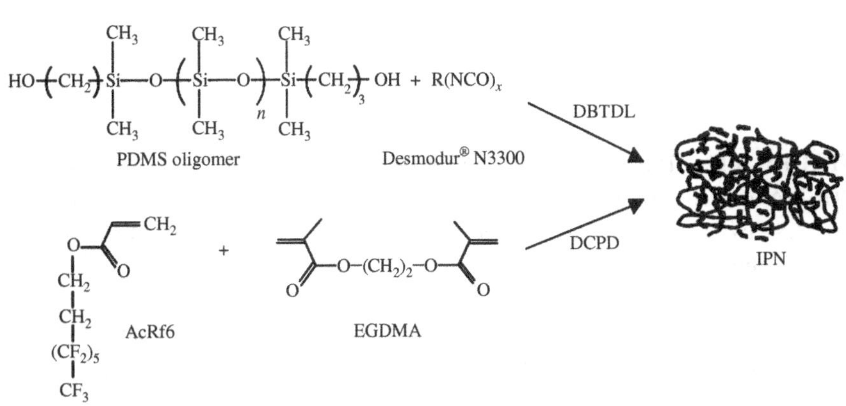

SCHEME 3.2 Synthesis pathway of PDMS/polyAcRf6 IPNs. Source: Darras *et al.* [34]. Reproduced with permission of Elsevier

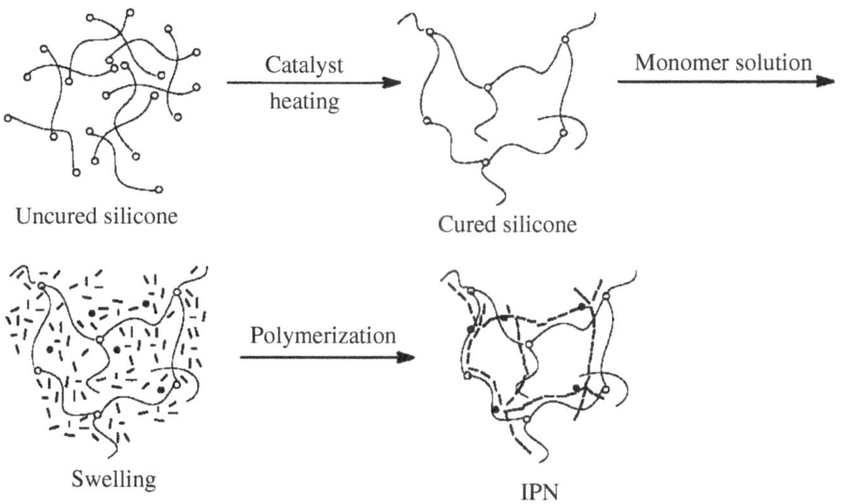

SCHEME 3.3 Schematic diagram of the synthesis of silicone rubber/poly(HEMA) sequential IPN. Source: Abbasi, Mirzadeh, and Katbab [36]. Reproduced with permission of John Wiley & Sons

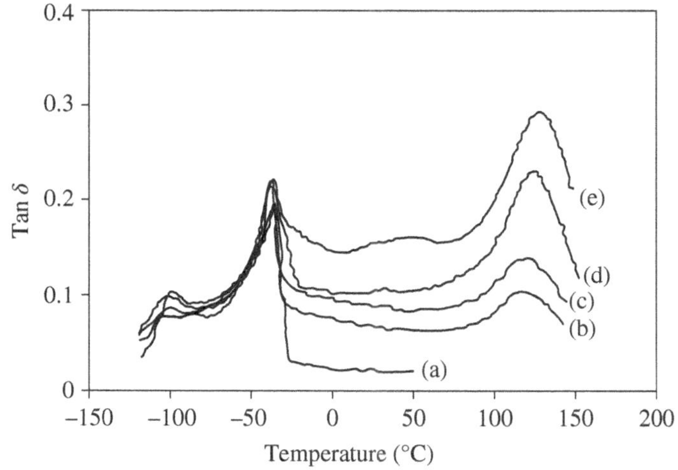

FIGURE 3.15 The tan δ versus temperature for the PDMS/PHEMA IPNs having PHEMA contents of (a) 0 wt%, (b) 10.8 wt%, (c) 19.5 wt%, (d) 30.4 wt%, and (e) 38.8 wt%. Source: Abbasi, Mirzadeh, and Katbab [37]. Reproduced with permission of John Wiley & Sons

physically blended systems. As can be seen from the curves (Figure 3.15) of temperature dependence of the tan δ for the IPNs (and physical blends) with a large content of PHEMA show a pronounced maximum in the region of α-transition of PHEMA, and for the IPNs with a small content of PHEMA, the curves show a low and wide maximum (Figure 3.15). The shape of the temperature dependencies of the tan δ for both systems is typical for two-phase polymer systems with incompatible components. The investigated systems are two-phase systems, but the shift of the maximum tan δ for the PHEMA in the IPN on the temperature scale from its position for pure polymer, as well as its broadening, indicates an incomplete phase separation in the IPN formation. Furthermore, the tan δ curve for the IPN having 40 wt% hydrogel content shows another maximum occurring at approximately 50°C. For tensile properties, the IPNs have higher tensile strength and higher elongation at break values than those for physical blends. Wang *et al.* [38] prepared silicone rubber/polyacrylate sequential IPNs by silicone rubber sheet dipped into the solution composed of different acrylate monomers and BPOs for different times at room temperature and then acrylate polymerized at 80°C for 2 h. Similarly results that the loss factor of the IPNs is improved after interpenetrating with polyacrylate ether were obtained.

Silicone Rubber/EP IPNs Kumar and coworkers [39,40] developed a novel siliconized epoxy interpenetrating coating system using EP as base, hydroxyl-terminated polydimethylsiloxane (HTPDMS) as modifier, γ-aminopropyltriethoxysilane (γ-APS) as crosslinking agent, and dibutyltin dilaurate (DBTDL) as catalyst. It is observed that the thermal stability of epoxy coating systems is enhanced when siloxane is incorporated to them, for the siloxane bond energy is significantly greater than those of carbon–carbon and carbon–oxygen bonds. There is a slight decrease in the glass transition temperature observed for silicone-modified epoxy coatings, and all siliconized epoxy coating systems exhibit single glass transition temperature (T_g), which in turn confirms the presence of intercrosslinked network structure. The heterogeneous morphology shown in Figure 3.16 also confirms the existence of intercrosslinking network structure in siliconized epoxy coating systems. The siliconized epoxy coatings can be effectively used as thermal barrier coatings.

Dhoke *et al.* [41] also studied the heat-resistant IPN prepared from epoxy and silicones using a crosslinking agent and a catalyst. For heat-resistant property, the silicone resins withstood up to 533 K, whereas the epoxy coating performed well up to 423 K. But the IPNs remained stable up to 478 K. The heat-resistance property of the EP was improved by the silicone resin through the formation of the IPN. To further enhance the heat-resistance property and other properties of the IPNs, modifiers such as black pigments like graphite, silicon carbide, and carbon black can be used for their shape, size, conducting nature, and heat stability property [42].

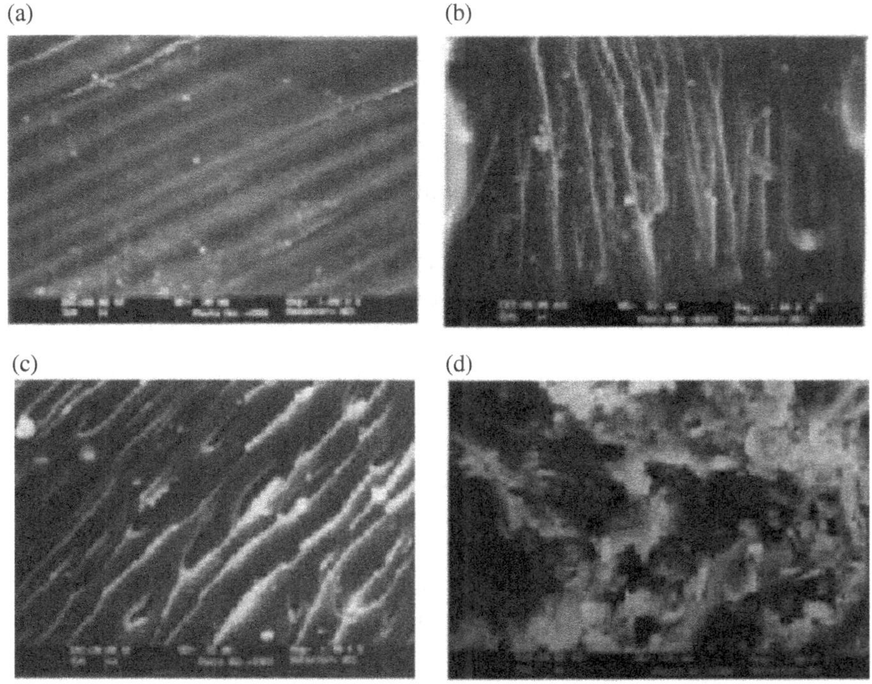

FIGURE 3.16 SEM of micrographs of the fractured surface of the unmodified epoxy (a) and siliconized epoxy coating systems having 5% (b), 10% (c), and 15% (d) siloxane content. Source: Kumar and Narayanan [39]. Reproduced with permission of Elsevier

The Formation of Silicone Rubber-Based IPNs with Other Polymers Miyata *et al.* [43] prepared PDMS/PS IPN membranes by the bulk copolymerization of styrene and DVB in the PDMS networks. All the PDMS/PS IPN membranes had a microphase-separated structure in which PS domains exist in a continuous PDMS phase (Figure 3.17). When an aqueous ethanol solution was permeated through the PDMS/PSt IPN membranes by pervaporation, the PDMS/PS IPN membranes exhibited ethanol permselectivity, regardless of the PDMS content. And the mechanical properties of the PDMS/PS IPN membranes could be improved by the introduction of PS without lowering the ethanol permselectivity of the PDMS membrane. A functional membrane was obtained.

An IPN combining a hydrophobic polymer (PDMS) and a hydrophilic polymer (polyvinylpyrrolidone or PVP) was synthesized in different solvents by Hillerstrom *et al.* [44]. The choice of solvent during polymerization of the IPN was affecting both the wettability and transparency of the PVP/PDMS IPN. The PVP/PDMS IPNs turned hydrophilic in all the solvents used in their study, but the transition from a hydrophobic to a hydrophilic PVP/PDMS IPN

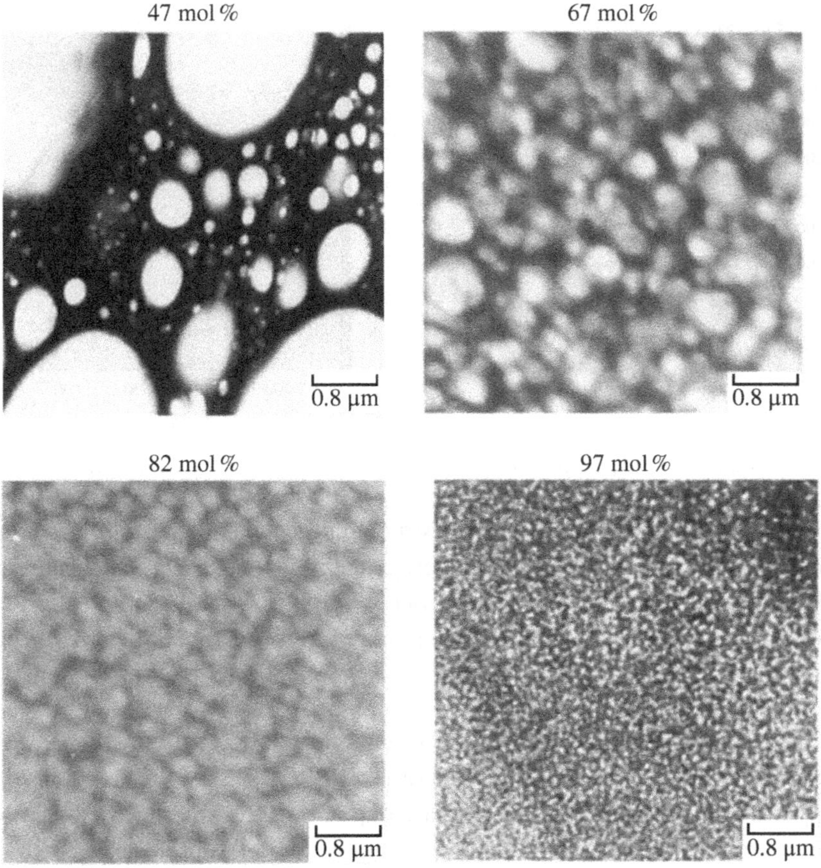

FIGURE 3.17 Transmission electron micrograph of the PDMS/PS IPN membranes with different PDMS contents. Source: Miyata *et al.* [43]. Reproduced with permission of John Wiley & Sons

occurred at lower PVP concentration if a solvent with similar solubility parameter as PVP was chosen. The properties of the PVP/PDMS IPN can be tuned by the selection of the solvent used during polymerization.

Kirubaharan *et al.* [45] reported on high-temperature-resistant IPNs prepared by blending silicone resin polymer with the organic polymers such as polypyrrole (PPy) and polythiophene (PTh). The IPNs with a molar ratio of silicone–PPy/PTh of 0.346 : 0.569 exhibited excellent heat-resistance properties. The thermal stability of silicone–PTh IPN was found to be significantly higher than that of the Silicone–PPy IPN. The atomic force microscopy (AFM) study (Figure 3.18) shows that surface profile of silicone–PTh is much more regular than that of silicone–PPy, indicating that the silicone-containing

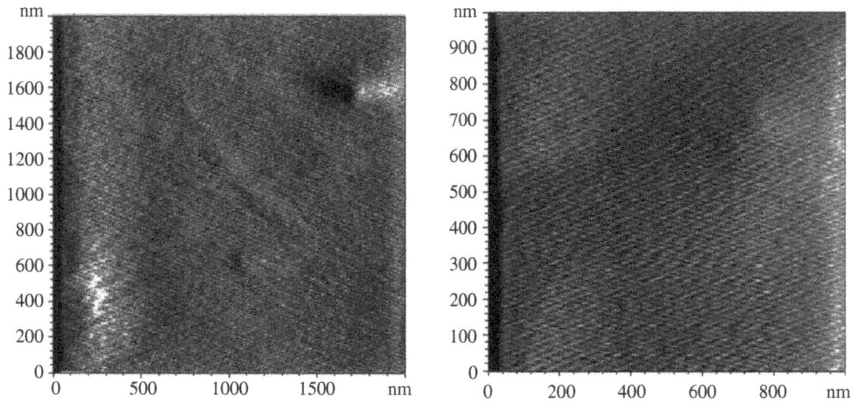

FIGURE 3.18 AFM images of silicone–PPy and silicone–PTh IPN. Source: Kirubaharan *et al.* [45]. Reproduced with permission of John Wiley & Sons

polymer matrices form new hybrid structured polymer systems when compared with the silicone resin. The heterogeneity of silicone–PTh IPN is higher when compared with that of the silicone–PPy IPN. It is the confirmation for the influence of PPy/PTh on silicone polymer to form smooth heterogeneous microstructures.

3.2.3.3 PU-Based IPNs PUs are organic polymers that contain the urethane group in the structure. They are typically made by the reaction of a polyol with a diisocyanate. They are a group of organic polymers that are normally grouped with rubbers. They are flexible and elastic polymers that are widely used in coatings, adhesives, thermoplastic elastomers, and composites. Nevertheless, pure PU has some disadvantages, such as low mechanical strength and poor heat resistance, and its applications in structural materials are thus limited [46]. Like the rubbers mentioned earlier, PUs often are combined with high mechanical plastic polymers such as epoxy, PS, and PMA in blending and especially in IPN form to improve their properties.

PU/EP IPNs EP has been employed in many kinds of structural composites for their high modulus and strength. However, such resins suffer from brittle behavior, which restricts their application in some applications [47]. PU/EP IPNs that integrated the advantages of PU and EP were first reported by Frisch *et al.* in 1974, and then their properties were extensively researched.

Hsieh and Han [48, 49] introduced PU based on poly(butylene adipate) (PU(PBA)) and poly(oxypropylene) (PU(PPG)) polyols into the diglycidyl ether of bisphenol A (DGEBA) (epoxy) to form graft IPNs. The tensile strength in PU(PPG)/epoxy and PU(PBA)/epoxy systems increases with increasing PU content. Because the IPN formation leads to more intimate interpenetration between the PU and epoxy, the tensile strength increased. Maximum values emerge at PU/epoxy ratios between 19/81 and 27/73. The toughness of epoxy resin can be improved. These graft-IPNs can have a morphology ranging from rubber particles dispersed in the rigid matrix (extensive phase separation) or a more homogeneous solution of PU and epoxy. The impact strength of the IPNs improves much more with homogeneous morphology than that of with the phase-separation system.

Wang and coworkers [47, 50] prepared a series of poly(tetramethylene glycol) (PTMG) and polycaprolactone (PCL)-based PU/EP graft IPNs and their mechanical properties, damping properties, and thermal stability were systematically studied in terms of composition and the value of the PU isocyanate index (R). For PU(PTMG)/EP IPNs [51], when PU(PTMG) was incorporated, the tensile strength decreased, but the elongation of the IPN increased with increasing PU content. With the value of PU isocyanate index increasing, the tensile strength was improved while the elongation was impaired. The impact strength of the IPNs was relatively good when the IPN is formed. The research highlight is the damping properties. The damping properties of the EP matrix can be improved by forming a graft IPN with PU(PTMG). The damping temperature range became broader and the glass transition temperature T_g became lower with the increase of PU content (Figure 3.19), and the maximum value of tan δ increased with the increase of the value of the isocyanate (Figure 3.20). Investigation of the morphologies revealed that a heterogeneous microstructure existed in the PU (PTMG)/EP IPNs. And, the microstructure became more heterogeneous with the increase

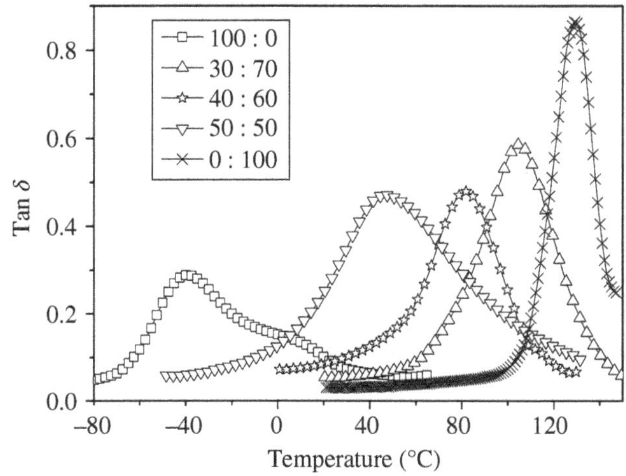

FIGURE 3.19 DMA traces of PU/EP IPNs with different component ratios at 10 Hz (R = 2). Source: Wang *et al.* [47]. Reproduced with permission of Springer

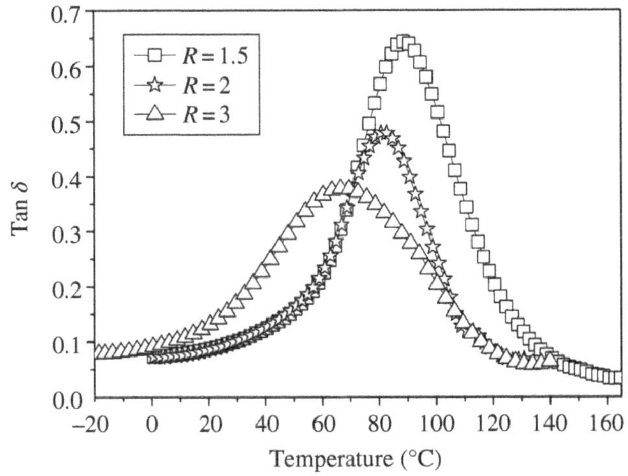

FIGURE 3.20 DMA traces of 40 : 60 PU/EP IPNs with different values of isocyanate index at 10 Hz. Source: Wang *et al.* [47]. Reproduced with permission of Springer

of PU content, which was the elucidation of the variation of the properties of the PU/EP IPNs. The formation of PU (PCL)/EP IPNs showed similar results in mechanical properties and better damping properties because the more hydrogen bonds exist in them [50].

The mechanical, thermal, and damping properties of poly diols-based PU/EP IPNs were also studied by Chen *et al.* [52], Bakar *et al.* [51], Cristea *et al.* [53], and others. The overall performance of the materials is improved when the IPNs are formed. Besides the poly diol-based PU rubbers, the poly triol-based PU rubbers are studied especially the CO-based PU and the PU(CO)/EP IPNs. CO is one of the more widely used natural products in the polymer field. It contains hydroxyl groups as well as carbon–carbon double bonds: it is especially well suited to a wide range of chemical modifications [54]. The PUs obtained from CO are typically water resistant and flexible due to the presence of a long fatty acid chain and lend themselves to thermosetting-type structures due to their trifunctional nature [55].

Raymond and Bui [2] prepared graft IPNs from DGEBA cured with 1.5% 2,4,6-tris(dimethylaminomethyl) phenol (DMP-30) and CO cured with 2,4-toluene diisocyanate (TDI) at NCO/OH = 1.50, following a sequential method (Scheme 3.4). Good interpenetration existed between the two networks and some degree of grafting existed between the two phases. Therefore, graft IPNs of 20% PU provided better heat resistance than that of cured epoxy as well as large synergistic effects on the tensile properties. However, their impact energy obtained from instrumented impact testing did not show much enhancement except for a narrow PU composition gap between 25 and 30%. Similar mechanical results were obtained by Chen *et al.* [56], and the damping properties of the IPNs were further studied extensively. The damping properties of EP matrix can be improved by forming graft IPNs with PU. With the increase of the PU content, the damping properties of the IPNs increased, whereas the glass transition temperature decreased (Figure 3.21). Further, the

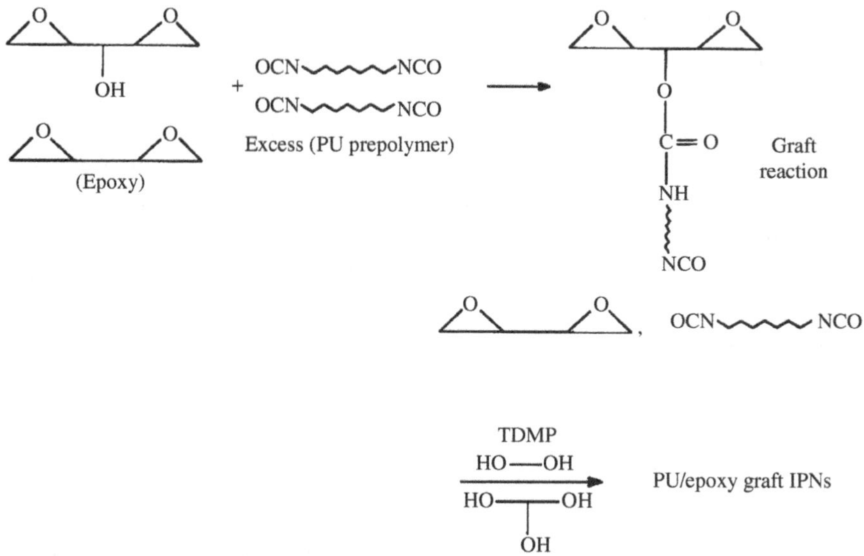

SCHEME 3.4 Scheme of the grafting process between epoxy and PU prepolymer. Source: Raymond and Bui [2]. Reproduced with permission of John Wiley & Sons

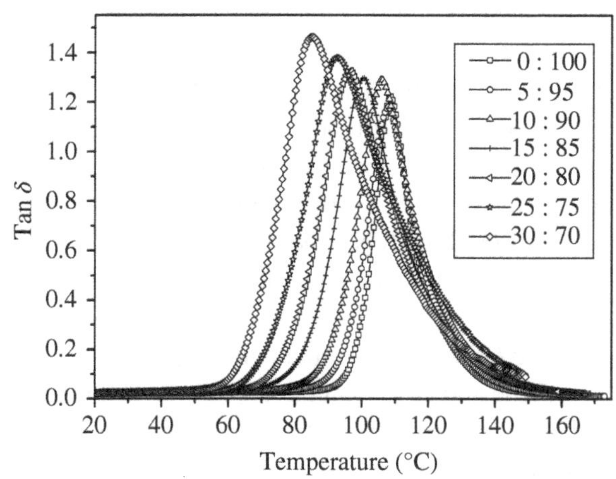

FIGURE 3.21 DMA curves of PU/EP IPNs with component ratios from 0 : 100 to 30 : 70 at 10 Hz. Source: Chen *et al.* [56]. Reproduced with permission of John Wiley & Sons

damping properties can be improved by modifiers–both organic [57, 58] and inorganic [59]. In addition, the PU/EP IPN composites showed better damping properties under higher frequencies.

PU/PS IPNs As early as 1975, Kim *et al.* [60] prepared the SIN-type IPNs composed of a PS network (crosslinked with DVB) and a PCL glycol-PU network (crosslinked with trimethylolpropane (TMP)). Phase separation was observed in PU/PS IPNs due to the incompatibility of each network. The continuous phase is PS at PU compositions of less than 75%. Enhanced tensile strength was observed because of the reinforcing effect of the rigid dispersed phase and the effect of increased permanent chain entanglement due to interpenetration. Valero *et al.* [55] also prepared the SIN from pentaerythritol-modified CO-based PU and PS. Due to permanent chain entanglements, a maximum in mechanical properties was observed at around 20% of PS (Figure 3.22). At concentrations above the 20% PS content, phase separation occurred, which presents a broad glass transition temperature area. Compositions (90% PU/10% PS) that are rich in PU show a continuous matrix with dispersed PS. Intermediate compositions (60% PU/40% PS) show signs of the existence of two continuous phases (Figure 3.23).

Gomza *et al.* [61] synthesized PU/PS sequential semi-IPNs (SIPNs) based on network PUs with various M_c (the molecular masses of chain segments between crosslinks) values and the linear PS. The network PU was synthesized from poly(oxypropylene glycol) (POPG) with $M = 500$ (PU-500), 1000 (PU-1000), or 2000 (PU-2000), triisocyanate adduct of TDI, and TMP in butyl acetate at 60°C. Then, the PU network was allowed to swell to equilibrium in a monomer (styrene) containing the dissolved radical initiator (AIBN), and then polymerization was carried out at 60°C. The structural features of them were studied by DSC and small-angle X-ray scattering. The studied SIPNs are composed of microregions of two basic types. The first type includes microregions, which conserve the features of the microphase-separated structure of PU, and the second type involves domains enriched in PS. Further, with an increase in the *Mc* of the PU constituent (the PU-2000–PS SIPN), the interference maximum appears. The research is beneficial for understanding of the microstructure of PU/PS SIPNs.

PU/Polymethacrylate IPNs For complementary properties of polymethacrylates with polyurethane and as incompatible pair, the combination of polyurethane and poly methacrylates with IPN technology has become a research focus in the IPN study. Xiao *et al.* [62] prepared two types of IPNs by simultaneous polymerization. The first type consisted of PU(PTMG)/PMMA, and the second one consisted of PU/poly(methyl methacrylate-*co*-methacrylic

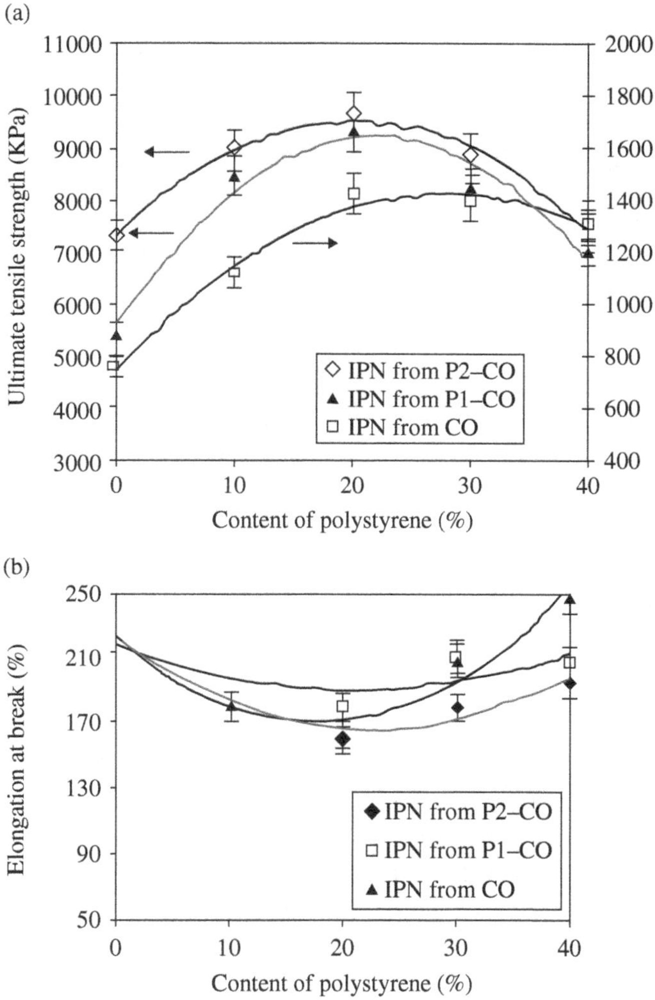

FIGURE 3.22 Mechanical properties of IPNs. (a) Ultimate tensile strength and (b) elongation % at break (P1-CO (molar ratio pentaerythritol/ castor oil = 0.2), P2[0.4]). Source: Valero *et al.* [55]. Reproduced with permission of Springer

acid) (P(MMA-MAA)) of constant composition (90/10) and (80/20) by weight, respectively. The mechanical properties, particularly the modulus of both types of IPN, increase with increasing NCO/OH ratio and decrease with increasing MW of the polyol in the PU. Phase compatibility improves and phase separation decrease in both types of IPN with increasing NCO/OH ratio and decreasing MW of the polyol used in the PU. Increasing the NCO/OH ratio and decreasing the MW of the polyol in the PU component could improve the interpenetration which is the explanation for the results.

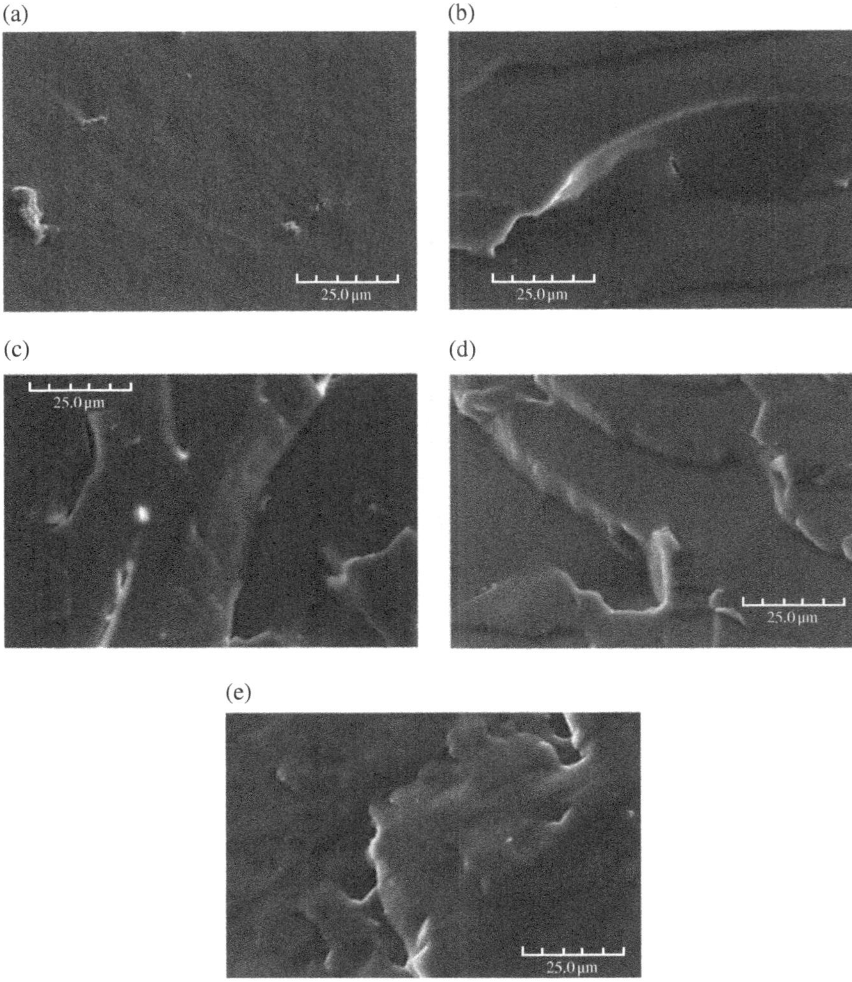

FIGURE 3.23 SEM of SIN-P1-CO, with various levels of PS at NCO/OH = 1: (a) PS0PU100, (b) PS10PU90, (c) PS20PU80, (d) PS30PU70, and (e) PS40PU60. Source: Valero *et al.* [55]. Reproduced with permission of Springer

Akay and Rollins [63] evaluated the damping properties of a variety of simultaneous and sequential IPNs based on a PU(POPG) network and PMMA in linear and network forms. The IPNs produced broad transitions of various types depending on the composition shown in Figures 3.24. The plots clearly show that at the extreme IPN compositions, that is PU/PMMA ratios of 80/20 and 20/80, single relatively sharp tan δ curves were obtained and at the intermediate IPN compositions broad tan δ curves were obtained. The elastic modulus, hardness and the tensile strength of the IPNs increase with increasing

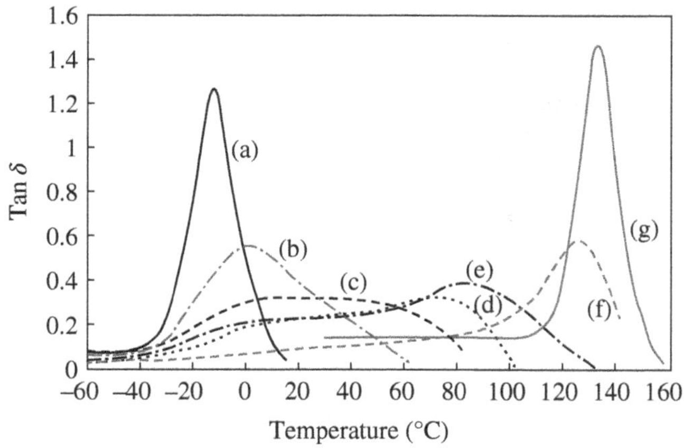

FIGURE 3.24 Dynamic mechanical properties for simultaneous full-IPNs with PU/PMMA compositions: (a) 100/0, (b) 80/20, (c) 70/30, (d) 60/40, (e) 50/50, (f) 20/80, and (g) 0/100. Source: Akay and Rollins [63]. Reproduced with permission of Elsevier

PMMA content. The distribution of the component polymers in the IPNs appears to be more homogeneous than in the sequential IPNs.

Kong *et al.* [64, 65] prepared semi- and full-IPNs using PU produced from a canola oil-based polyol with primary terminal functional groups and PMMA. For both types of IPNs, the reaction rates increase with the increases of PU content. With the same components, the reaction rates of full IPNs are faster than those of semi-IPNs. Both types of IPNs show two-phase systems with incomplete phase separation. However, the process of phase separation is more developed in the semi-IPNs compared with the full-IPNs as full-IPNs exhibit better compatibility because of the enhanced interpenetration shown in Figure 3.25. The original PMMA was rather brittle, and the PU was rubbery but fairly weak, while the mechanical properties of the IPNs are superior to those of its constituent polymers. The combination of PU with PMMA is useful in extending the elastomeric properties of PU and mitigating the brittleness of PMMA.

The Formation of PU-Based IPNs with Other Polymers The PU either can form IPNs with many other polymers including other synthetic rubbers to improve their properties such as PU/poly(cis-1,4-polyisoprene) IPNs [9], and CO-based PU/CR IPNs [16] in the preceding narrative. In addition, Parthiban *et al.* [66] prepared semi-IPNs using CO-based PU and chlorinated rubber. The toughness of chlorinated rubber can be improved by interpenetrating it with CO-based PU. These systems may be used in the coating industry, and

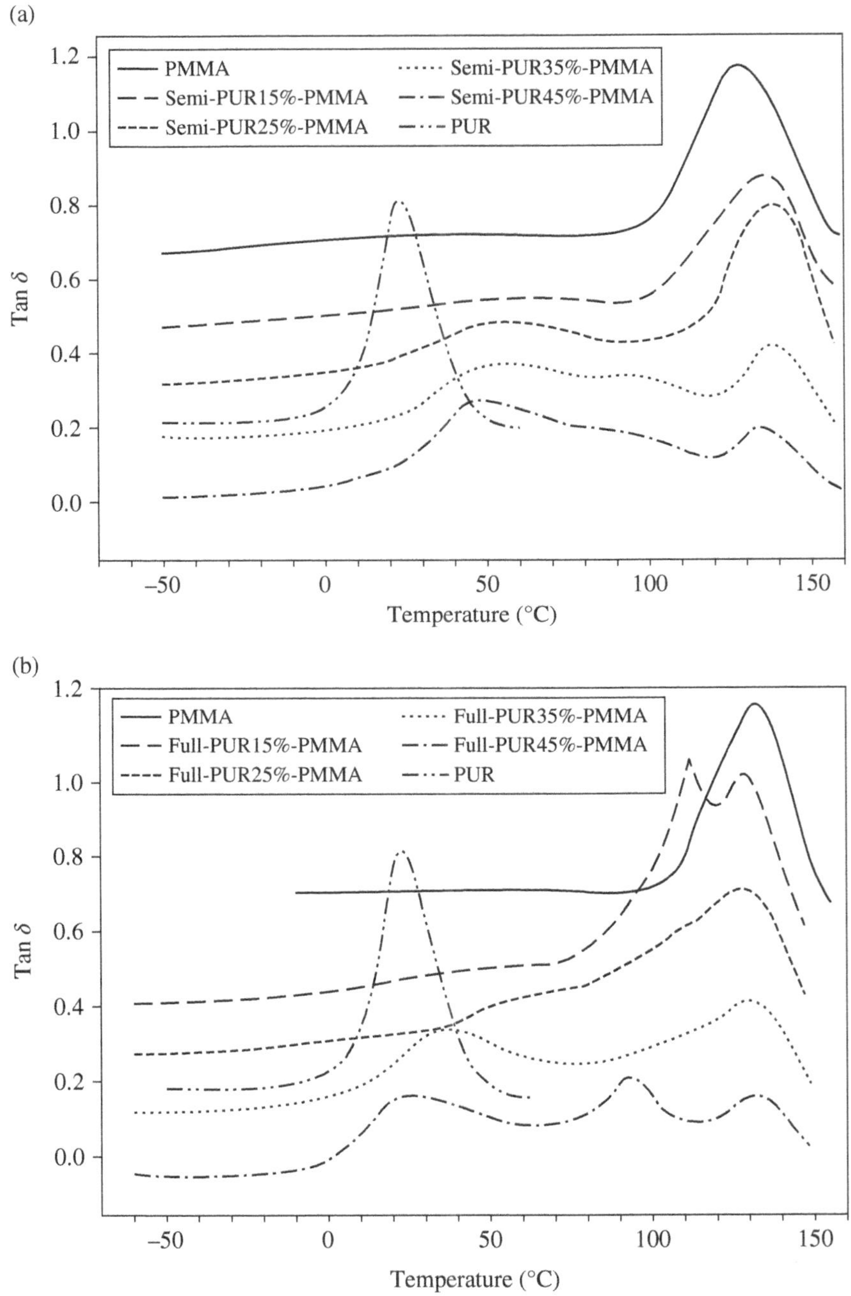

FIGURE 3.25 The dependences of tan δ on temperature for IPNs with different PU content (a) semi-IPNs and (b) full-IPNs. Source: Kong, Tan, and Narine [65]. Reproduced with permission of John Wiley & Sons

further the flame resistance of the PUs would be improved by the presence of chlorinated rubber. Semi-IPNs of acrylamide-based PUs were synthesized from different NCO-terminated polyurethane prepolymers derived from PTMG by Merlin and Sivasankar [67]. The IPNs showed better thermal stability, compatibility, tensile, and elongation properties compared to the individual polyacrylamide network.

3.3 SUMMARY AND CONCLUSIONS

The IPNs prepared from commonly used synthetic rubbers such as SBR, BR, CR, NBR, silicone rubber, PU, and so on, and other rigid plastic polymers (high mechanical strength but brittle) such as PS, PMMA, EP, and so on, represent one of the fastest-growing fields in polymer science. The morphology, phase separation, mechanical properties, damping properties, thermal properties, and other properties of the IPNs are extensively researched. Synthetic rubbers usually are incompatible pairs with rigid plastic polymers. The IPN technology can combine them together with intimate interpenetration. The toughness of rigid plastic polymers can be attained by the intimate interpenetration. In contrast, the strength of the rubber improves. Thus, the strength of the IPNs is better than the individual rubber networks, and the toughness of the IPNs is better than the individual rigid polymer networks. Varying degrees of phase separation will occur resulting in a two-phase polymer system according to the nature and composition of component polymers. They can have a morphology ranging from rubber particles dispersed in the rigid matrix (extensive phase separation) or a more homogeneous solution. The impact strength of the IPNs improves much more with homogeneous morphology than that of with the phase-separation system. However, the broad glass transition region comes from the micro heterogeneous phase separation. Proper phase separation in the IPNs leads to good damping properties. The synthetic rubber-based IPNs conquer many drawbacks of the rubbers and exhibit better overall mechanical, thermal, and damping properties, which can be used in many applications. However, the research about many synthetic rubbers needs further study, for example, the IPNs prepared from the rubbers IR, EPDM, IIR, and FKM that also have excellent properties, are scarcely reported. Moreover, the morphology control, damping optimization, and so on for specific applications when the IPNs are formed should be further studied. The research of synthetic rubber-based IPNs will continue to receive attention from both academia and industry.

ACKNOWLEDGMENTS

Financial supports from the National Science Found for Distinguished Young Scholars of China (grant no. 51025517) and the National Science Found of China (grant no. 51303189) are duly acknowledged.

REFERENCES

[1] Morton, M. (1981) History of synthetic rubber. *Journal of Macromolecular Science—Chemistry*, **A 15** (7), 1289–1302.

[2] Raymond, M.P. and Bui, V.T. (1998) Epoxy/castor oil graft interpenetrating polymer networks. *Journal of Applied Polymer Science*, **70** (9), 1649–1659.

[3] John, J., Suriyakala, R., Thomas, S. *et al.* (2010) Morphology, mechanical and thermal properties of nano-structured full-IPNs based on polyisoprene and PMMA. *Journal of Materials Science*, **45** (11), 2892–2901.

[4] Sperling, L.H. and Mishra, V. (1996) The current status of interpenetrating polymer networks. *Polymers for Advanced Technologies*, **7** (4), 197–208.

[5] Burford, R.P., Chaplin, R.P., Pittolo, M. and Mai, Y.W. (1986) Determination of k_c for an interpenetrating polymer network based on SBR and polystyrene. *Journal of Materials Science Letters*, **5** (1), 57–59.

[6] Ramaraj, B., Rajalingam, P. and Radhakrishnan, G. (1992) Semiinterpenetrating polymer networks based on styrene-butadiene rubber latex – thermal, mechanical and morphological properties. *Journal of Elastomers and Plastics*, **24** (4), 337–349.

[7] Patri, M., Reddy, C.V., Narasimhan, C. and Samui, A.B. (2007) Sequential interpenetrating polymer network based on styrene-butadiene rubber and polyalkyl methacrylates. *Journal of Applied Polymer Science*, **103** (2), 1120–1126.

[8] Ghosh, P. and Ray, P. (1991) Studies on polybutadiene rubber (PBR) polystyrene (PS) interpenetrating polymer networks. *Journal of Materials Science*, **26** (22), 6104–6112.

[9] Mishra, V., Murphy, C.J. and Sperling, L.H. (1994) Interpenetrating polymer networks based on thermoplastic polyurethanes (tpus) and cis-1,4-polyisoprene. *Journal of Applied Polymer Science*, **53** (11), 1425–1434.

[10] Donatelli, A.A., Sperling, L.H. and Thomas, D.A. (1976) Interpenetrating polymer networks based on SBR-PS. 1: Control of morphology by level of cross-linking. *Macromolecules*, **9** (4), 671–675.

[11] Donatelli, A.A., Sperling, L.H. and Thomas, D.A. (1976) Interpenetrating polymer networks based on SBR-PS. 2: Influence of synthetic detail and morphology on mechanical-behavior. *Macromolecules*, **9** (4), 676–680.

[12] Kim, M.S., Kim, G.H. and Chowdhury, S.R. (2007) Polybutadiene rubber/organoclay nanocomposites: effect of organoclay with various modifier concentrations on the vulcanization behavior and mechanical properties. *Polymer Engineering and Science*, **47** (3), 308–313.

[13] Xue, Y.P., Chen, Z.F. and Frisch, H.L. (1994) Pseudo-interpenetrating polymer networks of high-molecular-weight polychloroprene and cis-1,4-polybutadiene. *Journal of Applied Polymer Science*, **52** (12), 1833–1836.

[14] Yeh, M.H., Hwang, W.S. and Cheng, L.R. (2007) Microstructure and mechanical properties of neoprene-montmorillonite nanocomposites. *Applied Surface Science*, **253** (10), 4777–4781.

[15] Xue, Y.P. and Frisch, H.L. (1993) Interpenetrating polymer networks based on poly(chloroprene) and poly(carbonate-urethane). *Journal of Polymer Science Part A—Polymer Chemistry*, **31** (8), 2165–2171.

[16] Rajalingam, P. and Radhakrishnan, G. (1991) Poly(chloroprene) castor oil-based polyurethane semiinterpenetrating polymer network as an adhesive. *Polymer International*, **25** (2), 87–90.

[17] Burford, R.P. and Vo, C.D. (1999) Morphology development in polychloroprene-polystyrene latex interpenetrating polymer networks. *Journal of Applied Polymer Science*, **74** (3), 629–638.

[18] Ping, D. and Wang, Y.B. (2010) The dynamic mechanical properties of chlorobutyl rubber/polybutyl methacrylate sequential interpenetrating networks. *Polymer-Plastics Technology and Engineering*, **49** (13), 1310–1314.

[19] Zhang, Z.X., Chen, C.H., Gao, X.W. *et al.* (2011) A Study on the compatibility and physical properties of chlorinated polyethylene rubber/nitrile rubber blends. *Journal of Applied Polymer Science*, **120** (2), 1180–1185.

[20] Sriram, V., Subramani, S. and Radhakrishnan, G. (2001) AB-type crosslinked polymer networks from vinyl terminated polyurethane and poly (methyl methacrylate). *Polymer International*, **50** (10), 1124–1131.

[21] Adachi, H. and Kotaka, T. (1983) Structure and mechanical-properties of sequential interpenetrating polymer networks. 3: Poly(acrylonitrile-co-butadiene)/poly(methyl methacrylate) system. *Polymer Journal*, **15** (4), 285–292.

[22] Patri, M., Samui, A.B., Chakraborty, B.C. and Deb, P.C. (1997) Studies on IPNS based on nitrile rubber and polyalkyl methacrylates. *Journal of Applied Polymer Science*, **65** (3), 549–554.

[23] Manoj, N.R., Ratna, D., Dalvi, V. *et al.* (2002) Interpenetrating polymer networks based on carboxylated nitrile rubber and poly(alkyl methacrylates). *Polymer Engineering and Science*, **42** (8), 1748–1755.

[24] Manoj, N.R., Chandrasekhar, L., Patri, M. *et al.* (2002) Vibration damping materials based on interpenetrating polymer networks of carboxylated nitrile rubber and poly(methyl methacrylate). *Polymers for Advanced Technologies*, **13** (9), 644–648.

[25] Manoj, N.R., Raut, R.D., Sivaraman, P. *et al.* (2005) Sequential interpenetrating polymer network of poly(ethyl methacrylate) and carboxylated nitrile rubber: dynamic mechanical analysis and morphology. *Journal of Applied Polymer Science*, **96** (5), 1487–1491.

[26] Samui, A.B., Dalvi, V.G., Patri, M. *et al.* (2004) Studies on semi-interpenetrating polymer network based on nitrile rubber and poly(methyl methacrylate). *Journal of Applied Polymer Science*, **91** (1), 354–360.

[27] Samui, A.B., Dalvi, V.G., Chandrasekhar, L. *et al.* (2006) Interpenetrating polymer networks based on nitrile rubber and metal methacrylates. *Journal of Applied Polymer Science*, **99** (5), 2542–2548.

[28] Patri, M., Samui, A.B. and Deb, P.C. (1993) Studies on sequential interpenetrating polymer network (ipn) based on nitrile rubber and poly(vinyl acetate). *Journal of Applied Polymer Science*, **48** (10), 1709–1716.

[29] Chen, C.M., Xiong, C.X., Yang, J. and Dong, L.J. (2008) Microstructure and properties of fluoroelastomer/butadiene-acrylonitrile rubber interpenetrating polymer networks. *Journal of Wuhan University of Technology-Materials Science Edition*, **23** (1), 50–53.

[30] Bischoff, R. and Cray, S.E. (1999) Polysiloxanes in macromolecular architecture. *Progress in Polymer Science*, **24**, 185–190.

[31] He, X.W., Widmaier, J.M., Herz, J.E. and Meyer, G.C. (1989) Polydimethylsiloxane/poly(methylmethacrylate) interpenetrating polymer networks. 1: Efficiency of stannous octoate as catalyst in the formation of polydimethylsiloxane networks in methyl methacrylate. *Polymer*, **30** (2), 364–368.

[32] He, X.W., Widmaier, J.M., Herz, J.E. and Meyer, G.C. (1992) Polydimethylsiloxane poly(methyl methacrylate) interpenetrating polymer networks. 2:Synthesis and properties. *Polymer*, **33** (4), 866–871.

[33] Hou, S.S. and Kuo, P.L. (2001) Morphological, thermal and solid-state NMR study on a novel PMMA/crosslinked silicone semi-IPN. *Polymer*, **42** (23), 9505–9511.

[34] Darras, V., Fichet, O., Perrot, F. *et al.* (2007) Polysiloxane-poly(fluorinated acrylate) interpenetrating polymer networks: synthesis and characterization. *Polymer*, **48** (3), 687–695.

[35] Vuillequez, A., Moreau, J., Garda, M.R. *et al.* (2008) Polyurethane methacrylate/silicone interpenetrating polymer networks synthesis, thermal and mechanical properties. *Journal of Polymer Research*, **15** (2), 89–96.

[36] Abbasi, F., Mirzadeh, H. and Katbab, A.A. (2002) Sequential interpenetrating polymer networks of poly (2-hydroxyethyl methacrylate) and polydimethylsiloxane. *Journal of Applied Polymer Science*, **85** (9), 1825–1831.

[37] Abbasi, F., Mirzadeh, H. and Katbab, A.A. (2002) Comparison of viscoelastic properties of polydimethylsiloxane/poly(2-hydroxyethyl methacrylate) IPNs with their physical blends. *Journal of Applied Polymer Science*, **86** (14), 3480–3485.

[38] Wang, Y.B., Huang, Z.X. and Zhang, L.M. (2006) Damping properties of silicone rubber/polyacrylate sequential interpenetrating networks. *Transactions of Nonferrous Metals Society of China*, **16**, S517–S520.

[39] Kumar, S.A. and Narayanan, T. (2002) Thermal properties of siliconized epoxy interpenetrating coatings. *Progress in Organic Coatings*, **45** (4), 323–330.

[40] Kumar, S.A., Alagar, M. and Mohan, V. (2002) Studies on corrosion-resistant behavior of siliconized epoxy interpenetrating coatings over mild steel surface by electrochemical methods. *Journal of Materials Engineering and Performance*, **11** (2), 123–129.

[41] Dhoke, S.K., Palraj, S., Maruthan, K. and Selvaraj, M. (2007) Preparation and characterization of heat-resistant interpenetrating polymer network (IPN). *Progress in Organic Coatings*, **59** (1), 21–27.

[42] Dhoke, S.K., Maruthan, K., Palraj, S. and Selvaraj, M. (2006) Performance of black pigments incorporated in interpenetrating polymer network (IPN). *Progress in Organic Coatings*, **56** (1), 53–58.

[43] Miyata, T., Higuchi, J., Okuno, H. and Uragami, T. (1996) Preparation of polydimethylsiloxane/polystyrene interpenetrating polymer network membranes and permeation of aqueous ethanol solutions through the membranes by pervaporation. *Journal of Applied Polymer Science*, **61** (8), 1315–1324.

[44] Hillerstrom, A., Andersson, M., Pedersen, J.S. *et al.* (2009) Transparency and wettability of PVP/PDMS-IPN synthesized in different organic solvents. *Journal of Applied Polymer Science*, **114** (3), 1828–1839.

[45] Kirubaharan, A.M.K., Palraj, S., Selvaraj, M. and Rajagopal, G. (2011) Synthesis and characterization of high temperature resistance interpenetrating polymer network. *Journal of Applied Polymer Science*, **119** (4), 2339–2344.

[46] Clemitson, I.R. (2008) *Castable Polyurethane Elastomers*, CRC Press, New York.

[47] Wang, Q., Chen, S., Wang, T. and Zhang, X. (2011) Damping, thermal, and mechanical properties of polyurethane based on poly(tetramethylene glycol)/epoxy interpenetrating polymer networks: effects of composition and isocyanate index. *Applied Physics A: Materials Science & Processing*, **104** (1), 375–382.

[48] Hsieh, K.H. and Han, J.L. (1990) Graft interpenetrating polymer networks of polyurethane and epoxy. 1: Mechanical behavior. *Journal of Polymer Science Part B: Polymer Physics*, **28** (5), 623–630.

[49] Hsieh, K.H. and Han, J.L. (1990) Graft interpenetrating polymer networks of polyurethane and epoxy. 2: Toughening mechanism. *Journal of Polymer Science Part B: Polymer Physics*, **28** (6), 783–794.

[50] Chen, S., Wang, Q. and Wang, T. (2012) Damping, thermal and mechanical properties of polycaprolactone-based pu/ep graft ipns: effects of composition and isocyanate index. *Journal of Macromolecular Science, Part B*, **51**, 83–95.

[51] Bakar, M., Duk, R., Przybylek, M. and Kostrzewa, M. (2009) Mechanical and thermal properties of epoxy resin modified with polyurethane. *Journal of Reinforced Plastics and Composites*, **28** (17), 2107–2118.

[52] Chen, C.H. and Chen, M.H. (2006) Synthesis, thermal properties, and morphology of blocked polyurethane/epoxy full-interpenetrating polymer network. *Journal of Applied Polymer Science*, **100** (1), 323–328.

[53] Cristea, M., Ibanescu, S., Cascaval, C.N. and Rosu, D. (2009) Dynamic mechanical analysis of polyurethane-epoxy interpenetrating polymer networks. *High Performance Polymers*, **21** (5), 608–623.

[54] Liu, T.M. and Bui, V.T. (1995) Instrumented impact testing of castor oil-based polyurethanes. *Journal Applied Polymer Science*, **56** (3), 345–354.

[55] Valero, M.F., Pulido, J.E., Ramirez, A. and Cheng, Z.D. (2009) Simultaneous interpenetrating polymer networks of polyurethane from pentaerythritol-modified castor oil and polystyrene: structure-property relationships. *Journal of the American Oil Chemists Society*, **86** (4), 383–392.

[56] Chen, S., Wang, Q., Pei, X. and Wang, T. (2010) Dynamic mechanical properties of castor oil-based polyurethane/epoxy graft interpenetrating polymer network composites. *Journal of Applied Polymer Science*, **118** (2), 1144–1151.

[57] Chen, S., Wang, Q. and Wang, T. (2011) Physical properties of a high molecular weight hydroxyl-terminated polydimethylsiloxane modified castor oil based polyurethane/epoxy interpenetrating polymer network composites. *Applied Physics A: Materials Science & Processing*, **103** (4), 1047–1052.

[58] Chen, S., Wang, Q. and Wang, T. (2011) Hydroxy-terminated liquid nitrile rubber modified castor oil based polyurethane/epoxy IPN composites: damping, thermal and mechanical properties. *Polymer Testing*, **30** (7), 726–731.

[59] Chen, S., Wang, Q. and Wang, T. (2011) Damping, thermal, and mechanical properties of montmorillonite modified castor oil-based polyurethane/epoxy graft IPN composites. *Materials Chemistry and Physics*, **130** (1–2), 691–695.

[60] Kim, S.C., Klempner, D., Frisch, K.C. *et al.* (1975) Polyurethane-polystyrene interpenetrating polymer networks. *Polymer Engineering and Science*, **15** (5), 339–342.

[61] Gomza, Y.P., Klepko, V.V., Lipatov, Y.S. *et al.* (2008) Sequential semi-interpenetrating polymer networks based on polyurethane and polystyrene. *Polymer Science Series A*, **50** (9), 956–964.

[62] Xiao, H.X., Frisch, K.C. and Frisch, H.L. (1983) Interpenetrating polymer networks from polyurethanes and methacrylate polymers. 1: Effect of molecular-weight of polyols and NCO OH ratio of urethane prepolymers on properties and morphology of IPNs. *Journal of Polymer Science Part A—Polymer Chemistry*, **21** (8), 2547–2557.

[63] Akay, M. and Rollins, S.N. (1993) Polyurethane poly(methyl methacrylate) interpenetrating polymer networks. *Polymer*, **34** (9), 1865–1873.

[64] Kong, X. and Narine, S.S. (2008) Physical properties of sequential interpenetrating polymer networks produced from canola oil-based polyurethane and poly(methyl methacrylate). *Biomacromolecules*, **9** (5), 1424–1433.

[65] Kong, X.H., Tan, S.S. and Narine, S.S. (2009) Semi- and full-interpenetrating polymer networks based on polyurethane produced from canola oil and poly(methyl methacrylate). *Journal of Applied Polymer Science*, **114** (1), 139–148.

[66] Parthiban, S., Rajalingam, P. and Radhakrishnan, G. (1992) Semi-interpenetrating polymer networks from castor oil-based polyurethanes and chlorinated rubber. *Polymer International*, **29** (4), 289–292.

[67] Merlin, D.L. and Sivasankar, B. (2009) Synthesis and characterization of semi-interpenetrating polymer networks using biocompatible polyurethane and acrylamide monomer. *European Polymer Journal*, **45** (1), 165–170.

4

MICRO- AND NANOSTRUCTURED IPNs BASED ON THERMOSETTING RESINS

SANJA MARINOVIĆ[1], IVANKA POPOVIĆ[2] AND BRANKO DUNJIĆ[2]

[1] *Department of Catalysis and Chemical Engineering, Institute of Chemistry, Technology and Metallurgy, University of Belgrade, Belgrade, Serbia*

[2] *Faculty of Technology and Metallurgy, University of Belgrade, Belgrade, Serbia*

4.1 INTRODUCTION

An interpenetrating polymer network (IPN) is an intimate combination of two polymers both in network form [1]. Materials known as IPNs involve the preparation of crosslinked polymer I, subsequent swelling in a network of monomer II plus crosslinker, and the *in situ* polymerization of monomer II [2, 3]. Because they combine the properties of the components forming the networks, IPNs have become the subject of extensive studies [4–7].

Hyperbranched polymers (HBP) are built from monomers of AB_x type, that is, monomers containing two different functional groups. In order to control molecular weight and polydispersity, AB_x monomers are usually reacting with B_y-type molecules (core). These polymers are characterized by highly branched, treelike structure and large number of end groups, which affects their overall properties [8, 9]. The thermal properties of HBP depend on the nature of the end groups. The T_g of HBP decreases with decreasing

Micro- and Nano-structured Interpenetrating Polymer Networks: From Design to Applications, First Edition. Edited by Sabu Thomas, Daniel Grande, Uroš Cvelbar, K.V.S.N. Raju, Ramanuj Narayan, Selvin P. Thomas and Akhina H.

end group polarity [10]. The unique combination of low viscosity, high molecular weight, and big number of reactive end groups could be exploited in coatings [11–18].

The main interest of UV radiation lies in high polymerization rate, together with the advantage that it can be considered an environmentally friendly technique due to the solvent-free processes carried out at room temperature [19].

Urethane acrylates as resins for UV-curable coatings have the best mechanical and chemical properties among all UV-curable resins but suffer from the high viscosities [9, 20].

Although other authors have used the UV–thermal dual-cure process to produce IPNs [4, 7], to the best of our knowledge, the use of an acrylated HBP in IPNs obtained by dual curing has not been investigated so far. In the present study [21, 22], acrylated HBP was a part of the component of the IPN. The investigated systems were first subjected to UV-induced acrylate double bond conversion with the formation of a polyacrylic (PA) network, followed by the thermal radical polymerization of the urethane component. The properties of the obtained IPNs were analyzed by dynamic mechanical analysis (DMA), DSC, and thermogravimetric analysis (TGA). Film properties were also investigated with respect to hardness and flexibility. The aim was to explore the influence of different components on properties of IPNs in order to get superior properties with respect to scratch resistance.

4.2 EXPERIMENTAL DETAILS

4.2.1 Materials

Desmodur® N3390 (19.4×10^{-3} of NCO for 1 g of Desmodur), BAYER, Germany, was used as 90% solution in butyl acetate, MERCK, United States. Methyl methacrylate (MMA), BASF, Germany (purity ≥ 99.9%); butyl acrylate (BuA), BASF, Germany (purity ≥ 99.5%); and 2-hydroxypropyl methacrylate (HPMA), Cognis, were used with no further purification. 2,2′-Azobis (2-methylpropionitrile) (AIBN), DuPont, was used as the radical polymerization initiator.

2-Ethylhexyl acrylate (EHA), BASF, Germany (purity ≥ 99.6%), and hexanediol diacrylate (HDDA) were used as reactive diluents, while acrylated hyperbranched polyester HBP(A) was used as acrylated oligomer in the UV component of IPNs. 1-Hydroxycyclohexyl phenyl ketone (Irgacure 184), CIBA, United States, was used as the photoinitiator.

The structures of the employed monomers are presented in Table 4.1.

TABLE 4.1 Structure of Monomers Employed

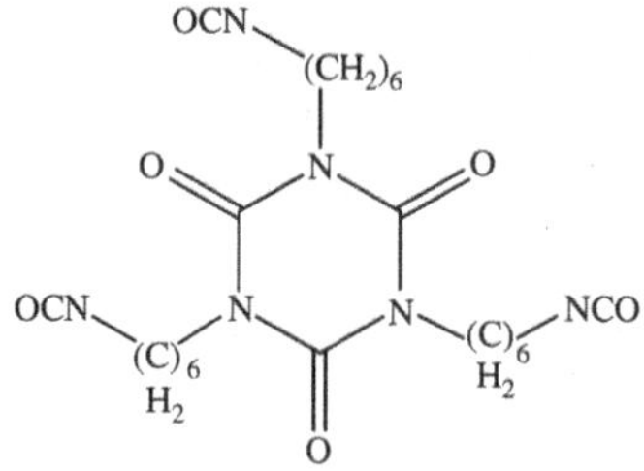

Desmodur N3390

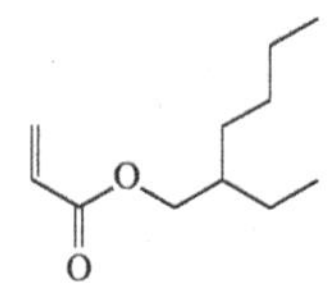

Acrylic acid 2-ethylhexyl escter (EHA)

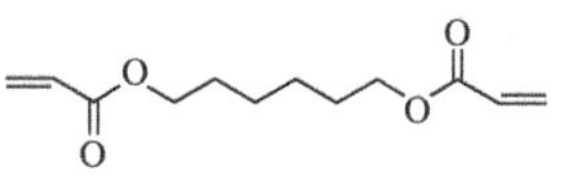

1,6 Hexanediol diacrylate (HDDA)

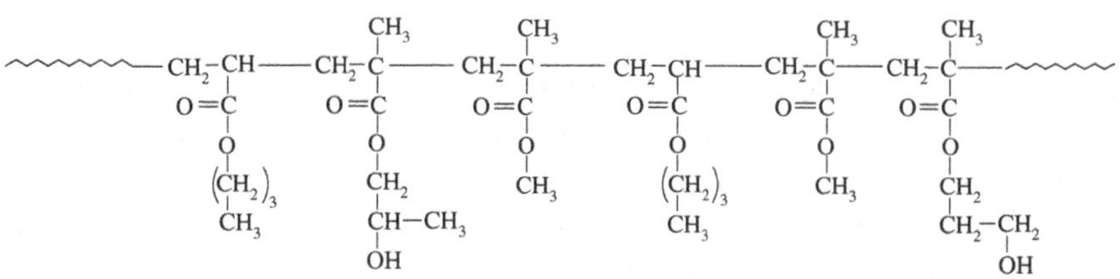

Hydroxyl functional acrylate copolymer (HA)

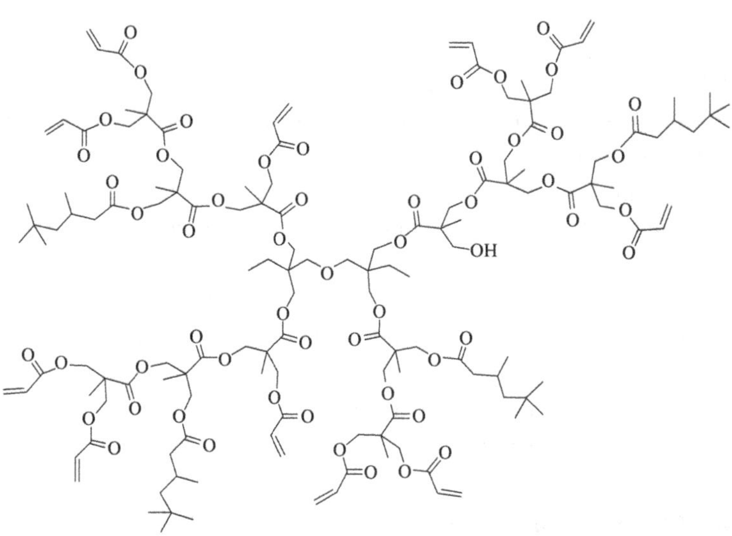

HBP (A)

Source: Simic *et al.* [21]. Reproduced with permission of Elsevier.

4.2.2 Synthesis of IPN Components and Sample Preparation

4.2.2.1 Synthesis of HA A four-necked round-bottom flask equipped with an inert gas (nitrogen) inlet, a temperature regulation lug for the contact thermometer, a mechanical impeller lug with a condenser, and dropping funnel was loaded with 1.76 mol MMA, 0.98 mol HPMA, 0.64 mol BuA, and 2.3 mol butyl acetate (BuAc). AIBN was used as the radical polymerization initiator (8.00 wt% diluted in monomer) and tertiary dodecyl mercaptan (tert-DDM) as the molar mass regulator (0.08 mol). The reaction went on for 9 h and was followed by determination of the solid material content (% SM) (the theoretical value of % SM was 60% in BuAc). An unusually high amount of initiator and molar mass regulator was used to obtain polymer with as low molar mass as possible.

4.2.2.2 Synthesis of HBP of the Second Generation A second-generation HBP (HBPG#2) was synthesized in a four-necked round-bottom flask equipped with a nitrogen inlet, a vacuum lug for water isolation, and mechanical stirring. Di-trimethylolpropane (DITMP, 0.36 mol), 2,2-dimethylolpropionic acid (bis-MPA, 4.32 mol), and the catalyst p-toluenesulfonic acid (p-TSA) were loaded in the flask. The bis-MPA required for the first generation of hyperbranched polyester was loaded first, and the reaction was led for an hour at 140°C, and the rest of the bis-MPA was then added with continuous heating at 140°C. The reaction was followed by measuring the acid number (AN). When it decreased to less than 12 (AN = 11.51), the reaction was stopped.

4.2.2.3 Synthesis of Hyperbranched Polyester with 25% of Hydroxyl Groups Modified by 3,5,5-Trimethyl Hexanoic Acid (Isononanoic Acid)-HBP (OH) HBPG#2 (0.272 mol) was loaded in a four-necked flask equipped with a nitrogen inlet, a contact thermometer lug, a mechanical impeller, and a Dean–Stark lug with a condenser and heated until it completely melted, whereupon 1.09 mol of isononanoic acid and 3 wt% of xylene for reflux were added. The temperature was increased gradually from 150 to 210°C, which was the esterification temperature. The reaction was followed by measuring the produced water and the AN. The reaction was stopped when the AN was 11.92. After that the Dean–Stark lug was changed to a vacuum lug, the nitrogen feed was stopped, and the mixture was subjected to vacuum, which led to a further decrease of the AN = 9.43.

4.2.2.4 Synthesis of HBP(OH) with 100% of Hydroxyl Groups of Acrylated HBP(A) HBP(OH) (0.045 mol) was loaded in a four-necked flask equipped with a contact thermometer lug, a mechanical impeller, and a Dean–Stark lug

TABLE 4.2 Composition of the IPN Samples (wt%)

	PU1	IPN1	IPN2	IPN3	IPN4	IPN5	IPN6
HA	65.2	43.7	43.7	43.7	43.7	43.7	43.7
N3390	34.8	23.3	23.3	23.3	23.3	23.3	23.3
EHA			8.3	16.5	24.7	16.5	
HDDA						16.5	16.5
HBP(A)		33	24.7	16.5	8.3		16.5
PU	100	67	67	67	67	67	67
PA	0	33	33	33	33	33	33

with a condenser and heated until it completely melted, whereupon 0.60 mol of acrylic acid, 1 wt% of hydroquinone (HQ), and 1.5 wt% of the catalyst p-TSA were added. Toluene (1.4 mol) for reflux was also added. To prevent the reaction of double bonds, the esterification was led with toluene reflux at 115°C. The reaction was followed by FTIR spectroscopy (decrease of the peak originating from hydroxyl groups) and the AN value. The reaction was stopped at AN = 19.7. Calcium hydroxide was added subsequently to neutralize the catalyst p-TSA, after which filtration under pressure was carried out and AN = 12.97 was attained. Precipitation was performed in petroleum ether to eliminate unreacted acrylic acid and to lower the AN, followed by precipitation in distilled water. HBP(A) was desiccated by $MgSO_4$ and filtered. The solid material content after drying was 92.5% and AN = 9.43. The hydroxyl number HB was 49.06 and the extent of modification of hydroxyl groups 88.01%.

4.2.2.5 Cured Sample Preparation Polyurethane–polyacrylic (PU/PA) IPNs were obtained by the UV–thermal dual-curing process that included the photoinitiated polymerization of double bonds of the acrylic components and the thermal polymerization of isocyanate (NCO) groups from Desmodur N3390 with the hydroxyl groups from HA.

All the IPNs had the same PU/PA weight ratio 67/33, as presented in Table 4.2.

Monomers were mixed with the photoinitiator Irgacure 184, and after stirring mixture was left to stabilize for few minutes, and subsequently a film was applied on a Teflon panel by a film applicator. The panel was left in air for a few minutes to provide evaporation of the solvent. UV curing of the acrylic component was performed by 20 passes on a moving ribbon placed in front of a UV lamp, at a rate of 6 m/min. A "PORTABLE 2" UV CURING SYSTEM" (UV Process Supply, Inc.) was used in the experiments. The sample on the panel was then placed in a heater for 2 h at 70°C

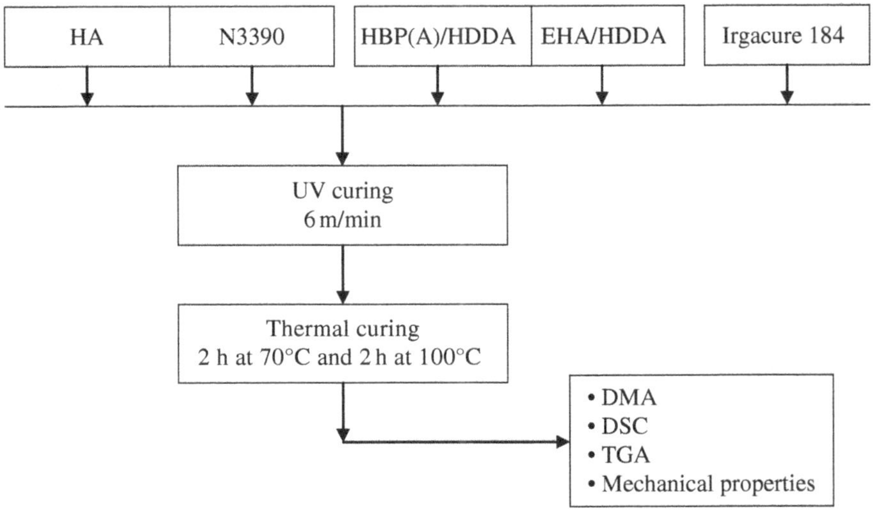

FIGURE 4.1 Sample preparation scheme. Source: Simic *et al.* [21]. Reproduced with permission of Elsevier

to achieve thermal curing and for two more hours at 100°C to complete the curing. The sample preparation scheme is shown in Figure 4.1.

4.2.3 IPN Characterization Techniques

During HBP(A) synthesis, the reaction was followed by FTIR spectroscopy employing a "BOMEM Michelson MB-102" instrument. The liquid formulations were coated, as thin films, onto a NaCl plate.

The IPN samples were characterized by the following methods DMA, DSC, and TGA and by determining the mechanical properties of the samples.

DMA was performed with a "TA Instruments Dynamic Mechanical Analyzer DMA 2980", at a frequency of 5 Hz in the temperature range 30–150°C, in tensile configuration.

DSC measurements were performed using a "PerkinElmer Pyris 6" instrument in the range 0–200°C, at a heating rate of 10°C/min.

Thermogravimetric measurements were performed on a "PerkinElmer TGA-2" instrument, under nitrogen (flow rate 22 cm^3/min), in the range between room temperature and 600°C at a heating rate of 20°C/min.

The mechanical properties measured included hardness and elasticity of the IPN samples. The hardness of the IPN samples was evaluated by monitoring the damping of the oscillations of a pendulum. The hardness values were expressed in Persoz seconds.

The elasticity of the IPN films applied on metal panels was investigated by the Erichsen cupping (DIN EN ISO 1250) method. The elasticity by Erichsen is expressed in millimeter.

Measurements of the tensile strength, maximal breaking force, and elongation at break were performed on an electromechanical pendulum INSTRON 1122.

4.3 INFLUENCE OF HBP(A) CONTENTS IN IPNs ON IPN MECHANICAL PROPERTIES

The intention was to investigate the influence of varying the weight ratio of HBP(A) and EHA on the PA component of the IPN by changing it from HBP(A)/EHA = 100/0 to 25/75 while keeping the composition of the PU component the same for all the IPNs (Table 4.2). IPNs were then compared to classical 2-pack polyurethane (PU1).

4.3.1 Dynamic Mechanical Analysis

Samples for DMA were prepared as films with the following dimensions: height ≈ 11 mm, width ≈ 5 mm, and thickness ≈ 0.5 mm. The values of the storage modulus (E′), loss modulus (E″), and loss tangent (tan δ) were determined at temperatures ranging from −30 to 150°C.

The dependence of E′ versus temperature is presented in Figure 4.2, while the data obtained for E′ at three different temperatures is shown in Table 4.3.

The temperatures of the maxima of the tan δ peaks are reported in the same table, as are the characteristics of the tan δ peaks (height-h, width-w, and surface area-S).

As may be seen in Figure 4.2, all the samples were in the glassy phase at room temperature, and the obtained storage modulus were about 10^9 Pa (~10^3 MPa). There is a slight increase of E′ in the rubbery plateau of IPN2, probably due to additional curing. According to the height of the rubbery plateau of E′, it is possible to obtain information on the crosslink density. Increase of the peak height is due to increasing crosslink density and decreasing flexibility of the IPN samples. It may be seen that the crosslink density decreases from IPN1 to IPN4 and, therefore, concluded that increasing the HBP(A) amount in UV-curable component makes IPNs more crosslinked and less flexible. Samples with 100 and 75 wt% of HBP(A) in UV-curable component have higher crosslink density than PU1, while those with 50 and 25 wt% of HBP(A) are less crosslinked, but with significantly higher T_gs than classical 2-pack polyurethane (PU1).

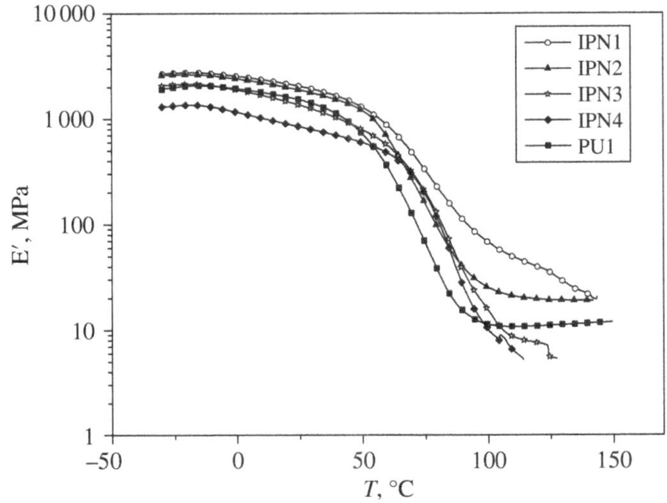

FIGURE 4.2 Storage modulus (E′) versus temperature curves for the investigated IPNs. Source: Simic *et al.* [21]. Reproduced with permission of Elsevier

TABLE 4.3 Damping Characteristics of the IPNs

	T_g, °C		E′, MPa			tan δ characteristics		
Sample	$(\tan \delta)_{max}$	$(E'')_{max}$	at 30°C	at 70°C	at 120°C	h	w, °C	S
IPN1	81.6	69.3	1.9×10^3	4.4×10^2	3.9×10^1	0.42	104	21.8
IPN2	80.6	68.4	1.8×10^3	2.6×10^2	2.0×10^1	0.64	74	23.8
IPN3	87.7	72.7	1.2×10^3	3.0×10^2	7.5×10^0	0.68	73	25.1
IPN4	88.4	74.2	7.9×10^2	2.9×10^2	—	0.78	72	28.1
PU1	79.9	59.5	1.4×10^3	1.4×10^2	1.1×10^1	0.81	60	24.2

Source: Simic *et al.* [21]. Reproduced with permission of Elsevier.

The tan δ versus temperature curves are reported in Figure 4.3. The maxima of the tan δ peaks define the glass transition temperatures (T_g) of the IPNs. IPN1 and IPN2 have similar T_gs, as do the T_gs for IPN3 and IPN4. It may be seen that T_g decreases with increasing HBP(A) content in the UV component. All dual-curing formulations upon hardening have higher T_gs than classical 2-pack polyurethane (PU1).

A single T_g of the IPNs indicates their homogeneity. It is also possible to assess their homogeneity according to the width of the tan δ peak. IPN4 has the narrowest tan δ peak, indicating the highest homogeneity of the sample. Comparison of the width of the tan δ peaks indicates that the homogeneity decreases with increasing HBP(A) content in the IPNs (Table 4.4).

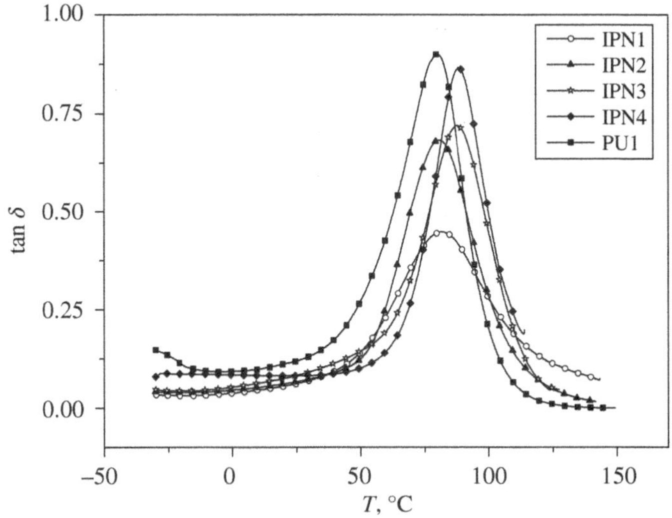

FIGURE 4.3 The dependence of tan δ on temperature for the investigated IPNs. Source: Simic *et al.* [21]. Reproduced with permission of Elsevier

TABLE 4.4 Glass Transition Temperatures of IPN Samples Obtained by DMA and DSC

Samples	T_g, °C (DMA-tg δ)	T_g, °C (DMA-E″)	T_g, °C (DSC)
IPN1	81.6	69.3	51
IPN2	80.6	68.4	50
IPN3	87.7	72.7	61
IPN4	88.4	74.2	62
PU1	79.9	59.5	52

Source: Simic *et al.* [21]. Reproduced with permission of Elsevier.

The width of the tan δ–T peak depends on the nature of the polymer–plasticizer interactions. If there is phase separation, the peak becomes wider. So, with increasing plasticizer content, the width of the tan δ peak increases, the height decreases, and the peak shifts toward lower temperatures. Comparison of the tan δ–T curves, shown in Figure 4.3, indicates that HBP(A) acts as a plasticizer, since the width of the tan δ peak increases, the height decreases, and the peak shifts toward lower temperatures with increasing HBP(A) content.

4.3.1.1 Thermogravimetric Analysis Thermogravimetric curves for IPN samples are presented in Figure 4.4, and the temperatures of 10, 50, and 90% mass loss are reported in Table 4.5.

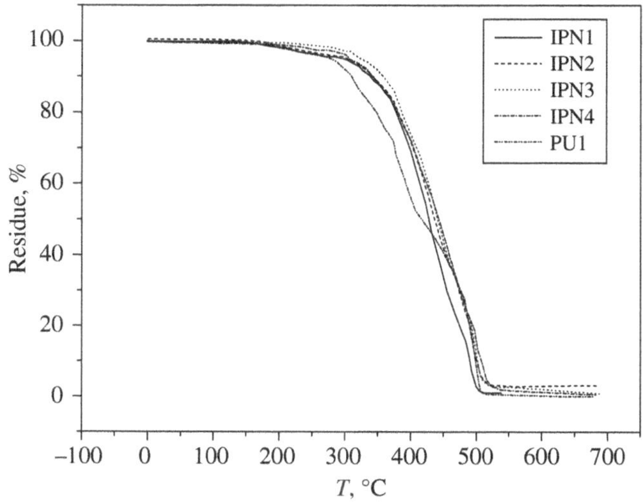

FIGURE 4.4 TG curves of the investigated IPN samples. Source: Simic *et al.* [21]. Reproduced with permission of Elsevier

TABLE 4.5 Temperatures of 10, 50, and 90% Mass Loss of the IPN Samples during Thermal Degradation, at a Heating Rate of 10°C/min

Sample	$T_{10\%}$	$T_{50\%}$	$T_{90\%}$
IPN1	338	428	491
IPN2	340	437	501
IPN3	358	441	502
IPN4	341	443	507
PU1	309	413	496

Source: Simic *et al.* [21]. Reproduced with permission of Elsevier.

It was observed that the sample IPN3 was the most thermally stable. All IPNs are more stable than PU1.

The mechanical properties of the IPNs are reported in Table 4.6: hardness, flexibility, tensile strength, maximal breaking force, and elongation at break. The hardness of the IPNs increases with increasing HBP(A) content in the UV component, while the flexibility remains the same for all the samples and is maximal, 8 mm. The hardness of classical PU coating (PU1) is increased by 8.5% by introduction of UV component with 100 and 75 wt% of HBP(A), while it is decreased when the content of HBP(A) in the UV component is less than 75%.

TABLE 4.6 Mechanical Properties of the IPN Samples

IPN Samples	Hardness, Persoz sec	Elasticity, mm	Max. Breaking Force, F_{max}, N	Tensile Strength, σ_z, MPa	Elongation, A, %
IPN1	295	8	155.0	32.5	5
IPN2	283	8	268.0	42.3	8.5
IPN3	252	8	132.0	30.1	5.5
IPN4	222	8	102.7	24.2	16.7
PU1	272	8	208.7	40.3	4.7

Source: Simic *et al.* [21]. Reproduced with permission of Elsevier.

IPN2 has the highest tensile strength, followed by PU1, IPN1, IPN3, and IPN4. On the basis of the obtained results, it may be concluded that the sample IPN2 has the best characteristics considering strength and stiffness because it has high tensile strength and elongation at break.

4.4 INFLUENCE OF REACTIVE DILUENT IN INTERPENETRATING POLYMER NETWORKS ON THEIR PROPERTIES

The intention was not only to investigate the influence of reactive diluent in IPNs on their behavior but also to see more clearly the role of the HBP component. The composition of the PU component was kept the same in all samples. In IPN3 the reactive diluent was EHA, while in IPN6 it was HDDA. In IPN5 the reactive diluent was EHA, but instead of HBP(A), HDDA was used. The IPN samples were also compared with classical 2-pack polyurethane (PU1).

All the IPN samples were prepared in the same way as described in Part 2.2. The dependence of E′ versus temperature is presented in Figure 4.5, while the data obtained for E′ at three different temperatures is shown in Table 4.7. The temperatures of the maxima of the tan δ peaks are reported in the same table, as are the characteristics of the tan δ peaks (height-h, width-w, and surface area-S).

All IPNs are in glassy state at room temperature, having E′ about 10^9 Pa. All the samples except IPN5 have one T_g which indicates good homogeneity between two phases. In IPN5 two T_g values indicate that phases are not homogeny mixed.

Firstly IPN3 and IPN5 were compared. Those samples differ in PA composition: in IPN5, instead of hyperbranched polyester with high acrylate functionality (theoretically 12), HDDA – a linear, low molecular weight compound was used.

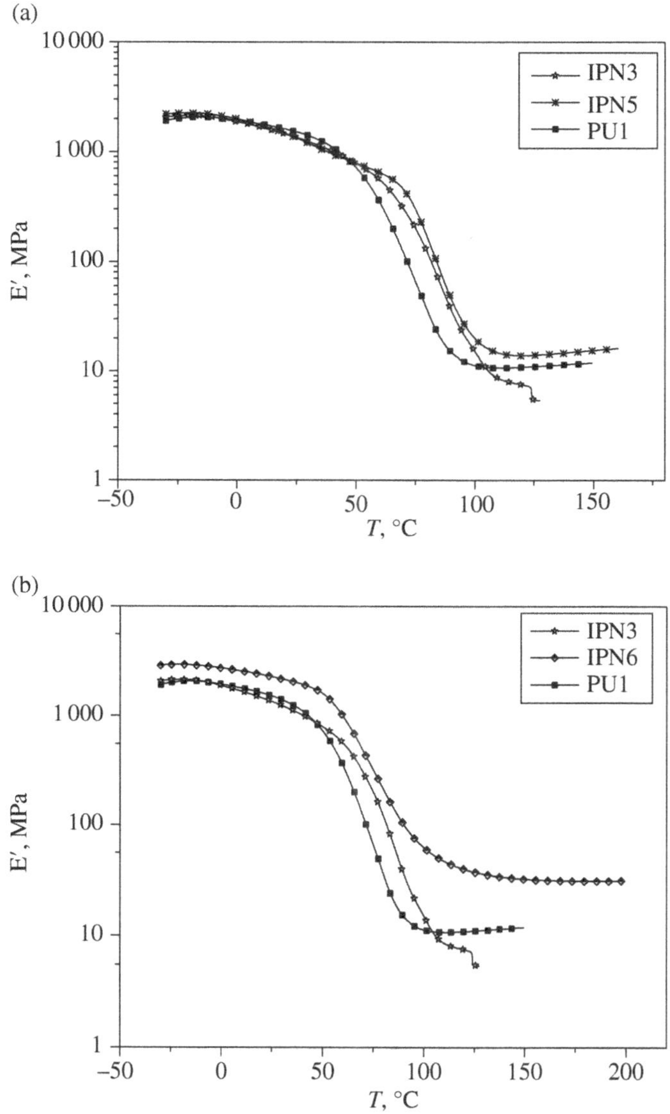

FIGURE 4.5 Storage modulus (E′) versus temperature curves for the investigated IPNs. (a) The influence of HBP(A)/HDDA in acrylic component of IPNs on storage modulus and (b) the influence of different reactive diluent in IPNs on storage modulus Source: Marinovic *et al.* [22]. Reproduced with permission of Elsevier

They are both in glassy state at room temperature. Surprisingly, the DMA spectra of both IPNs are very similar, indicating that much higher molecular weight and functionality of HBP(A) than HDDA do not have much influence on the mechanical properties of IPN. It can be seen that there is a slight increase of E′ in the rubbery plateau of IPN5, probably due to additional curing. The height of the rubbery plateau of E′ indicates the crosslink density of networks. Increase of height is due to increase of crosslink density and decrease of flexibility of the IPN samples. It can be seen that IPN5 has higher crosslink density than IPN3. It can be concluded that HBP(A) with EHA gives better phase homogeneity but lower crosslink density of networks in IPNs than HDDA.

The tan δ versus temperature curves are reported in Figure 4.6.

TABLE 4.7 Damping Characteristics of the IPNs

	T_g, °C		E′, MPa			tan δ Characteristics		
Sample	$(\tan\delta)_{max}$	$(E'')_{max}$	at 30°C	at 70°C	at 120°C	h	w, °C	S
IPN3	87.7	72.7	1.2×10^{3}	3.0×10^{2}	7.5×10^{0}	0.68	73	25.1
IPN5	36.8/88	24.6/76	1.2×10^{3}	4.6×10^{2c}	1.4×10^{1}	0.64	63	20.1
IPN6	82.4	62.7	2.2×10^{3}	4.7×10^{2}	3.9×10^{1}	0.48	83	20.0
PU1	79.9	59.5	1.4×10^{3}	1.4×10^{2}	1.1×10^{1}	0.81	60	24.2

Source: Marinovic *et al.* [22]. Reproduced with permission of Elsevier.

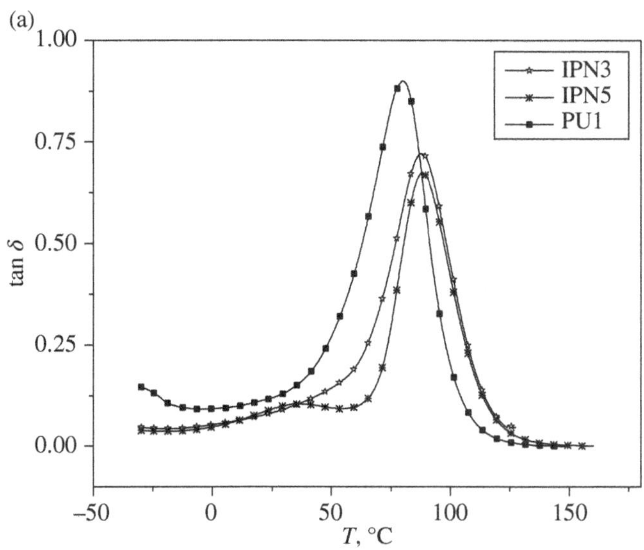

FIGURE 4.6 The dependence of tan δ on temperature for the investigated IPNs. (a) The influence of HBP(A)/HDDA in acrylic component of IPNs on tan δ, (b) the influence of different reactive diluent in IPNs on tan δ, and (c) the comparison of IPNs with different acrylic component composition with respect to tan δ. Source: Marinovic *et al.* [22]. Reproduced with permission of Elsevier

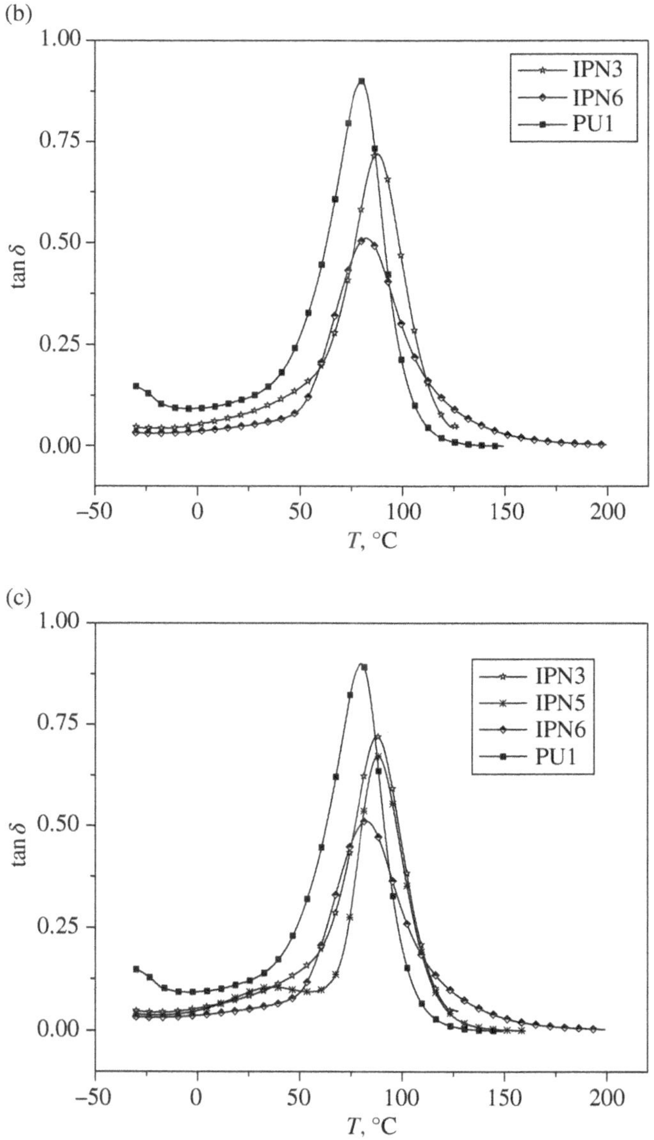

FIGURE 4.6 (*Continued*)

The maxima of the tan δ peaks define the glass transition temperatures (T_g) of the IPNs. IPN3 shows one T_g indicating homogenous phases of that samples, but IPN5 shows two T_g peaks indicating phase nonhomogeneity. Both dual-curing formulations upon hardening have higher T_gs than classical 2-pack polyurethane (PU1).

Comparing IPN3 and IPN6 which both have HBP(A) as one component of PA part of IPN, but have different reactive diluents (EHA in IPN3 and HDDA in IPN6), it can be seen that they both are in glassy state at room temperature. IPN6 has higher values of E′ in the whole temperature range, which is more prominent in the rubbery plateau showing much higher crosslink density of IPN5 than IPN3 or PU1.

The tan δ versus temperature curves show that both IPN samples have single T_g indicating homogeneity of phases. IPN5 has lower T_g than IPN3. The width of the tan δ peak indicates sample homogeneity. IPN3 has the lower width of the tan δ peak which means that its homogeneity is higher. This clearly shows the influence of reactive diluent.

If all the IPN samples were compared (Figure 4.6c), it can be seen that IPN5 has the lowest T_g. This can be attributed to the absence of high functionality HBP(A). All the IPNs have higher T_g than PU1.

Data for T_g values obtained by DMA and DSC are also given in Table 4.8. It can be seen that DSC gives lower values for T_g than DMA.

Thermogravimetric curves for Interpenetrating Polymer Networks samples are presented in Figure 4.7.

It was observed that the sample IPN3 was the most thermally stable, while IPN5 has the lowest thermal stability among IPNs. All IPNs are more stable than PU1. Higher thermal stability of samples with HBP(A) than classic 2-pack PU clearcoat composition can be explained by higher crosslink density. Also, samples with HBP(A) have higher thermal stability than the sample IPN5 that has HDDA instead of HBP(A).

Other properties of the IPNs were reported in Table 4.9: hardness, flexibility, tensile strength, maximal breaking force, and elongation at break.

It can be seen that hardness is highest with IPN6 and IPN6 also has tensile strength much higher than other samples. So it can be concluded that IPN6 has best characteristics considering strength and stiffness, indicating that combining high functionality [21] acrylated polyester with two-functional reactive diluent (HDDA) yields coating with superior properties.

TABLE 4.8 T_g Values of IPN Samples Obtained by DMA and DSC

Samples	T_g, °C (DMA-tg δ)	T_g, °C (DMA-E″)	T_g, °C (DSC)
IPN3	87.7	72.7	61
IPN5	36.7/88	24.6/76	59
IPN6	82.4	62.7	50.4
PU1	79.9	59.5	55

Source: Marinovic *et al.* [22]. Reproduced with permission of Elsevier.

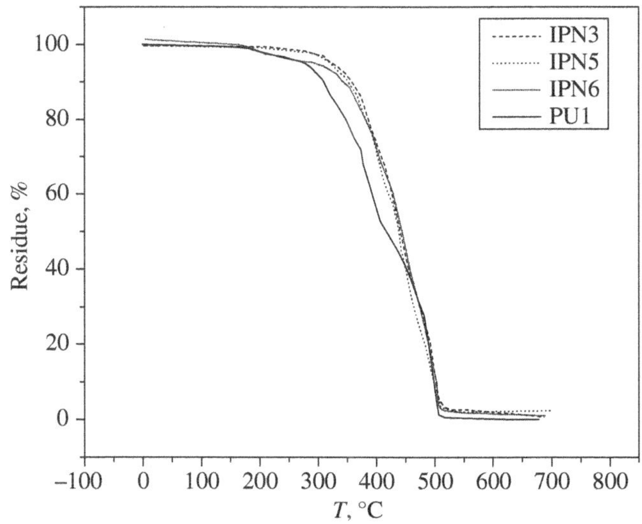

FIGURE 4.7 TG curves of the investigated IPN samples. Source: Marinovic *et al.* [22]. Reproduced with permission of Elsevier

TABLE 4.9 Mechanical Properties of the IPN Samples

IPN Samples	Hardness, Persoz s	Elasticity, mm	Max. Breaking Force, F_{max}, N	Tensile Strength, σ_z, MPa	Elongation, A, %
IPN3	252	8	132.0	30.15	5.5
IPN5	237	8	132.7	30.70	13.0
IPN6	285	7.8	282.0	53.20	8.5
PU1	272	8	208.7	40.30	4.7

Source: Marinovic *et al.* [22]. Reproduced with permission of Elsevier.

IPN6 sample has also the widest tan δ peak, meaning it has the highest flexibility and lowest homogeneity of the network of all IPNs.

4.5 CONCLUSIONS

IPNs were synthesized by the UV–thermal dual-cure process. Acrylic part was cured by UV radiation, and urethane part was crosslinked by heating at 70°C for 2 h and then at 100°C for another 2 h.

HBP(A) was used as part of the acrylic component. Apart from IPN5 all other samples are homogeneous materials with no phase separation, as evidenced by single T_g in DMA and DSC curves. These techniques have shown that HBP(A) lowers the IPN's T_g. HBP(A) plays the role of a plasticizer, as the T_g becomes lower as the weight ratio of HBP(A) increases in the IPNs.

The DMA results showed that the IPN crosslink density increases and flexibility decreases with increasing amount of HBP(A). Samples with 100 and 75 wt% of HBP(A) in UV-curable component have higher crosslink density than PU1, while those with 50 and 25 wt% of HBP(A) are less crosslinked, but with significantly higher T_gs than classical 2-pack polyurethane (PU1).

When it comes to thermal stability, all IPNs have higher stability than PU1. Samples with HBP(A) in PA part have higher thermal stability than sample with HDDA instead of HBP(A), indicating that hyperbranched structure combined with high functionality yields networks with fewer weak points.

ACKNOWLEDGMENT

The author is grateful to the Ministry of Education and Science of the Republic of Serbia for the support (Project No III45001).

REFERENCES

[1] Sperling, L.H. (1981) *Interpenetrating Polymer Networks and Related Materials*, Plenum Press, New York.

[2] Huelck, V., Thomas, D.A. and Sperling, L.H. (1972) *Macromolecules*, **5**, 340.

[3] Curtius, A.J., Covitch, M.J., Thomas, D.A. and Sperling, L.H. (1972) *Polymer Engineering & Science*, **12**, 101 reprinted in Rubber Chemistry and Technology, 45 (1972) 1582.

[4] Yang, J., Winnik, M.A., Ylitalo, D. and De Voe, R.J. (1996) *Macromolecules*, **29**, 7047.

[5] Ramis, X., Cadenato, A., Morancho, J.M. and Salla, J.M. (2001) *Polymer*, **24**, 9469.

[6] Chen, L. and Chen, S. (2004) *Progress in Organic Coatings*, **49**, 252.

[7] Sangermano, M., Malucelli, G., Priola, A. and Manea, M. (2006) *Progress in Organic Coatings*, **55**, 225.

[8] Kim, Y.H. (1998) *Journal of Polymer Science Part A: Polymer Chemistry*, **36**, 1685.

[9] Tasić, S., Božić, B. and Dunjić, B. (2004) *Progress in Organic Coatings*, **51**, 321.

[10] Hult, A., Johansson, M. and Malmström, E. (1999) *Advances in Polymer Science*, **143**, 1.

[11] Dzunuzovic, E., Tasić, S., Božič, B. *et al.* (2005) *Progress in Organic Coatings*, **52**, 136.

[12] Dzunuzovic, E., Tasić, S., Bozic, B. *et al.* (2004) *Journal of the Serbian Chemical Society*, **69**, 441.

[13] Dunjić, B., Tasić, S. and Božić, B. (2004) *European Coatings Journal*, **6**, 36.

[14] Kumari, S., Mishra, A.K., Chattopadhyay, D.K. and Raju, K.V.S.N. (2007) *Journal of Polymer Science, Part A: Polymer Chemistry*, **45**, 2673.

[15] Chen, Z. and Webster, D.C. (2007) *Polymer International*, **56**, 754.

[16] Xu, G. and Shi, W. (2005) *Progress in Organic Coatings*, **52**, 110.

[17] Johansson, M., Malmström, E., Hult, A. and Claesson, H. (2002) *European Coatings Journal*, **7–8**, 26.

[18] Van Benthem, R.A.T.M. (2000) *Progress in Organic Coatings*, **40**, 203.

[19] Fouassier, J.P. and Rabek, J.C. (1993) *Radiation Curing in Polymer Science and Technology*, vol. **I**, Elsevier, London.

[20] Schwalm, R., Häuβling, L., Reich, W. *et al.* (1997) *Progress in Organic Coatings*, **32**, 191.

[21] Simić, S., Dunjić, B., Tasić, S. *et al.* (2008) *Progress in Organic Coatings*, **63**, 43.

[22] Marinović, S., Popović, I., Dunjić, B. *et al.* (2010) *Progress in Organic Coatings*, **68**, 293.

5

MACRO-, MESO-, AND NANOPOROUS SYSTEMS DESIGNED FROM IPNs

Daniel Grande

Institut de Chimie et des Matériaux Paris-Est (ICMPE), UMR 7182 CNRS-Université Paris-Est Créteil, Thiais, France

5.1 INTRODUCTION

Over the past few decades, much attention has been paid to the design of porous materials because they find a large variety of applications in many areas, including advanced microelectronics, biomolecule immobilization, controlled drug delivery, tissue engineering, separation and purification techniques, gas or ion transport, as well as supported catalysis [1–6]. Organic porous materials have unique properties over their inorganic analogues, such as tunable mechanical properties, ease to be functionalized, higher biocompatibility, and especially lower production cost. The basic idea associated with the design of porous polymers relies on the introduction of various types of porogens (solvents, gases, small molecules, or macromolecules) within polymer structures, followed by their selective removal. In the early stages, porous polymers were synthesized using simple organic solvents and gases as porogens. For instance, the first porous polymer was prepared by free radical copolymerization of styrene and divinylbenzene (DVB) using an organic solvent as the porogen as described by

Micro- and Nano-structured Interpenetrating Polymer Networks: From Design to Applications, First Edition. Edited by Sabu Thomas, Daniel Grande, Uroš Cvelbar, K.V.S.N. Raju, Ramanuj Narayan, Selvin P. Thomas and Akhina H.

Staudinger in 1935 [7]. Solvents and gases used as porogens just create pores by physically or chemically induced phase separation, but the microstructure of the resulting materials is generally characterized by a broad pore size distribution.

Original approaches toward porous polymers with a defined porosity consist in using suitable templates that induce specific structural pores. Various types of simple organic molecules and specific macromolecules have been used as effective porogen templates for creating a wealth of porous materials with desired characteristics, in view of their promising applications from size/shape-selective nanoreactors to inner-layer dielectrics. Therefore, the design of diverse macromolecular precursors has become increasingly important for the generation of different kinds of porous materials that exhibit isotropic and anisotropic polymer morphologies with closed or open pore structures and a pore size range from the nanometer region to several micrometers (Table 5.1). Macro- (pore diameter > 50 nm), meso- (pore diameter between 2 and 50 nm), and microporous (pore diameter < 2 nm)—according to IUPAC classification [8]—materials have been prepared depending on the porogen type and/or the pore generation method. The major template-oriented methodologies include molecular imprinting [9, 10], removal of self-assembled molecules from supramolecular architectures [11], selective etching of one block in self-organized block copolymers [12–21], and selective removal of a porogen homogeneously blended within a thermostable polymer matrix [22–25]. Even though semidegradable block copolymers are arguably ideal precursors for the formation of ordered mesoporous systems, the relatively poor mechanical, thermal, and chemical stability of uncrosslinked polymer structures entails that crosslinking of the remaining block in the residual porous matrices generally appears to be crucial to get access to some of the aforementioned applications.

An alternative and original strategy for engineering porous crosslinked materials has been put forward only by a handful number of research teams via the use of (semi-)interpenetrating polymer networks (IPNs) as nanostructured precursors [26]. Semi-IPNs result from the formation of a network in the presence of an uncrosslinked polymer; the latter will act as a macromolecular porogen template and will afford the porosity in the material upon its selective degradation or solvent extraction. IPNs are generally defined as an intimate combination of two independently crosslinked polymers, at least one of which has been synthesized in the immediate presence of the other [27–29]. The network structure may result from either chemical crosslinking of polymer chains by covalent bonding or physical crosslinking of segments of multiple polymer chains containing crystalline or glassy phase domains. The most important types of IPNs are defined as follows: sequential IPNs are created by forming the two constituting subnetworks via two successive and independent crosslinking processes,

TABLE 5.1 Major Template-Oriented Methodologies Used for the Design of Porous Polymeric Materials

Methodology	Porogen Template	Porous Structure	Pore Size Range	Targeted Applications
Molecular imprinting	Small molecules or macromolecules, even cells	Crosslinked poly(meth) acrylates, polystyrenes, polyurethanes, and so on	1–50 nm	Separation, detection, solid-phase extraction, biosensors, biomedical and pharmaceutical systems
Selective destruction of supramolecular architectures	Self-assembled (macro) molecules	Polystyrene-*b*-poly (4-vinylpyridine), polyaniline, polyisoprene, phenolic resins, and so on	1 nm–10 μm	Smart membranes, confined chemistry (catalysis, nanomaterials design, etc.)
Selective etching of block copolymers	Sacrificial polymeric block or sequence	Polystyrenes, poly(meth) acrylates, and so on (uncrosslinked in general)	≥10 nm	Nanofiltration, confined chemistry (catalysis, nanomaterials design, etc.)
Selective removal of a porogen within thermostable matrices	Photo- or thermolabile thermoplastics, oligomers, plasticizers	Epoxy- and cyanurate-based thermosets, phenolic resol resins	≤1 μm	Interlayer dielectrics, membranes or adsorbents working under severe conditions
Selective removal of one partner from (semi-)IPNs	Uncrosslinked oligomers or polymers, polymer subnetwork	Crosslinked poly(meth) acrylates, polystyrenes, polyethylenes, and so on	10 nm–a few μm	Separation techniques, confined chemistry

while simultaneous IPNs are derived from two crosslinking processes carried out simultaneously. Both subnetworks must be generated by noninterfering reaction routes, namely, different polymerization mechanisms and/or distinct experimental conditions. Due to their interlocking framework, macrophase segregation may be prevented in IPNs, and the spatial scale of phase separation may be restricted to domain sizes of a few hundreds or even tens of nanometers. It is noteworthy that IPN morphology is mainly governed by the interplay between two types of competitive factors: thermodynamic ones, that is, intermolecular interactions between both constitutive partners, versus kinetic ones, that is, competition between crosslinking and phase separation. Hence, the microphase-separated morphologies developed in IPNs can be synthetically controlled by the proper selection of component nature and composition, as well as polymerization kinetics and mechanisms. Such complex polymer structures are of particular interest when they arise from the association of two components that exhibit a contrasted degradability under specific conditions. Indeed, (nano)porous networks can be generated from such IPNs by resorting to selective degradation methods [26]. Although detailed studies on porous polymer systems designed from (semi-)IPNs remain scarce, the most relevant literature data on such systems, and particularly our contributions in this research field, are addressed in the present chapter.

5.2 POROUS SYSTEMS DERIVED FROM SEMI-INTERPENETRATING POLYMER NETWORKS

5.2.1 Porous Networks by Selective Degradation of Uncrosslinked Chains

Polymeric hydrogels based on semi-IPNs composed of poly(2-hydroxyethyl methacrylate) (PHEMA) and poly((D,L-lactide)-*co*-(ε-caprolactam)) (poly(LA-*co*-CLM)) were reported by Baysal's group [30]. Linear poly(LA-*co*-CLM) chains were previously synthesized, and they were then interpenetrated into PHEMA subnetworks that were prepared by free radical copolymerization in the presence of uncrosslinked copolymer chains. The extraction results showed that the entrapping level between the two components was very high. However, the presence of main-chain ester linkages in the linear copolymer partner allowed its gradual and selective hydrolytic degradation under mild conditions (pH 7.4, T=37°C) that led to the formation of biocompatible porous PHEMA hydrogels. The scanning electron microscopy (SEM) analysis of semi-IPNs, before and after hydrolysis, was used to monitor the pore generation and structure.

5.2.2 Porous Networks by Solvent Extraction of Uncrosslinked Chains

Our group reported the synthesis and characterization of semi-IPNs based on various linear oligoesters and poly(methyl methacrylate) (PMMA) subnetworks, and we examined the scope and limitations of utilizing such reference systems as precursors to (nano)porous networks [31–34]. Porous polymer networks were readily obtained by mere extraction of uncrosslinked oligoesters from various oligoester/PMMA (50/50 wt%) semi-IPNs (Figure 5.1). Such an extraction was effected for 24 h in a Soxhlet apparatus at 40°C (CH_2Cl_2 reflux).

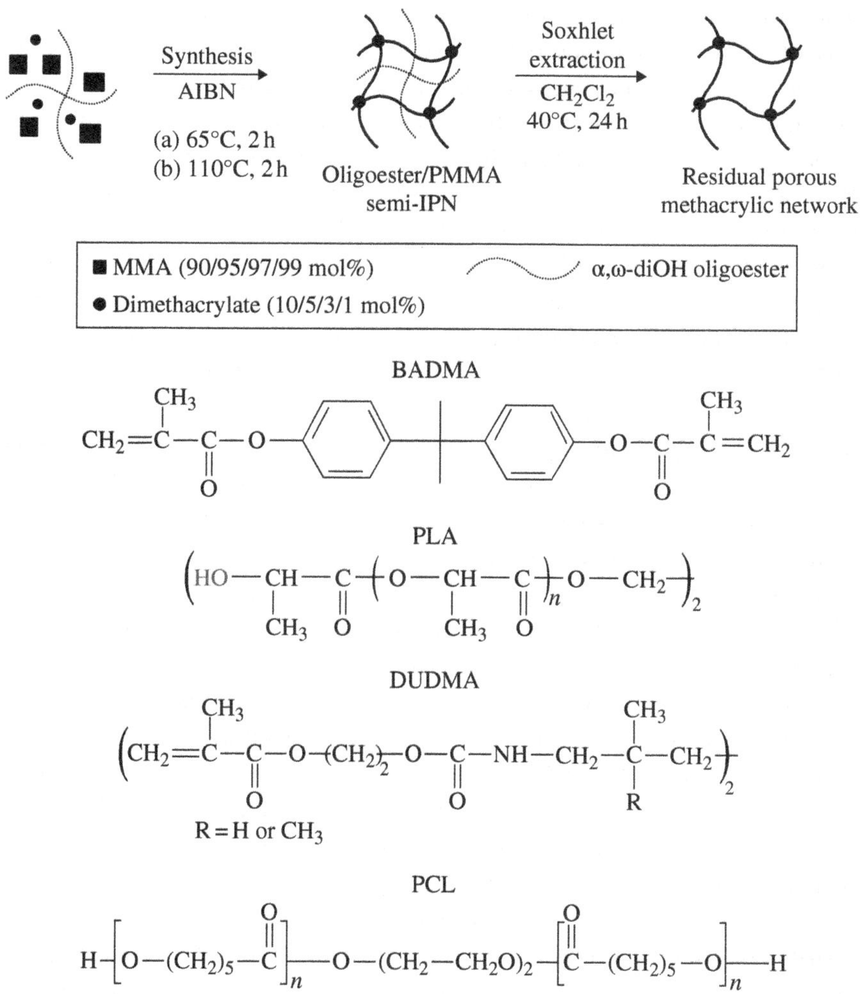

FIGURE 5.1 Design of porous networks from oligoester/PMMA semi-IPNs

Regardless of the dimethacrylate nature (bisphenol A dimethacrylate (BADMA) or diurethane dimethacrylate (DUDMA)) and the crosslink density of PMMA network, as well as the oligoester nature (poly(ε-caprolactone) (PCL) or PLA) and molar mass ($500 < M_n < 4000\,g{\cdot}mol^{-1}$), the extraction was quantitative, as indicated by values close to 50 wt% found for the sol fractions. The complete disappearance of the characteristic bands of oligoesters in the FTIR and solid-state ^{13}C NMR spectra of extracted semi-IPNs confirmed this assertion. Moreover, the T_g values of the residual porous networks clearly matched those of the corresponding PMMA single networks (110–160°C). It is noteworthy that such a quantitative extraction was performed with a good solvent of oligoesters at a temperature far below the T_g value of PMMA networks to avoid the collapse of the residual porous structures. The SEM analysis of semi-IPN samples before extraction displayed compact structures without any pore. The corresponding PMMA single networks, after extraction, did not exhibit any pore either, indicating a high compacity in this type of networks. In sharp contrast, the morphology exhibited by extracted semi-IPNs was dramatically different, despite identical chemical structures. The micrographs of the latter samples revealed highly porous structures, thus showing the effective role of oligoester subchains as template porogens (Figure 5.2 as typical examples). Pore sizes varied from a few micrometers to tens of nanometers, depending on the structural parameters used. Although SEM is a common and easy tool to observe porous materials, it is not a very precise technique to determine pore sizes smaller than a few tens of nanometers. Accordingly, we resorted to an alternative technique, namely, DSC-based thermoporometry, which is a quantitative and sensitive technique allowing for the precise determination of pore size distributions in a wide array of porous materials, provided pore diameters are smaller than about 200–300 nm. It relies on the melting temperature (T_m) depression and the Gibbs–Thomson effect shown by a solvent constrained within the pores [35–37]. When applying thermoporometry through DSC measurements using water as the penetrant solvent to the porous networks under investigation, two endothermic peaks were generally detected, that is, one related to the melting of bulk water with a maximum around 0°C and a second one, not fully resolved, at lower temperatures arising from the melting of water confined within the pores. The use of PLA (instead of PCL) and the lowest molar mass oligomers ($M_n \sim 500\,g{\cdot}mol^{-1}$) increased the degree of miscibility between both semi-IPN partners, thus leading to a significant decrease in pore sizes and a narrowing of pore size distributions of the corresponding porous networks [31, 34]. A decrease in the crosslinker content led to a decrease in pore sizes, and the pore size distributions became narrower. Moreover, because of the presence of highly favorable hydrogen

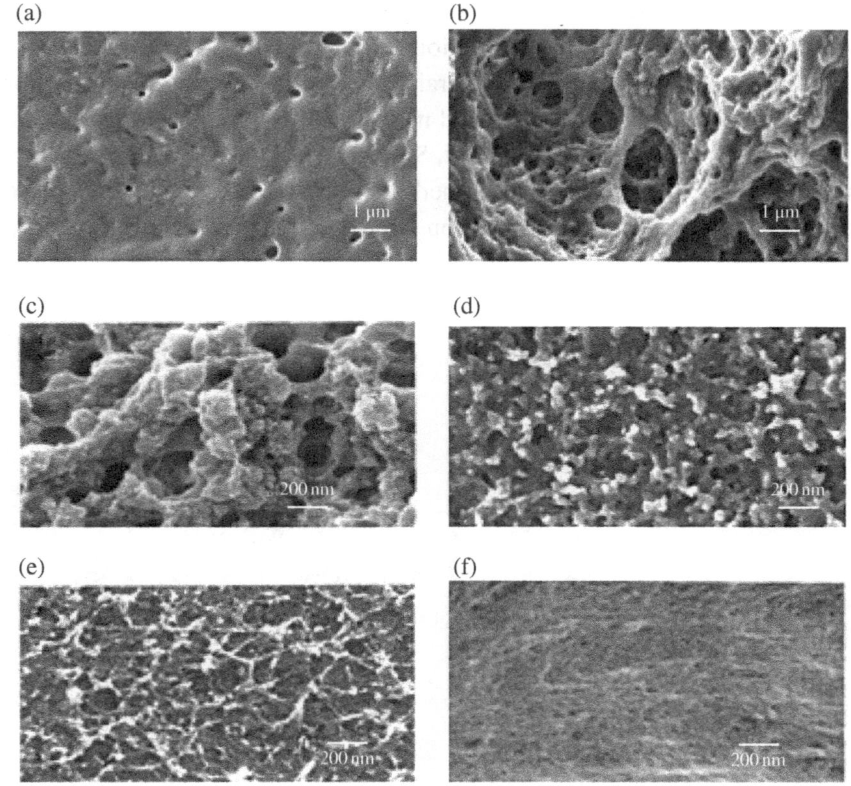

FIGURE 5.2 SEM micrographs of oligoester/PMMA (50/50 wt%) semi-IPNs: (a) Before extraction, PCL/PMMA (MMA/BADMA 90/10 mol%). After extraction, (b) PCL/PMMA (MMA/BADMA 90/10 mol%). (c) PLA/PMMA (MMA/BADMA 90/10 mol%). (d) PLA/PMMA (MMA/DUDMA 90/10 mol%). (e) PLA/PMMA (MMA/BADMA 99/1 mol%). (f) PLA/PMMA (MMA/DUDMA 90/1 mol%)

bonds in DUDMA-derivatized semi-IPNs, the extent of miscibility between both constitutive partners was enhanced, and pore sizes in the resulting porous networks were smaller than those in their BADMA-based homologues [33]. Actually, the dependence of pore sizes in such porous materials mirrored the differences observed in semi-IPN precursors from visual aspects, turbidity and DSC measurements, as well as polymer–polymer miscibility inferred from the calculation of oligoester/PMMA interaction parameters [33]. As a summary, a low molar mass ($M_n < 2000\,g{\cdot}mol^{-1}$) PLA oligomer and the lowest dimethacrylate content were the optimal parameters for preparing mesoporous networks with narrow pore size distributions.

More recently, semi-IPN systems in which PLA or PCL oligomers were embedded within a functional vinylic polymer network could be effectively used as nanostructured precursors to novel functionalized (meso)porous networks [38, 39]. Indeed, the extraction of uncrosslinked oligoesters from functional semi-IPNs constituted a straightforward and versatile strategy for engineering (meso)porous crosslinked methacrylic or styrenic polymers with miscellaneous functionalities (COOH, SO_3H, pyridyl groups). In this regard, the utilization of PCL/poly(methyl methacrylate-*co*-methacrylic acid)-based semi-IPNs afforded COOH-functionalized nanoporous networks [39].

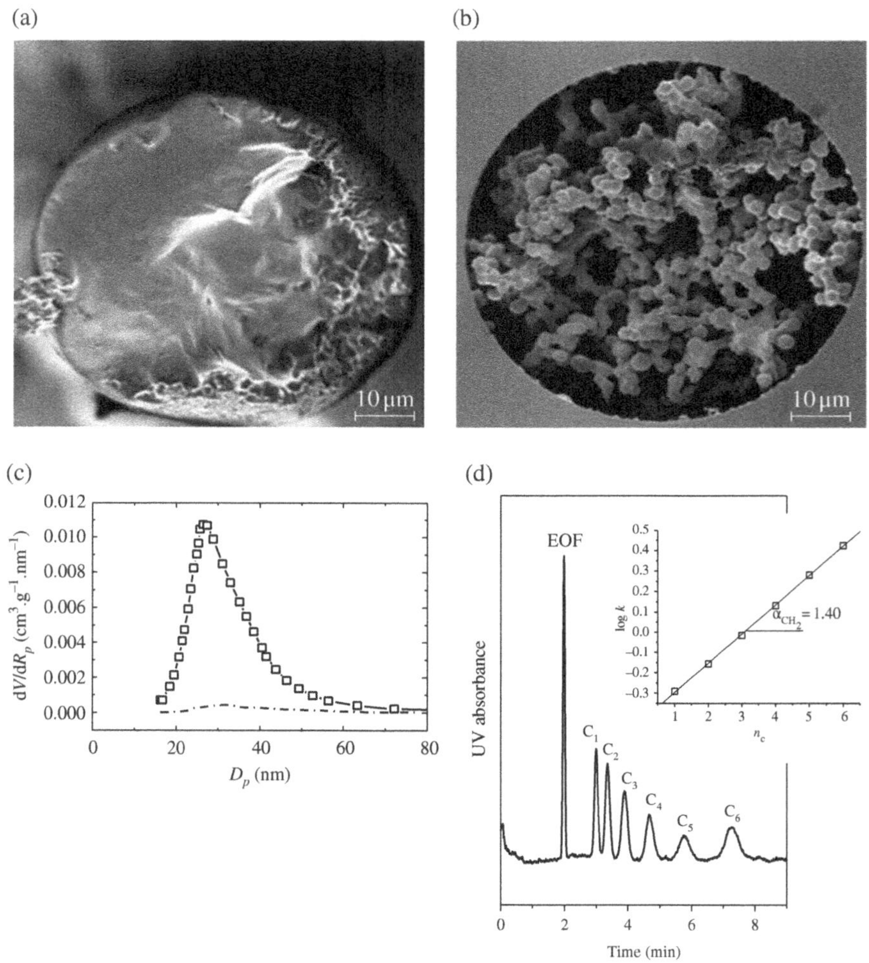

FIGURE 5.3 SEM micrographs of a PCL/PS (40/60 wt%) semi-IPN monolithic capillary column: (a) Before PCL extraction. (b) After PCL extraction. (c) Pore size distribution profiles as determined by DSC-based thermoporometry for corresponding bulk monolith (-·- before PCL extraction, □ after PCL extraction). (d) Electrochromatographic separation of alkyl benzenes injected on PS-based monolithic column

Interestingly, such functional frameworks could be used as cation exchange supports in ion exchange chromatography for the separation of proteins (α-chymotrypsin, cytochrome *c*, lysozyme), provided the structures contain a significant initial content of carboxylic acid functions ($[COOH]_0 > 2\,meq \cdot g^{-1}$). This investigation illustrated the major role played by the presence of the interconnected pores generated by the oligoester template in the protein retention. The resolution turned out to be better than that obtained with a classically prepared porous support using an organic solvent (THF) as a porogen.

Lately, PCL/polystyrene (PS) semi-IPNs were also designed for the development of porous capillary monoliths [40, 41]. Extraction of the uncrosslinked PCL oligomers from the PS matrix allowed for the generation of the corresponding porous monolithic materials that displayed a vast majority of macropores—ranging from 100 nm to a few μm—and a small amount of mesopores in the 30 nm range. Such as-obtained porous monoliths could be efficiently used as reversed-phase stationary phases for the rapid separation of alkyl benzene derivatives by capillary electrochromatography (Figure 5.3).

5.3 (NANO)POROUS SYSTEMS DERIVED FROM INTERPENETRATING POLYMER NETWORKS

5.3.1 Pioneering Studies

In pioneering works, the primary aim was less preparation and exploitation of porous materials rather than the development of an approach to get further insights into the cocontinuous morphologies of IPNs. The first porous polymers were derived from sequential IPNs based on poly(*n*-butyl acrylate) (PBA) and PS of various compositions in which one subnetwork contained labile crosslinks resulting from acrylic acid anhydride (AAA) and the other subnetwork contained stable crosslinks arising from DVB [42, 43]. De-crosslinking of the AAA-based network was accomplished by hydrolyzing selectively the IPNs in a 10% ammonium hydroxide aqueous solution, thus leading to the formation of porous networks. SEM was used to examine the morphology of the resulting polymers which revealed a continuous open pore structure, in the case of a PS-based matrix. The extent of pore continuity actually depended on the composition of components in IPN precursors. However, when the stable subnetwork was constituted of PBA and the PS subnetwork contained the hydrolyzable AAA-type crosslinks, no pores were observed in the resulting PBA-based network after hydrolysis. This might be attributed to a complete collapse of the latter flexible network, being well above its T_g during hydrolysis.

Goethals' group reported the synthesis of porous polymers from sequential IPNs prepared by copolymerization of an acrylate-terminated poly(1,3-dioxolane) bismacromonomer with methyl (meth)acrylate, followed by copolymerization of methyl (meth)acrylate and tetraethylene glycol diacrylate [44]. The poly(1,3-dioxolane) segments in IPNs were degraded using a trifluoromethanesulfonic acid solution or strong Lewis acid vapors [45], thus affording residual porous poly(meth)acrylate networks. The morphology of IPNs, before and after degradation, was studied by optical microscopy techniques.

5.3.2 Porous Networks by "Selective" Electron Beam Degradation

Electron beam irradiation is another interesting technique for pore generation, as reported by Pionteck for IPN-like systems based on polyethylene (PE) and poly(methyl methacrylate-*co*-butyl methacrylate) [46–48]. Such IPNs were synthesized by the *in situ* method starting from a homogeneous mixture of PE, methacrylate comonomers, a peroxide radical initiator, and a dimethacrylate as a crosslinker. During the simultaneous formation of both subnetworks, grafting of polymethacrylate chains onto the PE subnetwork occurred, so both subnetworks were linked through covalent bonding, and this type of IPN was referred to as an IPN-like system. Interestingly, electron beam irradiation acts differently on both IPN components: polymethacrylates decompose, whereas PE crosslinks. Partial decomposition of the polymethacrylate partner was achieved by using an irradiation dose of 800 kGy while further crosslinking the PE subnetwork. This "selective" degradation resulted in an open porous PE structure which provided a new possibility to get access to polymer membranes useful for micro- or ultrafiltration [48]. The same group also utilized PE/poly(dodecyl methacrylate-*co*-ethyl methacrylate) IPN-like materials as precursors for the design of porous materials with smaller pore size, as polymethacrylates with long aliphatic side chains degraded less than their homologues with short pendant groups [47, 48].

5.3.3 Nanoporous Networks by Selective Hydrolysis

Our group reported the synthesis and characterization of partially hydrolyzable IPNs and porous materials derived therefrom [31, 32, 34, 49–51]. In this regard, PCL/PMMA- or PLA/PMMA-based sequential IPNs were first synthesized by the *in situ* sequential method which consists in mixing initially all the components homogeneously and then forming both subnetworks via two successive and noninterfering crosslinking processes. Typically, the polyester subnetwork was first generated at room temperature by an end-linking reaction between *α,ω*-dihydroxy oligoesters ($500 < M_n < 4000\,g{\cdot}mol^{-1}$) and a triisocyanate (Desmodur® RU) as the crosslinker, whereas the PMMA subnetwork was then obtained by heating to 65°C through free radical copolymerization of MMA and a dimethacrylate (from 1 to 10 mol%) as the crosslinker.

The selective hydrolysis of the polyester subnetwork associated with polyester/PMMA (50/50 wt%) IPNs afforded nanoporous methacrylic networks, thus showing the effective role of crosslinked PLA subchains as porogen templates. The hydrolytic degradation was conducted at 60°C by using a mixture of methylamine and ethanol. It has to be stressed that one such degradation was performed at a temperature intermediate between the T_g value of polyester subnetwork and that of PMMA subnetwork to avoid the collapse of the residual porous methacrylic structures while allowing for an efficient degradation of polyesters. Ethanol is a good solvent of oligoesters, and its role was to increase the hydrophilic character of polyester/PMMA IPNs, thus favoring a better diffusion of the hydrolysis medium into the network structure and accelerating the degradation rate (Figure 5.4). Under the basic conditions selected, mass loss values close to 50 wt% were encountered after 24 h, thus suggesting a quantitative degradation of the polyester subnetwork in IPNs, regardless of the structural parameters used. This was confirmed by FTIR and solid-state ^{13}C NMR, as indicated by the total disappearance of the characteristic bands of polyesters in the spectra of IPNs after hydrolysis. SEM analyses clearly showed that nondegraded IPNs were compact and nonporous. After hydrolysis, the resulting networks exhibited nanoporous structures with relatively small pore sizes and narrow pore size distributions (ranging from 10 to 100 nm—at most) (Figure 5.5 as typical examples). The dependence of pore sizes on dimethacrylate nature and content was not very significant and was

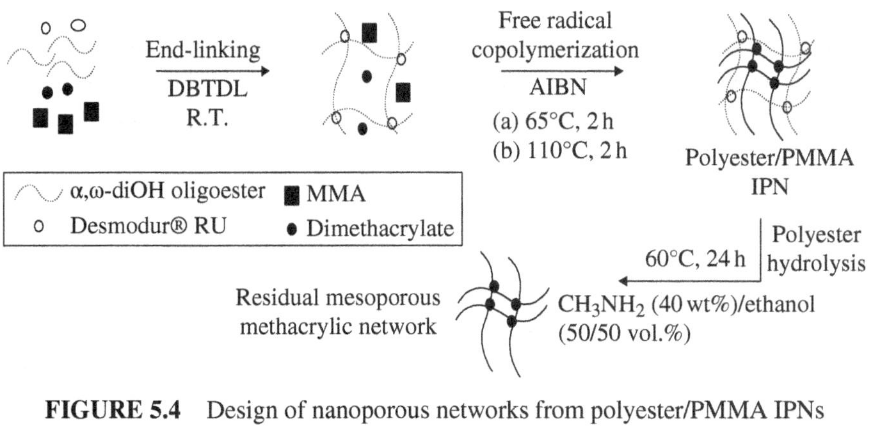

FIGURE 5.4 Design of nanoporous networks from polyester/PMMA IPNs

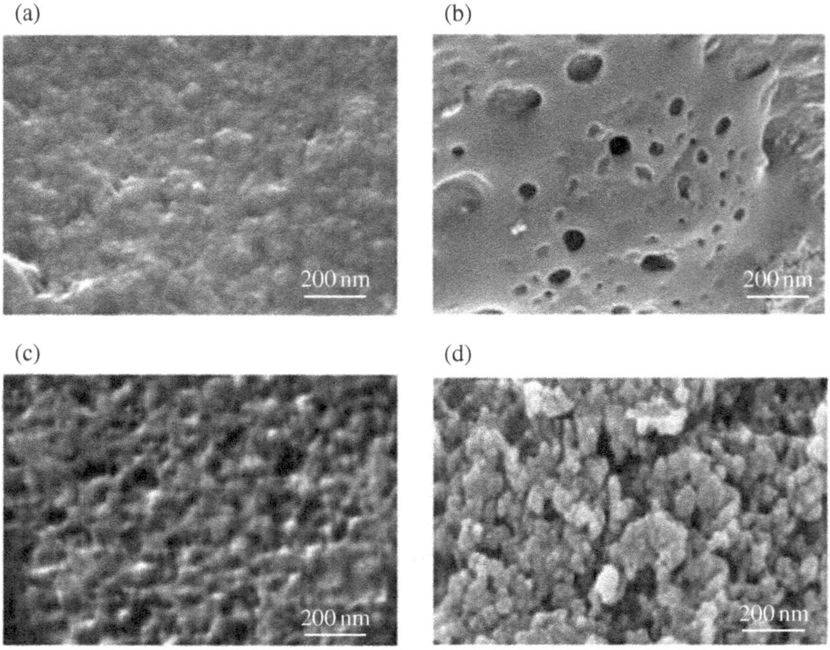

FIGURE 5.5 SEM micrographs of polyester/PMMA (50/50 wt%) IPNs: (a) Before selective polyester hydrolysis, PCL/PMMA (MMA/BADMA 90/10 mol%). After selective polyester hydrolysis, (b) PCL/PMMA (MMA/BADMA 90/10 mol%). (c) PLA/PMMA (MMA/DUDMA 90/10 mol%). (d) PLA/PMMA (MMA/BADMA 90/10 mol%)

essentially controlled by the good chain interpenetration of both constitutive subnetworks in the peculiar interlocking framework of IPN precursors. The transparency of all IPN samples and the DSC investigation of their microphase separation gave evidence of the latter assertion. Nevertheless, an increase in the molar mass of the oligoester could lead to a substantial increase in pore sizes and a broadening of pore size distributions (Figure 5.6). Overall, it is noteworthy that the pore sizes obtained in the materials derived from polyester/PMMA IPNs were much smaller than those encountered in the systems designed from corresponding semi-IPNs, due to the much smaller spatial scale of phase separation in IPN precursors than that developed in corresponding semi-IPN ones [50].

More recently, IPN systems based on PLA and a functional vinylic polymer could be effectively used as nanostructured precursors to novel functionalized (meso)porous networks [38, 52]. Indeed, the quantitative hydrolysis of the polyester subnetwork provided a straightforward and versatile strategy for

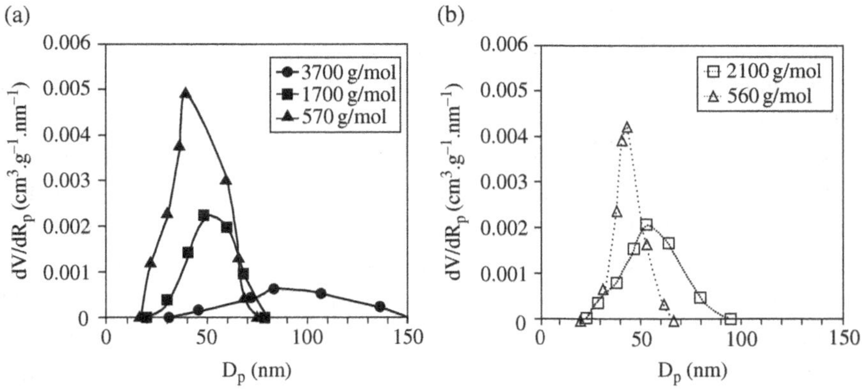

FIGURE 5.6 Pore size distribution profiles of porous methacrylic networks as determined by thermoporometry for 10mol% DUDMA-containing polyester/PMMA IPNs after selective polyester hydrolysis: (a) PLA- and (b) PCL-based systems

engineering nanoporous crosslinked methacrylic or styrenic polymers with miscellaneous functionalities (COOH, SO_3H, pyridyl groups). Values of specific surface areas as high as $80\,m^2{\cdot}g^{-1}$ could be obtained by nitrogen sorption porosimetry. The potential applications of such functional porous materials are mainly expected in the areas of separation techniques (chromatographic supports) and chemistry in confined media (nanoreactors, catalytic supports).

5.4 CONCLUSIONS

As mentioned in section "Introduction," the utilization of an organic solvent as a porogen in free radical crosslinking reactions provides an easy and practical methodology for preparing porous polymeric materials. However, the final materials generally display broad pore size distributions ranging from micro- to macropores, and the polymer properties and morphologies strongly depend on the experimental conditions used (stirring, temperature, time, suspension or bulk polymerization techniques, etc.). Moreover, original template-oriented routes have recently been developed to engineer nanoporous polymeric materials with controlled porosity. They generally entail the selective removal of single polymer domains acting as porogen templates from macromolecular architectures with controlled degradability.

This chapter has specifically reviewed the scope and limitations of the use of semi-IPNs and IPNs as nanostructured precursors to (nano)porous crosslinked materials. Straightforward and versatile synthetic strategies have been developed by a few research groups. On the one hand, porous networks with varying pore sizes can be engineered by either selective degradation of uncrosslinked subchains or solvent extraction of the latter from semi-IPN precursors. On the other hand, nanoporous networks can be effectively generated by resorting to selective degradation approaches (e.g., hydrolysis or irradiation processes) from IPNs of controlled degradability, that is, constituted of a degradable subnetwork and a stable one under specific conditions. Even though the use of semi-IPN precursors generally leads to the formation of macroporous networks, the semi-IPN approach may constitute a straightforward and effective route toward mesoporous networks, provided the extent of miscibility between both semi-IPN partners is high. Alternatively, the utilization of IPNs offers more versatility, since nanoporous networks are generally derived therefrom, regardless of the structural parameters associated with the stable constitutive subnetwork. The nanoporosity of residual porous networks can be attributed to the good degree of chain interpenetration associated with both subnetworks in IPN precursors arising from their peculiar interlocking framework.

Complementary approaches involving semi-IPN and IPN systems may provide an interesting means to tune the morphology and functionality associated with (nano)porous polymeric materials. The potential applications of such functional (nano)porous polymer systems are mainly expected in the areas of separation techniques (chromatographic supports, selective membranes) and chemistry in confined media (nanoreactors, catalytic supports).

ACKNOWLEDGMENTS

Financial support from the "Centre National de la Recherche Scientifique" (CNRS) and the "Université Paris-Est Créteil" (UPEC) is gratefully acknowledged. The author is especially indebted to Prof. Ph. Guérin (who passed away in December 2009), Prof. F. Lauprêtre, and Dr. S. Boileau for helpful discussions in the field of IPNs, as well as Dr. G. Rohman, Dr. T.-X. Lav, and Dr. R. Majdoub for their involvement in the area of porous polymeric materials designed from (semi-)IPNs.

REFERENCES

[1] Koros, W.J., Fleming, G.K., Jordan, S.M. *et al.* (1988) *Progress in Polymer Science*, **13**, 339.

[2] Odani, H. and Masuda, T. (1992) *Design of Polymer Membranes for Gas Separation*, Wiley-VCH, New York.

[3] Thomas, J.D.R. (1994) *Analyst*, **119**, 203.

[4] Dunleavy, M.J. (1996) *Medical Device Technology*, **5**, 14.

[5] Tavolaro, A. and Drioli, E. (1999) *Advanced Materials*, **11**, 975.

[6] Svec, F., Peters, E.C., Sykora, D. *et al.* (2000) *Journal of High Resolution Chromatography*, **23**, 3.

[7] Staudinger, H. and Husemann, E. (1935) *Chemische Berichte*, **68**, 1618.

[8] Sing, K.S.W., Everett, D.H., Haul, R.A.W. *et al.* (1985) *Pure and Applied Chemistry*, **57**, 603.

[9] Haupt, K. and Mosbach, K. (1998) *Trends in Biotechnology*, **16**, 468.

[10] Wulff, G. (2002) *Chemical Reviews*, **102**, 1.

[11] (a) Maki-Ontto, R., de Moel, K., de Odorico, W. *et al.* (2001) *Advanced Materials*, **13**, 117; (b) De Moel, K., Alberda van Ekenstein, G.O.R., Nijland, H. *et al.* (2001) *Chemistry of Materials*, **13**, 4583; (c) Ikkala, O. and ten Brinke, G. (2002) *Science*, **295**, 2407.

[12] (a) Lee, J.S., Hirao, A. and Nakahama, S. (1988) *Macromolecules*, **21**, 274; (b) Lee, J.S., Hirao, A. and Nakahama, S. (1989) *Macromolecules*, **22**, 2602.

[13] Hashimoto, T., Tsutsumi, K. and Funaki, Y. (1997) *Langmuir*, **13**, 6869.

[14] (a) Park, M., Harrison, C., Chaikin, P.M. *et al.* (1997) *Science*, **276**, 1401; (b) Li, R.R., Dapkus, P.D., Thompson, M.E. *et al.* (2000) *Applied Physics Letters*, **76**, 1689.

[15] (a) Hedrick, J.L., Carter, K.R., Richter, R. *et al.* (1998) *Chemistry of Materials*, **10**, 39; (b) Mansky, P., DeRouchey, J., Russell, T.P. *et al.* (1998) *Macromolecules*, **31**, 4399; (c) Thurn-Albrecht, T., DeRouchey, J., Russell, T.P. and Jaeger, H. (2000) *Macromolecules*, **33**, 3250; (d) Thurn-Albrecht, T., Steiner, R., DeRouchey, J. *et al.* (2000) *Advanced Materials*, **12**, 787; (e) Xu, T., Kim, H.C., DeRouchey, J. *et al.* (2001) *Polymer*, **42**, 9091; (f) Goldbach, J.T., Russell, T.P. and Penelle, J. (2002) *Macromolecules*, **35**, 4271; (g) Thurn-Albrecht, T., DeRouchey, J., Russell, T.P. and Kolb, R. (2002) *Macromolecules*, **35**, 8106; (h) Xu, T., Hawker, C.J. and Russell, T.P. (2003) *Macromolecules*, **36**, 6178; (i) Goldbach, J.T., Lavery, K.A., Penelle, J. and Russell, T.P. (2004) *Macromolecules*, **37**, 9639.

[16] (a) Liu, G.J., Ding, J.F., Guo, A. *et al.* (1997) *Macromolecules*, **30**, 1851; (b) Liu, G.J., Ding, J.F., Hashimoto, T. *et al.* (1999) *Chemistry of Materials*, **11**, 2233.

[17] Chan, V.Z.H., Hoffman, J., Lee, V.Y. *et al.* (1999) *Science*, **286**, 1716.

[18] (a) Zalusky, A.S., Olayo-Valles, R., Taylor, C.J. and Hillmyer, M.A. (2001) *Journal of the American Chemical Society*, **123**, 1519; (b) Zalusky, A.S., Olayo-Valles, R., Wolf, J.H. and Hillmyer, M.A. (2002) *Journal of the American Chemical Society*, **124**, 12761; (c) Cavicchi, K.A., Zalusky, A.S., Hillmyer, M.A. and Lodge, T.P. (2004) *Macromolecular Rapid Communications*, **25**, 704; (d) Rzayev, J. and Hillmyer, M.A. (2005) *Macromolecules*, **38**, 3; (e) Mao, H. and

Hillmyer, M.A. (2005) *Macromolecules*, **38**, 4038; (f) Hillmyer, M.A. (2005) *Advances in Polymer Science*, **190**, 137; (g) Mao, H. and Hillmyer, M.A. (2006) *Soft Matter*, **2**, 57.

[19] Li, M., Douki, K., Goto, K. *et al.* (2004) *Chemistry of Materials*, **16**, 3800.

[20] Uehara, H., Yoshida, T., Kakiage, M. *et al.* (2006) *Macromolecules*, **39**, 3971.

[21] (a) Grande, D., Penelle, J., Davidson, P. *et al.* (2011) *Microporous and Mesoporous Materials*, **140**, 34; (b) Majdoub, R., Antoun, T., Le Droumaguet, B. *et al.* (2012) *Reactive and Functional Polymers*, **72**, 495.

[22] (a) Hedrick, J.L., Miller, R.D., Hawker, C.J. *et al.* (1998) *Advanced Materials*, **10**, 1049; (b) Hedrick, J.L., Carter, K.R., Labadie, J.W. *et al.* (1999) *Advances in Polymer Science*, **141**, 1; (c) Nguyen, C., Hawker, C.J., Miller, R.D. *et al.* (2000) *Macromolecules*, **33**, 4281.

[23] Eigner, M., Voit, B., Estel, K. and Bartha, J.W. (2002) *e-Polymers*, 028.

[24] Loera, A.G., Cara, F., Dumon, M. and Pascault, J.P. (2002) *Macromolecules*, **35**, 6291.

[25] (a) Grande, D., Grigoryeva, O., Fainleib, A. *et al.* (2008) *European Polymer Journal*, **44**, 3588; (b) Grigoryeva, O., Gusakova, K., Fainleib, A. and Grande, D. (2011) *European Polymer Journal*, **47**, 1736; (c) Grande, D., Grigoryeva, O., Gusakova, K. and Fainleib, A. (2013) *European Polymer Journal*, **49**, 2162.

[26] Balaji, R., Boileau, S., Guérin, P. and Grande, D. (2004) *Polymer News*, **29**, 205.

[27] Sperling, L.H. (1981) *Interpenetrating Polymer Networks and Related Materials*, Plenum Press, New York.

[28] Sperling, L.H. and Mishra, V. (1996) *Polymers for Advanced Technologies*, **7**, 197.

[29] Kim, S.C. and Sperling, L.H. (eds) (1997) *IPNs around the World: Science and Engineering*, Wiley, Chichester.

[30] Kayaman-Apohan, N. and Baysal, B.M. (2001) *Macromolecular Chemistry and Physics*, **202**, 1182.

[31] Grande, D., Pastol, J.-L., Guérin, P. and Boileau, S. (2003) *Polymer Preprints (American Chemical Society, Division of Polymer Chemistry)*, **44** (1), 44.

[32] Grande, D., Lacoudre, N., Guérin, P. and Boileau, S. (2003) *Polymer Preprints (American Chemical Society, Division of Polymer Chemistry)*, **44** (1), 148.

[33] Rohman, G., Grande, D., Lauprêtre, F. *et al.* (2005) *Macromolecules*, **38**, 7274.

[34] Grande, D. and Rohman, G. (2007) *Polymeric Materials Science and Engineering*, **97**, 124.

[35] Brun, M., Lallemand, A., Quinson, J.F. and Eyraud, C. (1977) *Thermochimica Acta*, **21**, 59.

[36] Quinson, J.F., Mameri, N., Guihard, N. and Bariou, B. (1991) *Journal of Membrane Science*, **58**, 191.

[37] Hay, J.N. and Laity, P.R. (2000) *Polymer*, **41**, 6171.

[38] Grande, D., Beurroies, I. and Denoyel, R. (2010) *Macromolecular Symposia*, **291**, 168.

[39] Grande, D., Rohman, G. and Millot, M.-C. (2008) *Polymer Bulletin*, **61**, 129.

[40] Lav, T.-X., Carbonnier, B., Guerrouache, M. and Grande, D. (2010) *Polymer*, **51**, 5890.

[41] Lav, T.-X., Grande, D., Gaillet, C. *et al.* (2012) *Macromolecular Chemistry and Physics*, **213**, 64.

[42] Widmaier, J.M. and Sperling, L.H. (1982) *Macromolecules*, **15**, 625.

[43] Widmaier, J.M. and Sperling, L.H. (1984) *British Polymer Journal*, **16**, 46.

[44] Du Prez, F. and Goethals, E.J. (1995) *Macromolecular Chemistry and Physics*, **196**, 903.

[45] De Clercq, R.R. and Goethals, E.J. (1992) *Macromolecules*, **25**, 1109.

[46] Hu, J., Pompe, G., Schulze, U. and Pionteck, J. (1998) *Polymers for Advanced Technologies*, **9**, 746.

[47] Hu, J., Schulze, U. and Pionteck, J. (1999) *Polymer*, **40**, 5279.

[48] Pionteck, J., Hu, J. and Schulze, U. (2003) *Journal of Applied Polymer Science*, **89**, 1976.

[49] Balaji, R., Boileau, S., Guérin, P. and Grande, D. (2005) *Polymer Preprints (American Chemical Society, Division of Polymer Chemistry)*, **46** (1), 300.

[50] Rohman, G., Lauprêtre, F., Boileau, S. *et al.* (2007) *Polymer*, **48**, 7017.

[51] Rohman, G. and Grande, D. (2008) *Macromolecular Symposia*, **267**, 21.

[52] Grande, D., Gorzolnik, B. and Antoun, T. (2009) *Polymeric Materials: Science and Engineering*, **101**, 1108.

6

NATURAL RUBBER-BASED MICRO- AND NANOSTRUCTURED IPNs

SA-AD RIYAJAN

Department of Chemistry, Faculty of Science and Technology, Thammasat University, Pathum Thani, Thailand

6.1 INTRODUCTION

Industrial products made from elastomer materials especially natural rubber (NR) latex have found extensive applications such as in tire, tubings, surgical gloves, catheters, balloons, and other products. But the compatibility of devices made from NR is different when compared to those made of synthetic polymers due to rubbers' elasticity, flexibility, and resistance against splitting that provides NR products with more advantages than those of other polymers [1, 2]. In addition, the advantage of NR is renewable resource. In order to improve their properties, modification of NR polymers is essential. In the previous work with over the last fifth decades, there have been three main approaches utilized to modify the properties of NR polymers including changes to its chemical microstructure, addition of different material reagents, and blending with other polymers. Blending with other polymers is a technique to improve the properties of the polymer to get the new commercially useful materials having desired properties which cannot be derived from a

Micro- and Nano-structured Interpenetrating Polymer Networks: From Design to Applications, First Edition. Edited by Sabu Thomas, Daniel Grande, Uroš Cvelbar, K.V.S.N. Raju, Ramanuj Narayan, Selvin P. Thomas and Akhina H.

single polymer [3–9]. In addition, it is an easy, cheap, and economical method and can produce new materials with novel properties [10–12]. NR has been blended with many polymers such as polypropylene (PP), polyethylene (PE), and poly(vinyl alcohol) (PVA). Interpenetrating polymer network (IPN) is a one of the chemical blending and it consists of combination of two polymers, both in network form, at least one of which is synthesized and/or crosslinked in the immediate presence of the other. The network cannot be separated unless chemical bonds are broken. The two or more networks can be envisioned to be entangled in such a way that they are concatenated and cannot be pulled apart but not bonded to carbon–carbon chemical reaction. The IPN was divided into two types such as semi-IPNs and full IPNs. When one of the phases is crosslinked, pseudo- or semi-IPNs will be formed. When both phases are crosslinked, full IPNs are developed. IPNs are generally prepared by sequential, simultaneous, or latex polymerization techniques. The parameters affecting the properties of the IPN are dependent on phase behavior, chemical additive agent, blend ratio, and crosslinking levels. The aim of this chapter is to prepare and characterize the IPN obtained from NR and the synthetic polymer as well as crosslinking agent. In addition, the effect of filler on the properties of IPN and its application were reported in this chapter.

6.2 NATURAL RUBBER

6.2.1 Basic Information of NR

NR is made from Para rubber tree (*Hevea brasiliensis*, Müll. Arg.), a member of the spurge family *Euphorbiaceae* as shown in Figure 6.1, and it consists of *cis*-1,4-polyisoprene and nonrubber component such as carbohydrate, protein, and mineral substances. After tapping a milky colloid is produced by some plants. The plants would be "tapped," that is, an incision made into the bark of

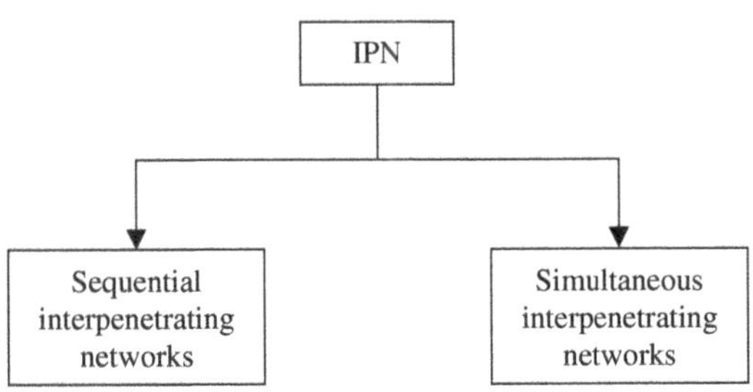

FIGURE 6.1 Two methods for preparing IPN, including sequential interpenetrating networks and simultaneous interpenetrating networks

the tree and the sticky, milk colored latex sap collected and refined into a usable rubber. NR is used extensively in many applications and products since it is normally very stretchy and flexible and extremely waterproof. This is largely because it responds to wounding by producing more latex; also this means that the tree is able to photosynthesize more.

6.2.2 Properties

Comparing other rubbers, NR shows outstanding in mechanical strength due to nonrubber component and it exhibits the stress crystalline. In addition, its ability crystallizes at large deformations resulting in an extraordinary toughness. The induced crystals of NR may arrest the possible crack development where large local strains occur. Table 6.1 shows the influence of nonrubber component on the properties of NR [13]. For example, some proteins in NR were reported to prevent the radical formation. Moreover, protein and free amino acid react with abnormal group in NR, resulting in storage hardening of NR.

6.2.3 Applications

Nowadays, NR is widely used ranging from household (such as glove and condom) to industrial products (such as tire, tubes, and balloon), entering the production stream at the intermediate stage or as final products. Among rubber

TABLE 6.1 Influence of Nonrubber Component on the Properties of NR

Properties	Nonrubber Component
Latex stability	Carbohydrate
Color	Yellow—beta-carotenes
	Black—reaction between enzymatic reaction and polyphenol oxidase
Vulcanization behavior	Function of phospholipid and protein as accelerators
	Function of fatty acid as activators
Oxidation	Tocotrienols as an antioxidant
	Cu^{2+}, Mn^{2+}, and Fe^{2+} as prooxidants
Storage hardening	Protein and free amino acid react with abnormal group in NR
Creep and stress relaxation	Protein and ash absorb humidity, leading to increasing creep and stress relaxation
Modulus	Protein
Tear strength	Protein
Dynamic crack growth	Protein

products, tires and tubes are the largest consumers of rubber. Tires and tubes are made from dried rubber, while gloves (medical, household, and industrial) and toy balloons are produced from the concentrated NR latex. In addition, NR is utilized in adhesives in many manufacturing industries and products such as rubber bands and pencil erasers as well as textile industry because of its excellent elongation and recovery properties. Many aircraft tires and inner tubes are still made of NR because of the high cost of certification for aircraft use of synthetic replacements. However, the limitations of using NR are poor thermal resistance, odor, occurrence of storage hardening, and unstable price. Therefore, the IPN technique is used for these problems.

6.3 SYNTHESIS OF POLYMER IPN

Firstly, NR has been used to improve the properties of PS to get a high impact PS (HIPS). The tensile strength, tear strength, and modulus of the full IPN NR/PS are higher than those of semi-IPN NR/PS [14]. But the semi-IPN NR/PS shows higher toughness and elongation than full IPN NR/PS. NR/poly(methyl methacrylate) (PMMA) [15] and IPN NR/PS have been synthesized in the latex state [16]. After synthesis of IPN was done, the IPN synthesis method leads to improved compatibility and mechanical properties. Semi- and full IPN NR/PMMA and NR/PS have been synthesized, and parameters affecting on their properties are the PMMA, or PS, content and the crosslinker. At present, PSf and natural rubber glove waste (NRG) needed to be managed after use [17]. The other biopolymers mostly used in packaging application are starch and cellulose. To prepare ecocomposites, a good dispersion of starch in NR matrix was obtained by cogrinding with NR [18]. Since the hydrophilic behavior of starch was reduced, the interaction between the initial hydrophobic matrix and hydrophilic filler was enhanced without using any compatibilizer [19]. Moreover, starch was used to improve the mechanical strength of NR [20–22]. Recently, the properties of IPN NR/PS were improved by adding fillers [20, 23, 24]. Two methods including sequential interpenetrating networks and simultaneous interpenetrating networks are used to produce the IPN as shown in Figure 6.1. In the first method or sequential interpenetrating networks, there are two steps as follows. The polymer A is firstly polymerized from monomer A in the presence of initiator and catalyst and then crosslinked with crosslinker to get the polymer network A as shown in Figure 6.2. In the second step, the monomer B and crosslinker are swollen into polymer network A and then polymerized *in situ*.

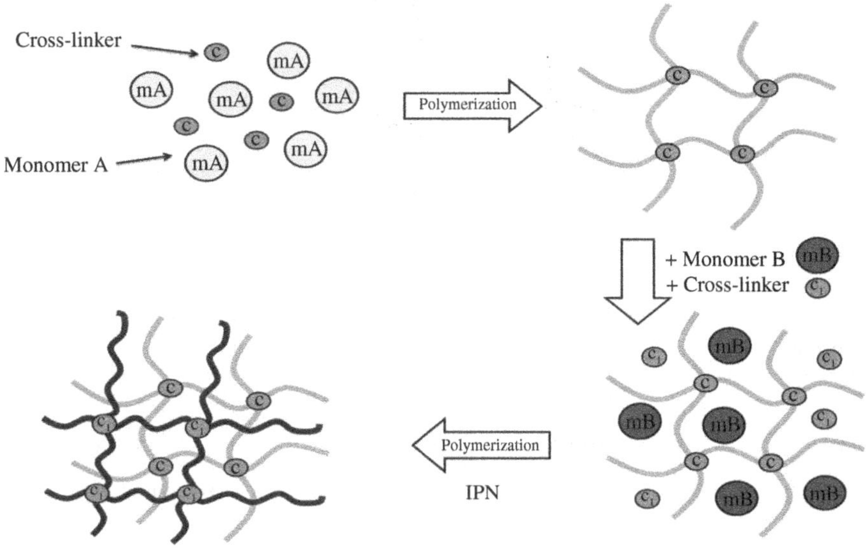

FIGURE 6.2 Possible model of IPN made from sequential interpenetrating networks (full IPNs)

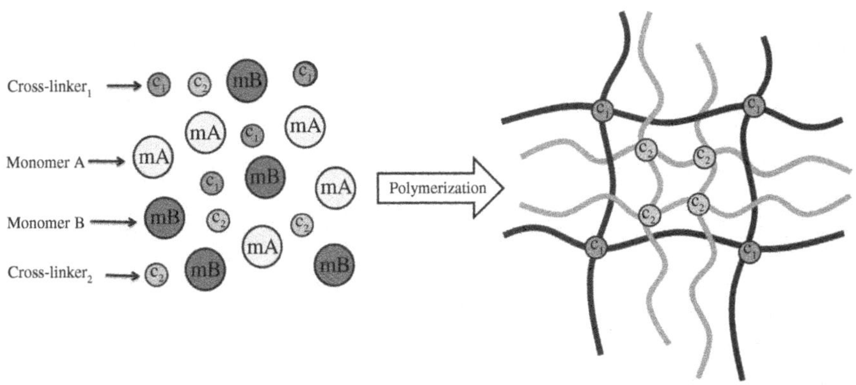

FIGURE 6.3 Possible model of IPN made from the simultaneous interpenetrating networks

For example, the NR was crosslinked first with vulcanizing agent such as dicumyl peroxide (DCP), and this was followed by swelling in styrene monomer and then polymerization and crosslinking of the PS phase. In the case of the simultaneous interpenetrating networks, the monomer A and monomer B in the presence of crosslinkers and activators of both networks are blended. At the same time, the reactions are carried out simultaneously but no interfering reactions as shown in Figure 6.3.

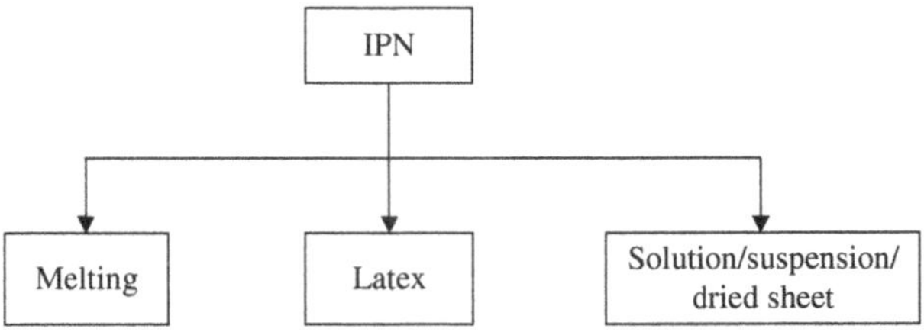

FIGURE 6.4 Techniques of preparing IPN from NR

TABLE 6.2 Examples for Technique of Preparing IPN from NR and Synthetic Polymer

	NR	Synthetic Polymer	
State	Crosslinkers	Crosslinker	References
Latex	No crosslinker/ sulfur	PS; divinylbenzene PMMA; ethylene glycol dimethacrylate	Schneider *et al.* [16]
Melting	Dicumyl peroxide	PS: divinylbenzene	Thomas *et al.* [28], [29], [34]
Solution suspension	Sulfur	PS: no crosslinker	Riyajan *et al.* [17]
Dried sheet	No crosslinker	PMMA/ethylene glycol dimethacrylate	Jayasuriya and Hourston [15]

In addition, IPN can be divided into three categories: latex, solution, and melting (Figure 6.4).

An example for preparing IPN is presented in Table 6.2. The IPN was prepared by the sequential technique. In the case of the sequential method, the NR is crosslinked first using DCP followed by the polymerization and crosslinking of PS phase as shown Figure 6.5. NRG waste from Sri Trang Company (Songkhla, Thailand) was masticated by using two-roll mill (Charoenchaikarnchang; Bangkok, Thailand) at 30–32°C for 30 min. The masticated NRG was immersed in toluene for 2 days at ambient temperature and then stirred for 5 h. After stirring with a mechanical stirrer for 1 h, the particle size and size distribution of the NRG suspended in toluene were measured by a laser particle size analyzer (LPSA; Beckman Coulter LS 230). The NRG suspension with various contents was blended with 10% w/v of PSf in toluene at room temperature by using a homogenizer (PL Trading International Co. Ltd; Bangkok, Thailand): 0.5% BPO, 5% glycerol, and different MA concentrations were added into the blend under stirring. In order to form a sheet of 0.45 mm in thickness, the mixture was poured into a

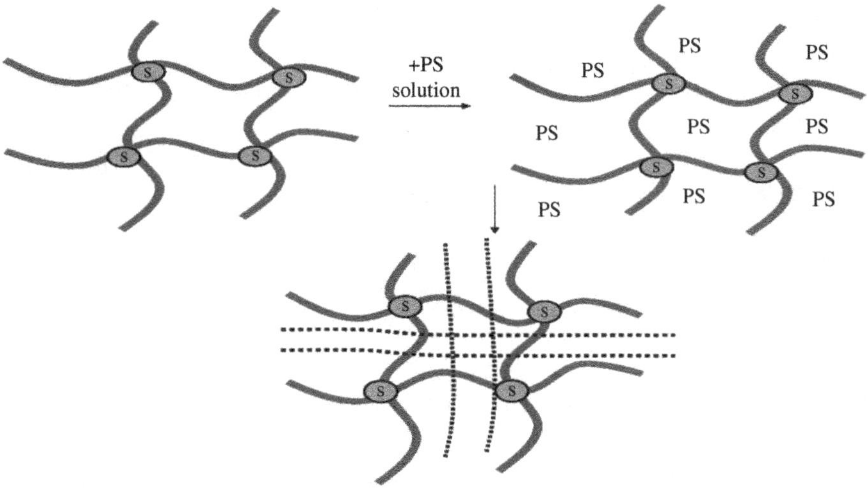

FIGURE 6.5 Possible model of IPN made from NRG and PSf or PS

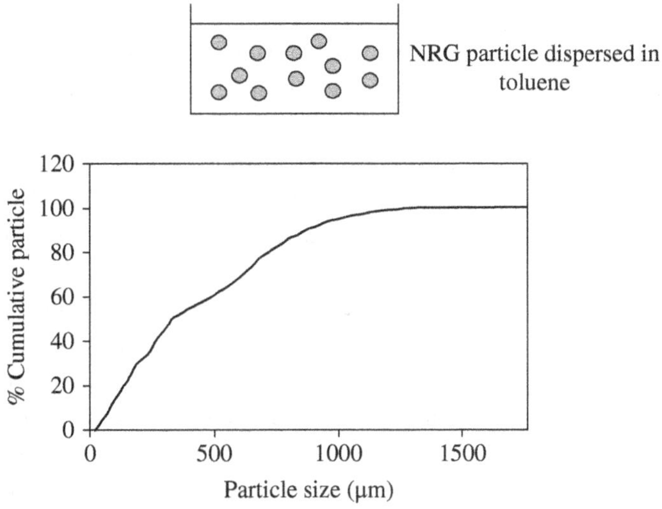

FIGURE 6.6 Relationship between % cumulative particle and particle size of NRG in toluene (author's experiment)

glass plate ($10 \times 10 \times 0.5\,cm^3$) at ambient temperature. Under a hot-press, the sample was preheated at 150°C for 6 min and then compressed for 4 min before being cooled under 3 MPa at 30°C for 4 min.

Results indicated that 90% cumulative particle size of the dispersed NRG in toluene was less than 1000 μm as shown Figure 6.6.

After NR was crosslinked with sulfur, the PS solution was blended with different sulfur prevulcanized NR (SPNR) latex (60% w/v) contents at room

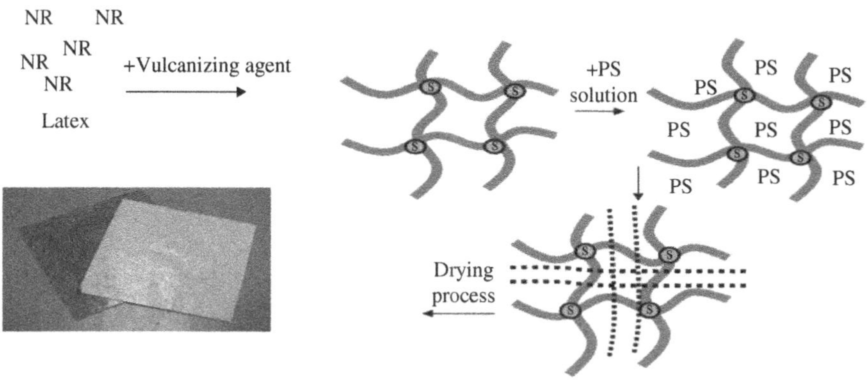

FIGURE 6.7 Possible model of IPN made from NR latex and PSf (PS)

temperature while being continuously stirred by homogenizer as shown in Figure 6.7. After that, PS penetrated into NR network. After dyeing process, we got IPN sheet as shown in Figure 6.7.

In the case of latex form, Schneider *et al.* [16] studied the preparation of IPN from prevulcanized and a noncrosslinked NR latex which were coated with a shell of crosslinked PMMA or PS through batch emulsion polymerization. The effect of NR/styrene swelling ratio, initiator, the crosslinking degree, and the polymerization temperature and distinct differences of the phase arrangement of polymers in the latex particles were investigated by transmission electron microscopy (TEM) in work. They found that the different staining methods allowed us to enhance the contrast between the NR phase and the secondary polymers in the composite latex particles. In addition, a semicontinuous feeding process decreased the PS subinclusion size. By using azobisisobutyronitrile (AIBN) initiation or the hydrophobic redox, the morphology of IPN showed the rigid PS subinclusions within the NR particles during batch emulsion polymerization. The effect of higher monomer/NR swelling ratio on the morphology of IPN was investigated that the larger-sized crosslinked PS subinclusion phase within the NR was appeased. When the temperature for preparing IPN increased to near the glass transition temperature of PS, the morphology of IPN allowed showed the phase separation in large-sized PS subinclusions. In the case of PMMA, the morphology of IPN obtained from PMMA was more core–shell than that of PS due to the hydrophilicity of PMMA observed from TEM. After NR cured with crosslinker, the penetration of styrene monomer in NR particle was lower than that of uncured NR (Table 6.3).

A thermoplastic blend was obtained from blending between epoxidized natural rubber (ENR) and PS [31]. ENRs with different amounts of epoxide

TABLE 6.3 Examples of Preparing Polymer IPN Made from NR and Other Polymers

Type	Polymer	Crosslinker	Topic	References
Semi-IPN	Condensed tannin	Enzyme mimetic	Preparation and properties	Kadokawa *et al.* [25]
	Poly(acrylic acid)	Ethylene glycol	Preparation of membranes for the pervaporation of water–ethanol mixtures	Amnuaypanich and Kongchana [26]
	PVA	Maleic acid	Preparation and properties Good mechanical properties	Riyajan *et al.* [27]
	PS	Dicumyl peroxide oxide	Thermal stability	Mathew *et al.* [34]
			Morphology, mechanical properties, and failure	Mathew *et al.* [28]
Full IPN	PS	Dicumyl peroxide (NR)	Kinetics of diffusion	Mathew *et al.* [29]
		Divinylbenzene (PS)	Parameter affecting on properties	Mathew *et al.* [34]
			Thermal stability	Mathew *et al.* [30]
			Viscoelastic properties	Mathew *et al.* [34]

groups were prepared via *in situ* epoxidation method. PS was prepared by emulsion polymerization. The obtained ENRs were blended with commercial and prepared PS at room temperature in the latex state at the amounts of dry PS of 0, 20, 40, 60, 80, and 100 phr. It was found that the tensile strength and elongation at break increased as the amount of PS increased and then decreased later at higher amounts of PS. However, these values decreased as the amount of epoxide groups increased. The hardness of the specimens increased with the increase of epoxide groups and PS content. In addition, the ENR/PS blends could be described as partially miscible blends. At present, the preparation of chemical blend was made from NR or ENR and hydrophilic polymer such as PVA, polyacrylic acid, chitosan, and polyacrylamide. For example, a new semi-IPN based on ENR and PVA containing maleic acid (MA) as a crosslinking reagent was synthesized. The effect of curing time and dose of MA on the properties of the polymer blend was investigated. In addition, the crystalline content of PVA dramatically decreased after adding MA in the semi-IPN, as observed from its XRD.

6.3.1 Preparation of Semi-IPN ENR and PVA

The PVA solution (20% w/v) was prepared in Millipore water by heating it on a water bath at 70°C while being continuously stirred. The desired quantity of MA (dose varied from 10 to 60% w/w with respect to the weight of PVA) was mixed with the solution at ambient temperature with the help of a mechanical stirrer. The mixture was poured in the form of a sheet on a glass surface as shown in Figure 6.8. The sheet with a 0.55 mm thickness was dried at ambient temperature and subsequently cured in an oven at different curing times (e.g., 60 min) at 120°C. Preparation of ENR latex was followed by epoxidation of NR latex: 96.3 g of high ammonia natural rubber (HANR) latex with 20% DRC was mixed with a nonionic surfactant, Terric 16 A16, at room temperature while stirring for 1 h. The mixture was neutralized followed by acid modification with 22 g of 85% formic acid: 88 ml of a 30% aqueous solution of hydrogen peroxide was added into the mixture at 50°C within 15–20 min. The epoxidation was then carried out at 50°C for 2.5 h and the ENR latex obtained having 20% of epoxidized content was used in the study. The preparation of the ENR latex and the PVA blend was similar to the case of forming the semi-IPN between NR and PVA.

The model of the polymer blend based on NR/PVA and ENR/PVA is presented in Figure 6.9. The PVA was crosslinked first with MA and this was followed by adding NR. Then, NR particle was dispersed in PVA as shown in Figure 6.9a. In the case of ENR, ENR was dispersed in crosslinked PVA with MA as shown in Figure 6.9b.

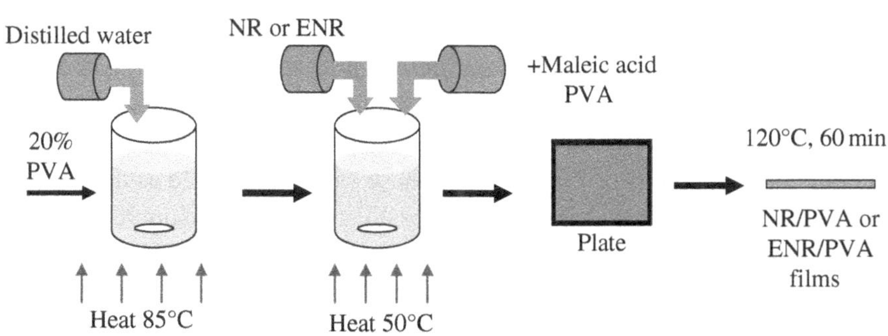

FIGURE 6.8 Preparation of IPN made from NR and ENR with PVA in the presence of MA as a crosslinker

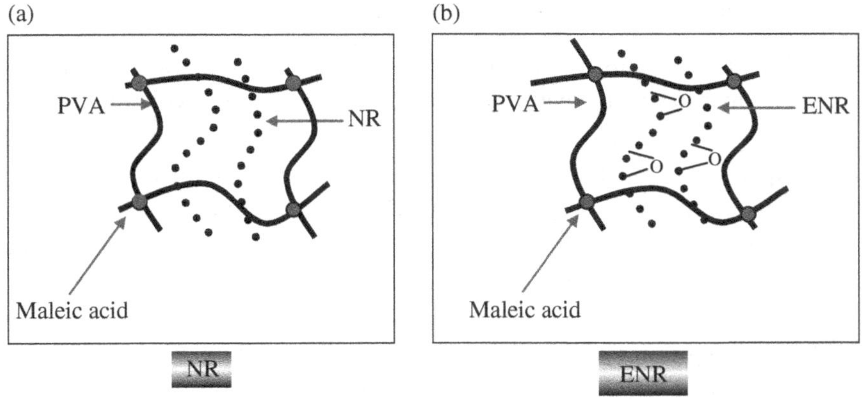

FIGURE 6.9 Model of the polymer blend based on (a) NR and PVA and (b) ENR and PVA

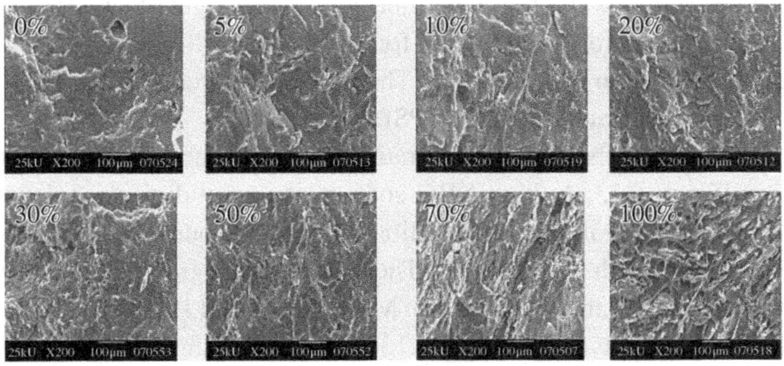

FIGURE 6.10 SEM images of 40/60 NR/PSf 6% MA with different Cfiber (author's experiment)

6.3.2 Characterization of Morphology and Chemical Interaction IPN

Synthesis, morphology, structure, and properties will be discussed in detail.

The phase morphology IPN of NR/PS depends on blend ratio, crosslinker level, and initiating systems (Figure 6.10).

6.4 PROPERTIES OF IPN MADE FROM NR AND PLASTICS

The properties of IPN depend on the crosslink density of the elastomer, the crosslink density of the plastic, and the elastomer/plastic ratio as well as filler, as will be discussed in detail in the following section.

6.4.1 Swelling Behavior and Solvent Resistance

The diffusion mechanism in rubbers is essentially connected with the ability of the polymer to provide pathways for the solvent to progress in the form of randomly generated voids. The swelling ratio of IPN sample was estimated using Equation 6.1. As the void formation decreases with Cfiber addition, the solvent uptake also decreases as shown in Figure 6.11. But when adding over 70 phr Cfiber, the swelling ratio of the polymer blend slightly decreased due to excess void. The presence of bonding agent binds the alkali-treated fiber and rubber so that swelling is highly restricted in the composite.

The swelling index which is a measure of the swelling resistance of the rubber compound is calculated using Equation 6.1:

$$\text{Swelling ratio} = \frac{W_2 - W_1}{W_1} \quad (6.1)$$

where W_1 = original weight of the sample and W_2 = weight of swollen sample.

The influence factors of SPNR/PS foam ratio on swelling ratio of IPN were investigated as shown in Figure 6.12. The lowest swelling ratio of the polymer blend was observed at 40/60 SPNR/PSf.

The influence of NRG/PS blend ratio on the toluene resistance was presented in Figure 6.13. At high NRG content, the good dimensional stability was recorded due to more NRG portion in blend. In our previous work, the NRG was filled with Cfiber [17]). The optical micrographs of PS blended with 5% Cfiber and various amounts of MA (0, 5, 10, and 15%) after immersion in toluene are displayed in Figure 6.13e. The good dimensional stability of PSf/Cfiber added with MA especially of 15% was observed in Figure 6.13e.

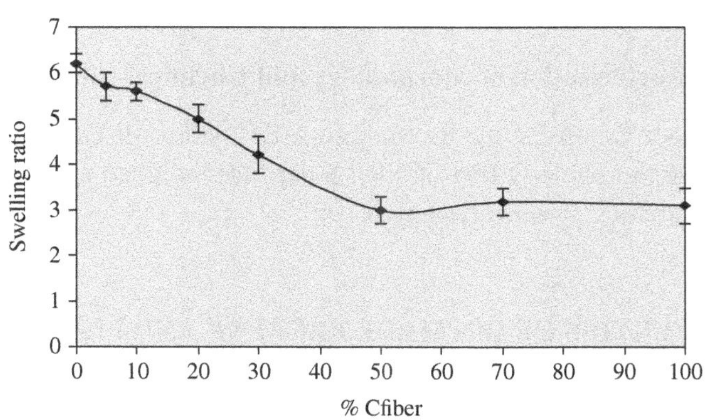

FIGURE 6.11 Effect of Cfiber on the swelling ratio of the polymer composite at different Cfiber contents (author's experiment)

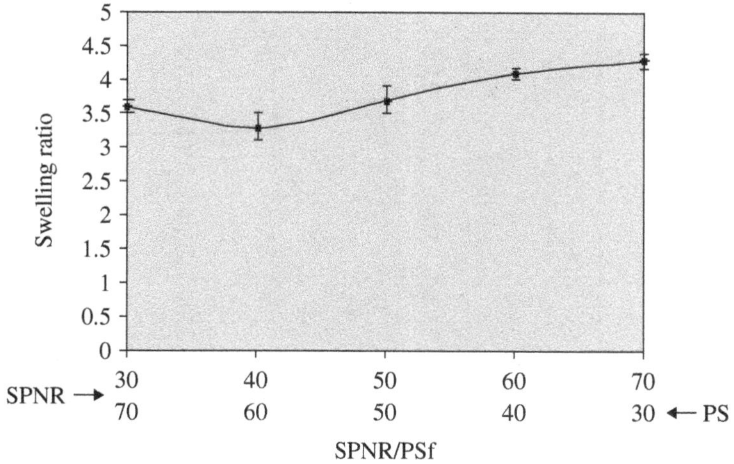

FIGURE 6.12 Effect of blend ratio on swelling ratio of the polymer composite at 50 % Cfiber

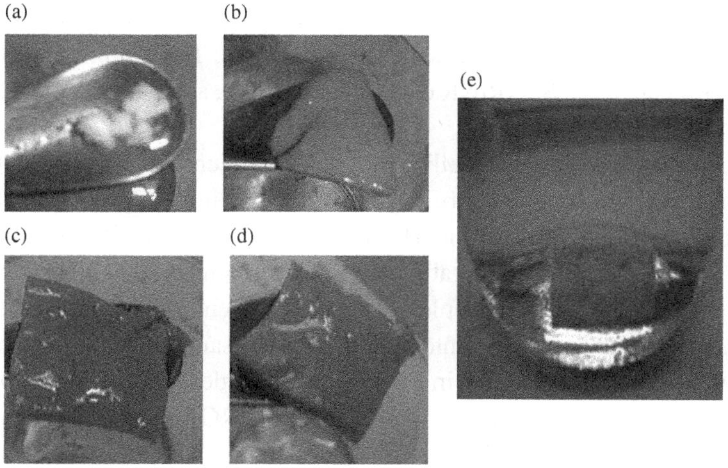

FIGURE 6.13 Photograph of IPN sample resistances in organic solvent (a) 20 : 80, (b) 30/70, (c) 50/50, and (d) 30/70 NRG/PS blend as well as (e) 50/50 NRG/PS in the presence of 40% Cfiber (author's experiment)

This result indicated the occurrence of chemical interaction between PSf and MA or between cellulose and MA during blending process. The swelling ratio of the polymer composite film immersed in toluene at ambient temperature for 5 days was used for the confirmation of the chemical reaction between MA and SPNR/PS as well as Cfiber. The swelling ratio of the polymer composite dramatically decreased as a function of the Cfiber. The effect of MA contents

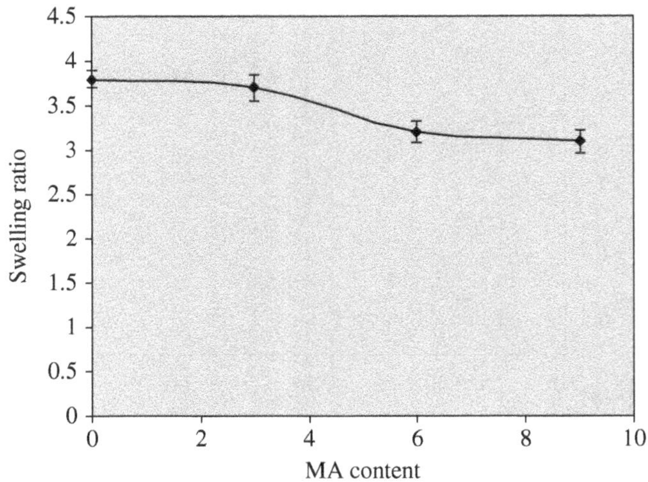

FIGURE 6.14 Influence of the MA contents on the swelling ratio of the polymer composite at 50/50 SPNR/PSf (author's experiment)

on the swelling of the polymer blend was presented in Figure 6.14. The swelling ratio of the polymer decreased with increasing MA. The swelling ratio of the polymer blend 50/50 SPNR/NR containing 0, 2.5, 6, and 9 MA% was 3.8, 3.6, 3.4, and 3.2, respectively.

The possible model of swelling for polymer blend and IPN sample was presented in Figure 6.15a and b, respectively. Swelling is a volume expansion caused by absorption of solvent. The swelling behaviors of IPN depend on its chemical structure, polymer ratio, crosslinking density, and filler. The stability dimension of IPN sample in organic solvent is higher than that of polymer blend due to higher interaction. An **IPN** sample swells but does not dissolve in **solvents** as shown in Figure 6.15. In order to improve the biodegradation, the cassava starch was added into PS/NRG IPN sample. The effect of cassava starch (Cs) or gelatinized Cs (GCs) and its content on swelling ratio of the polymer blend was investigated. With increasing the starch contents from 0, 5, 10, 15 to 20%, the swelling ratio of the polymer blend having Cs decreased from 2.7, 2.4, 2.3, 1.8 to 1.7, whereas that containing GCs decreased from 2.2, 2.1, 1.6 to 1.4, respectively, as shown in Figure 6.16. The swelling ratio of GCs was lower than that of Cs because the gelatinization process broke down intermolecular bonds of Cs granules containing both linear amylose and branched amylopectin. Hence, the hydrogen bonding sites (hydroxyl hydrogen and oxygen) engaged more water and irreversibly dissolves Cs granules. Penetration of water at high temperature increased randomness in the granule structure and decreased the number and size of crystalline regions.

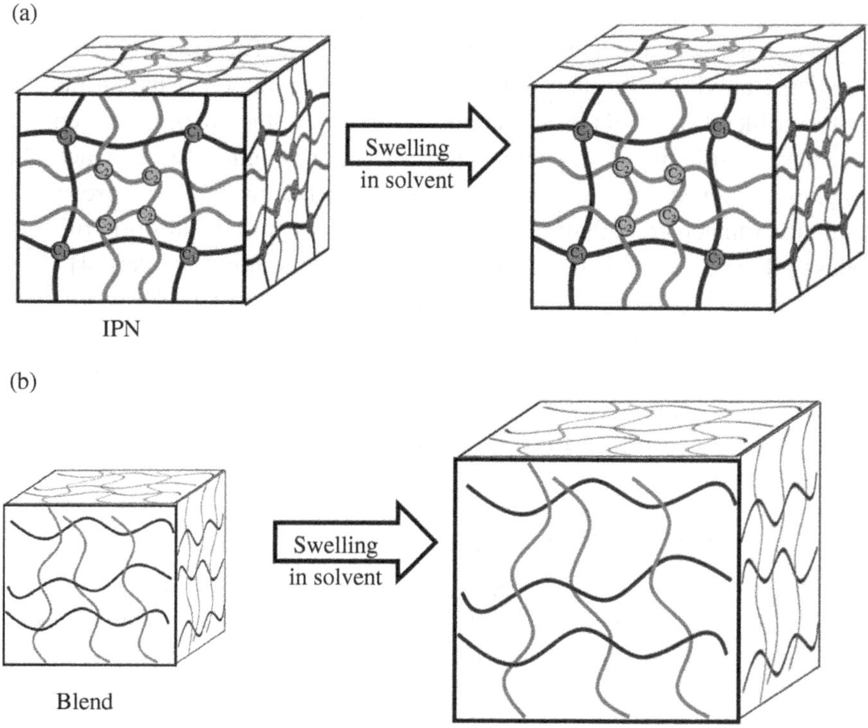

FIGURE 6.15 Model of swelling (a) IPN sample and (b) polymer blend

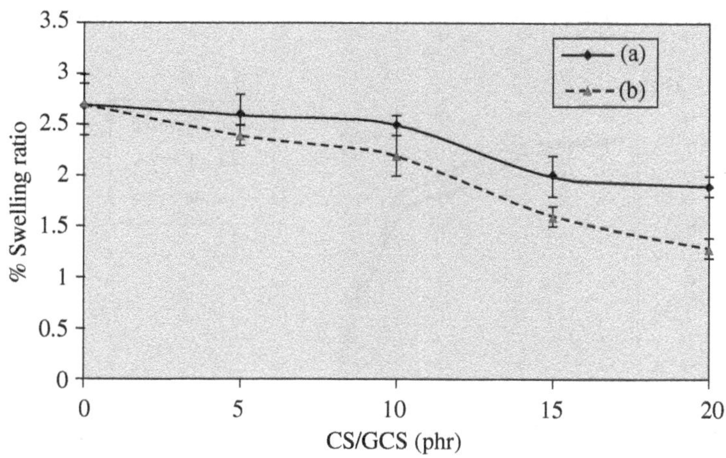

FIGURE 6.16 Influence of PS on the elongation at break of the SPNR/PS blend (author's experiment)

Therefore, the swelling ratio of the polymer blend in the presence of GCs in toluene was lower than that of sample containing Cs. The linear amylose and branched amylopectin chains from starch might migrate to the surface of polymer blend sheet and prevented the adsorption of the nonpolar solvent.

Phase continuity and phase interpenetration might also enhance the impact resistance. It was reported that the cocontinuous IPN structure exhibited high impact strength of polymer composite because the uniform rubbery phase allowed the energy to be dissipated into the whole sample [14]. This effective dissipation of energy was the principal role of rubbery phase in impact-resistant materials. When adding SPNR was over 50% in IPN, the impact strength of polymer blend decreased. These results are referred that the poor miscibility between SPNR and PSf was found. The highest impact strength was observed in 40/60 SPNR/PSf blend.

6.4.2 Thermal Properties

The thermal stability of SPNR/PSf was studied by thermogravimetric analysis (TGA) and compared to that of SPNR as shown in Figure 6.17. Results showed that thermal decomposition of SPNR in nitrogen atmosphere started at about 380°C followed by two major losses of weight during the main volatilization and pyrolysis, which are essentially completed by about 550°C [32]. The thermal degradation of SPNR containing 10, 25, and 50 wt% Cfiber was slightly lower onset degradation temperature (265°C) than SPNR. The thermal degradation of NR-g-cellulose changed in its weight loss with four steps. The first step occurred from 195 to 292°C which indicates water and the second step appeared from 292 to 350°C referring to the main chain of Cfiber. When

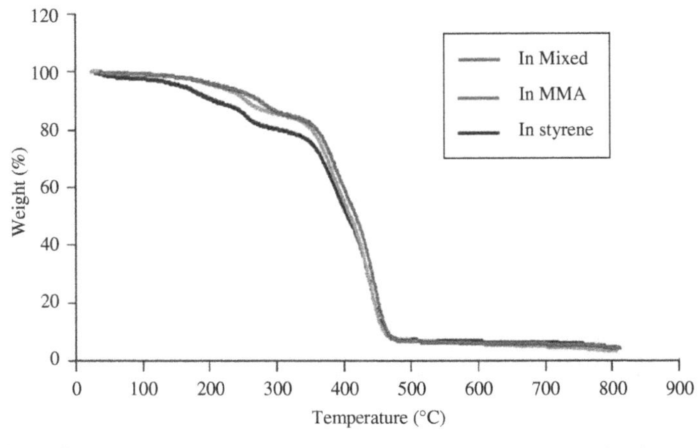

FIGURE 6.17 TGA thermograms of 50/50 SPNR/PSf blend (author's experiment)

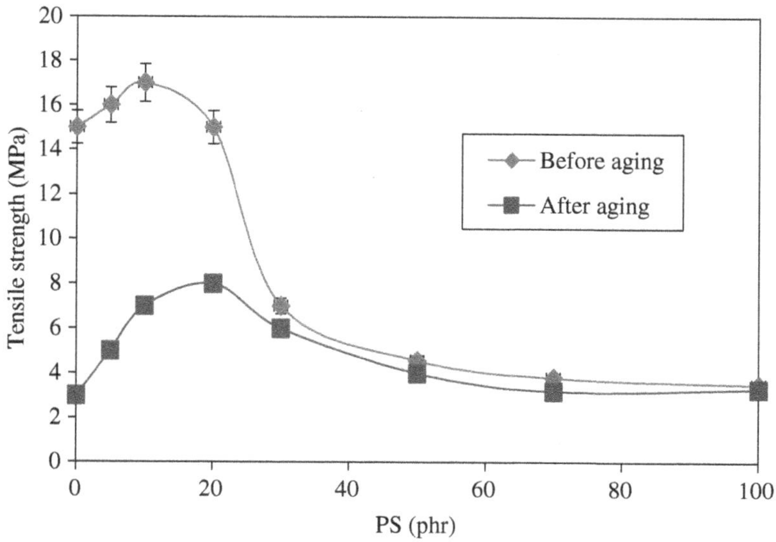

FIGURE 6.18 (a) Effect of aging at 90°C on the tensile strength of SPNR/PS blend with different PS contents

temperature continued from 350 to 400°C as shown in step 3, the degradation of main SPNR molecules was destroyed. The final step occurred at over 400°C which referred to the SPNR-g-Cfiber. This confirmed grafting between SPNR and Cfiber. In addition, the Cfiber in the graft copolymer show more sensitivity to thermal degradation temperature than rubber. The onset degradation temperature of Cfiber was found to be at about 170°C as shown in Figure 6.34. Starting from about 390°C, both SPNR and SPNR-g-Cfiber followed almost the same thermal degradation behavior.

In order to improve the thermal resistance, when the SPNR was blended with the PS, the results were reported in Figure 6.18.

6.4.3 Mechanical Strength

6.4.3.1 Tensile Strength The tensile strength of SPNR/PS IPN is illustrated in Figure 6.19. The tensile strength of SPNR increased as a function of PS content and then decreased with increasing PS content. The highest tensile strength of the SPNR/PS blend was found at 10 phr due to its reinforcement filler in the polymer blend. In addition, the tensile strength of IPN increased due to the interaction and interlocking of SPNR and PS networks, which result in the reinforcement of the SPNR phase. However, the tensile strength of the IPN sample dramatically decreased after adding over 10 phr due to phase separation.

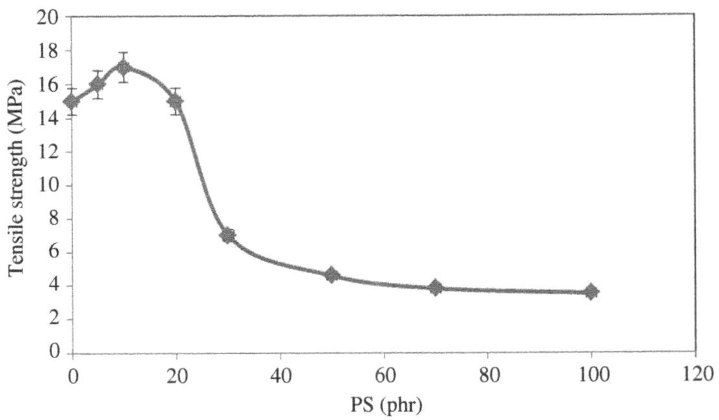

FIGURE 6.19 Effect of PS on the tensile strength of SPNR/PS IPN (author's experiment)

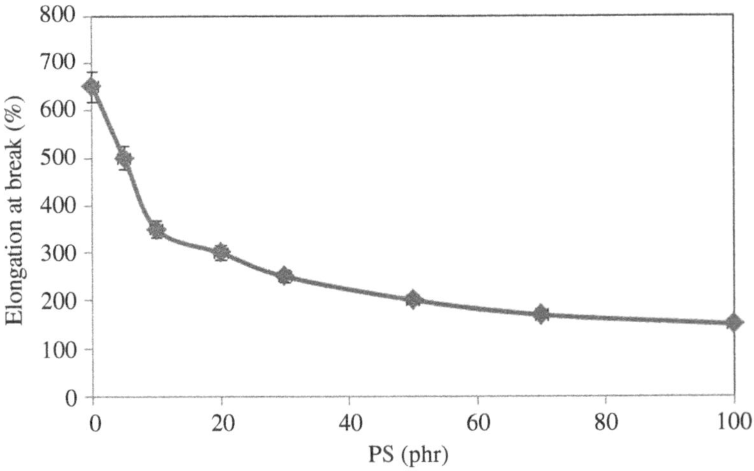

FIGURE 6.20 Influence of Cfiber on the elongation at break of the 50/50 SPNR/PS blend (author's experiment)

The tensile strength of the IPN having 5, 20, 30, 50, 70, and 100 was 16, 15, 7, 4.5, 3.7, and 3.5 MPa, respectively, while that of the SPNR was 15 MPa.

The effect of PS on the elongation at break of SPNR/PS IPN is presented in Figure 6.20. It was clear that the elongation at break of SPNR/PS IPN decreased as a function of PS content in the blend.

The reduction in elongation at break due to its phase separation between SPNR and PS as shown in Figure 6.20. In addition, this is due to the plastic nature and lower elongation of PS in comparison with SPNR.

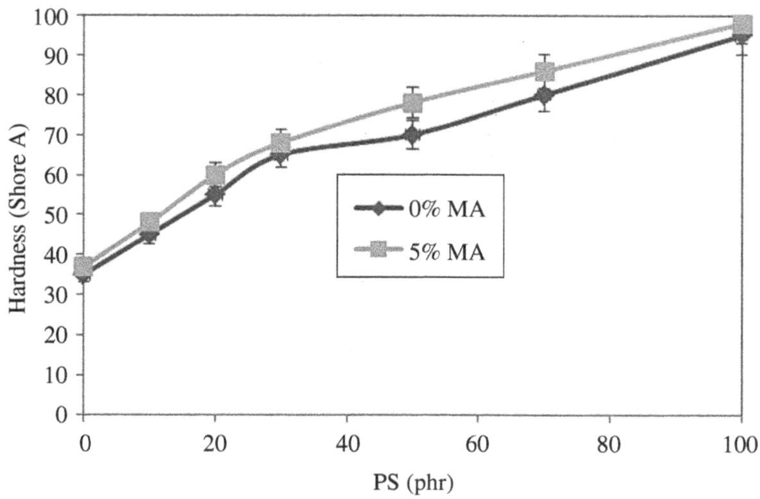

FIGURE 6.21 Influence of MA and PS content on the hardness of SPNR/PS with different PS contents (author's experiment)

6.4.3.2 Hardness The hardness of SPNR/PS IPN has been investigated by means of the intention technique. The effect of PS content and MA on the hardness of SPNR/PS IPN was presented in Figure 6.21. It was clear that after inclusion of PS into the SPNR, the hardness of SPNR was improved. The hardness of SPNR/PS IPNR increased as a function of PS content. The highest hardness was found at 100 phr PS. In addition, the hardness of SPNR/PS IPN was also raised after the addition of MA. It was referred that the SPNR was occurred the chemical reaction between the SPNR and PS by help of MA. Hardness changes are correlated with the variation of the glass transition temperature.

6.4.3.3 Tear Strength The effect of sulfur content on the tear strength of SPNR/PS blend is illustrated in Figure 6.22. It was found that the tear strength of the SPNR/PS blend was 50 N/mm. When the sulfur was added in the SPNR/PS blend, the tear strength of the SPNR/PS blend was enhanced and was proportional to sulfur content. The maximum tear strength of SPNR/PS blend was found at 3 (g) and its value was 65 N/mm.

The effect of cellulose fiber on the tear strength of SPNR/PS IPN is presented in Figure 6.23.

Results show that the tear strength of SPNR/PS IPN increased with increasing cellulose fiber contents and then decreased with increasing cellulose content. The highest tear strength was found at 50 phr and its value was 70 N/mm.

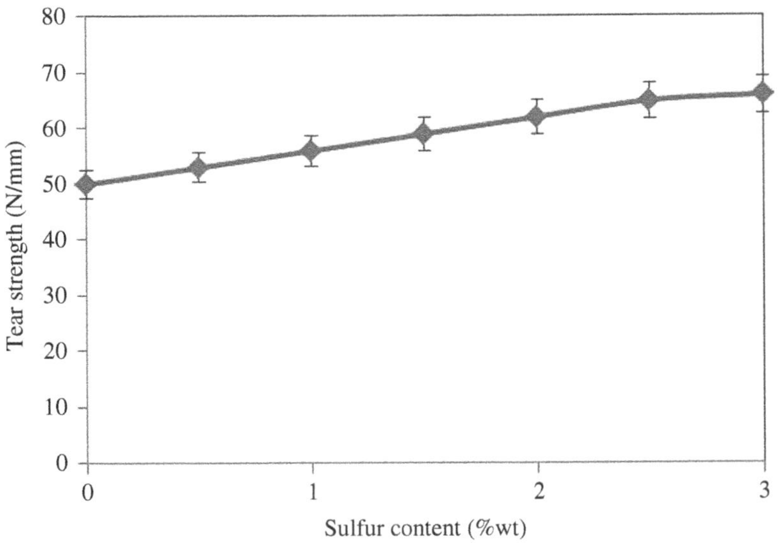

FIGURE 6.22 Influence of sulfur on the tear strength of SPNR/PS IPN (author's experiment)

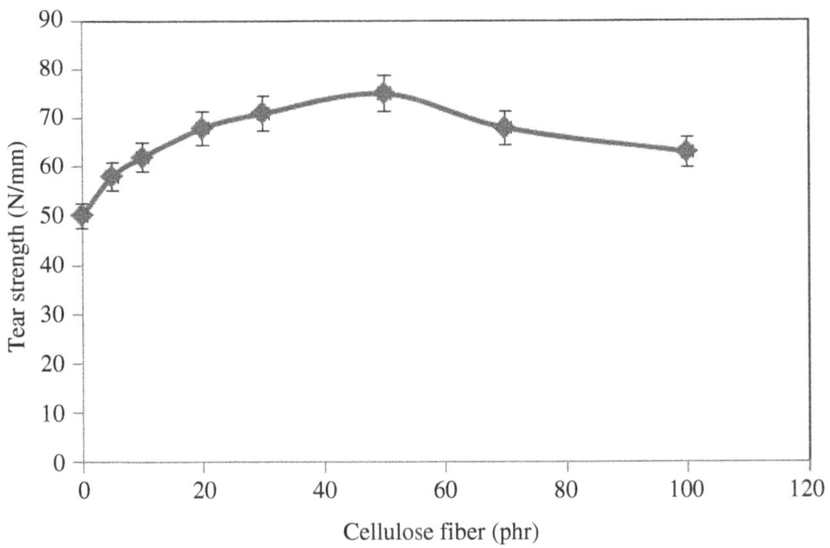

FIGURE 6.23 Influence of cellulose fiber content on the tear strength of SPNR/PS IPN (author's experiment)

6.4.3.4 Impact Strength The impact strength of SPNR/PS foam with different PS contents is presented in Figure 6.24. After the incorporation of PS into the SPNR, the impact strength of SPNR decreased. The impact strength of SPNR/PS IPN at 10 phr PS was 37 J/m^2, while the impact strength of SPNR/PS IPN at 50 phr was 32. Moreover, the minimum impact strength was 25 J/m^2.

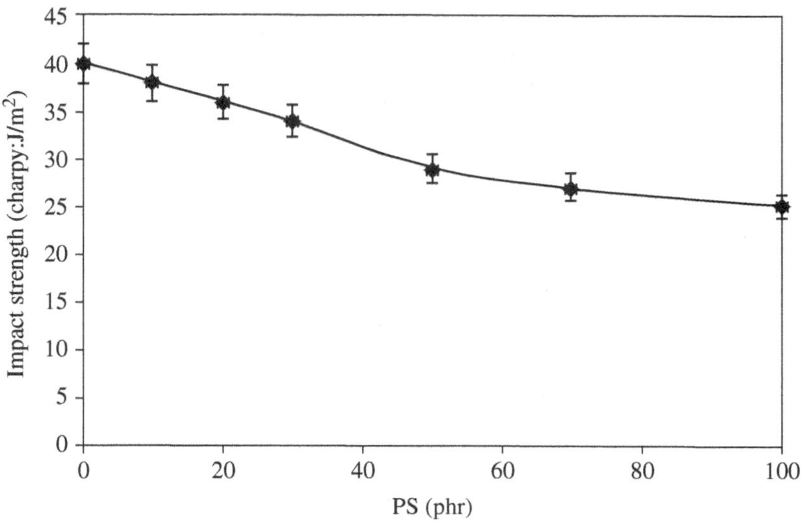

FIGURE 6.24 Influence of PS content on the impact strength of SPNR/PS IPN (author's experiment)

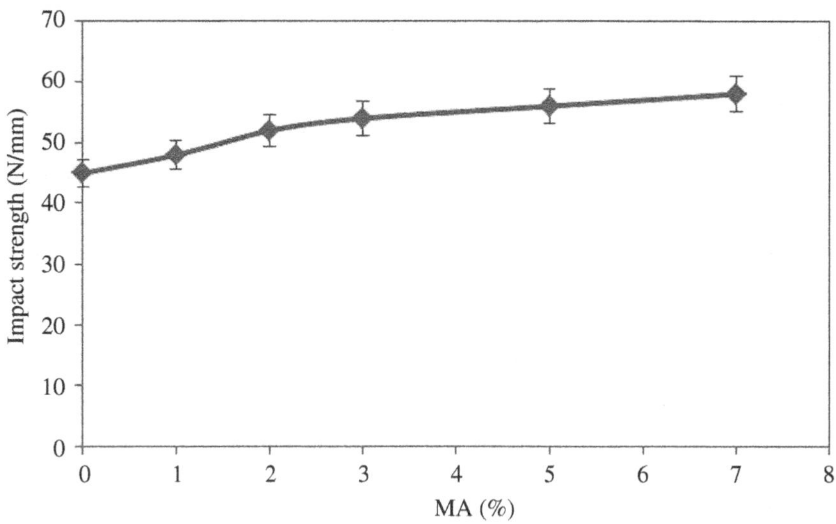

FIGURE 6.25 Influence of MA content on the impact strength of SPNR/PS blend IPN (author's experiment)

After the incorporation of PS into the SPNR, the impact strength of SPNR decreased. The impact strength of SPNR/PS IPN at 10 phr PS was 37 J/m^2. Moreover, the minimum impact strength was 25 J/m^2. The effect of MA on the impact strength of SPNR/PS blend IPN is presented in Figure 6.25. The impact strength of SPNR/PS blend was 45 N/mm. After the addition of MA in the SPNR/PS blend, the impact strength of SPNR/PS blend IPN was increased as a function

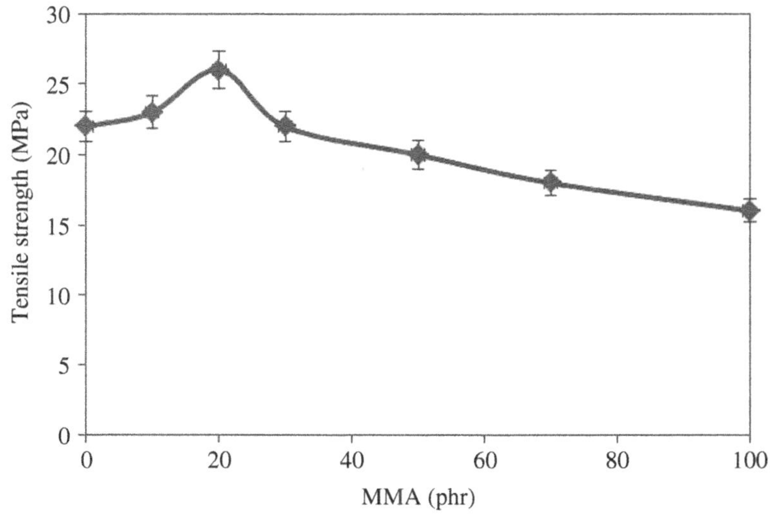

FIGURE 6.26 Influence of MMA content on the tensile strength of ENR/MMA blend (author's experiment)

of MA content. This might be due to the chemical reaction between polymer matrix and MA. The highest impact strength was found at 7% (Figure 6.26).

The tensile strength of ENR/MMA blend increased with increasing MMA followed by the reduction in the tensile strength. The highest tensile strength was found at 20 phr MMA in the blend. This result might be due to reinforcement of MMA in the blend. In addition, the higher chemical interaction between the MMA and ENR was expected. However, when the amount of MMA was more than 20, the tensile strength of the blend decreased due to its phase separation between the ENR and MMA. The effect of MMA on the elongation at break of ENR/MMA is illustrated in Figure 6.27. It was clear that the elongation at break of ENR/MMA blend dramatically decreased. The lowest elongation at break was observed at 100 MMA and its value was 50%. Because the characteristic behavior of MMA is brittle.

After the incorporation of plaster content into the 50/50 SPNR/PS blend, the toluene resistance of 50/50 SPNR/PS blend was clearly improved and their results are presented in Figure 6.28. The swelling ratio of the 50/50 SPNR/PS blend without plaster was 2.7, while the swelling ratio of 50/50 SPNR/PS blend in the presence of 10 phr plaster was 2.3. The minimum swelling ratio of SPNR/PS blend was observed at 100 phr plaster and its value was about 1.9. This result was explained that explained that the toluene penetrate difficultly in the polymer matrix due to the addition of plaster in the 50/50SPNR/PS blend restricting to toluene penetration.

The effect of plaster contents on the tensile strength and elongation at break of SPNR/PS blend is presented in Figure 6.29. The influence of plaster content

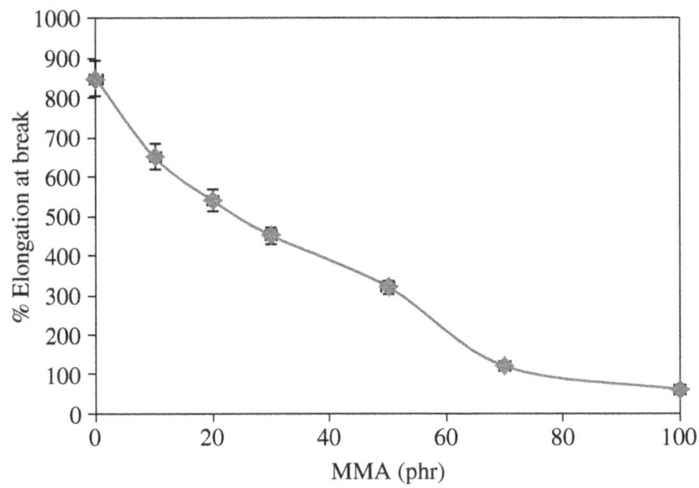

FIGURE 6.27 Effect of MMA on the elongation at break of ENNR/MMA blend (author's experiment)

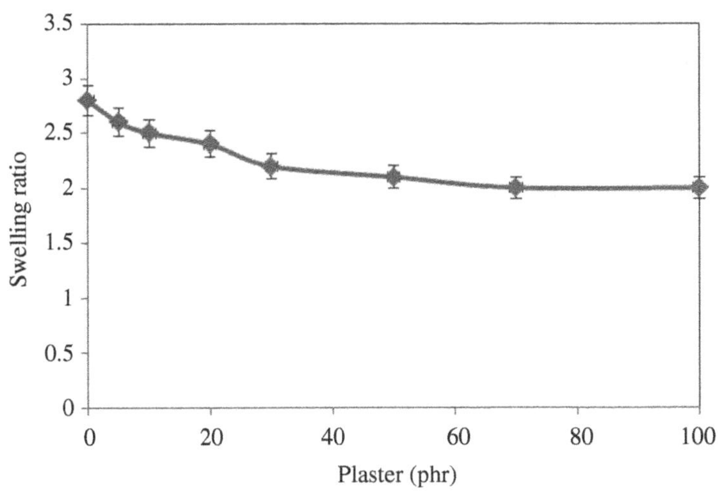

FIGURE 6.28 Influence of plaster content on the swelling ratio of SPNR/PS blend in toluene (author's experiment)

on the hardness of the SPNR/PS blend is presented in Figure 6.29. It was clear that the hardness of the SPNR/PS blend was increasing as a function of plaster content due to plaster nature. The maximum hardness of the SPNR/PS blend was observed at 100 phr plaster and its value was about 100 Shore A.

The effect of plaster content on the impact strength of the polymer blend is presented in Figure 6.30. Results showed that the hardness of SPNR/PS blend was increased as a function of plaster content and then decreased with

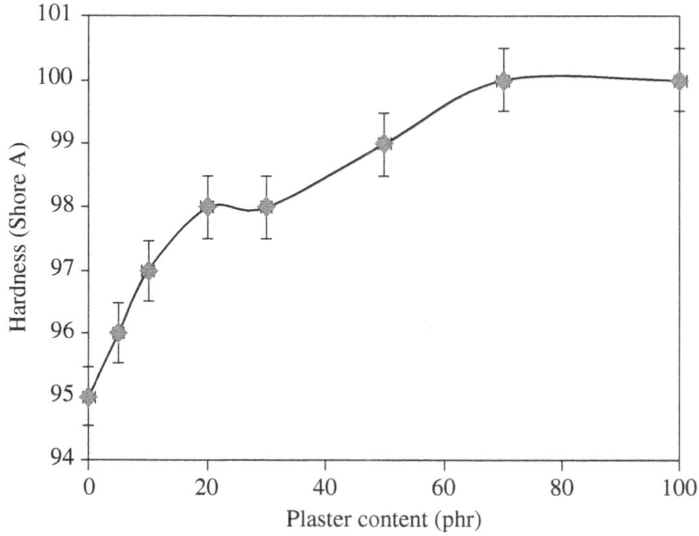

FIGURE 6.29 Influence of plaster content on the hardness of the SPNR/PS blend (author's experiment)

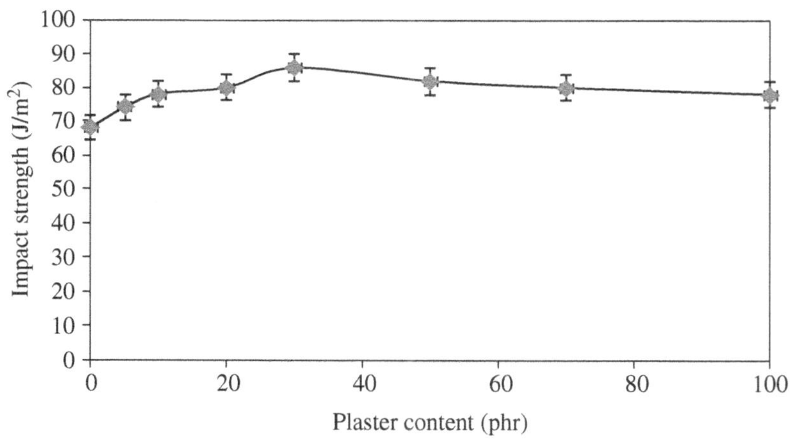

FIGURE 6.30 Effect of plaster content on the impact strength of SPNR/PS blend (author's experiment)

increasing plaster content. The highest impact strength of the SPNR/PS was found at 30 phr plaster content. When the amount of plaster was more than 30 phr, the impact strength of SPNR/PS blend decreased. This might be due to its poor adhesion between the SPNR and the PS matrix [33, 34].

The effect of plaster on the tensile strength, elongation at break, and modulus of the SPNR/PS blend is presented in Figure 6.31a, b, and c, respectively. The tensile strength of SPNR/PS blend increased with increasing plaster

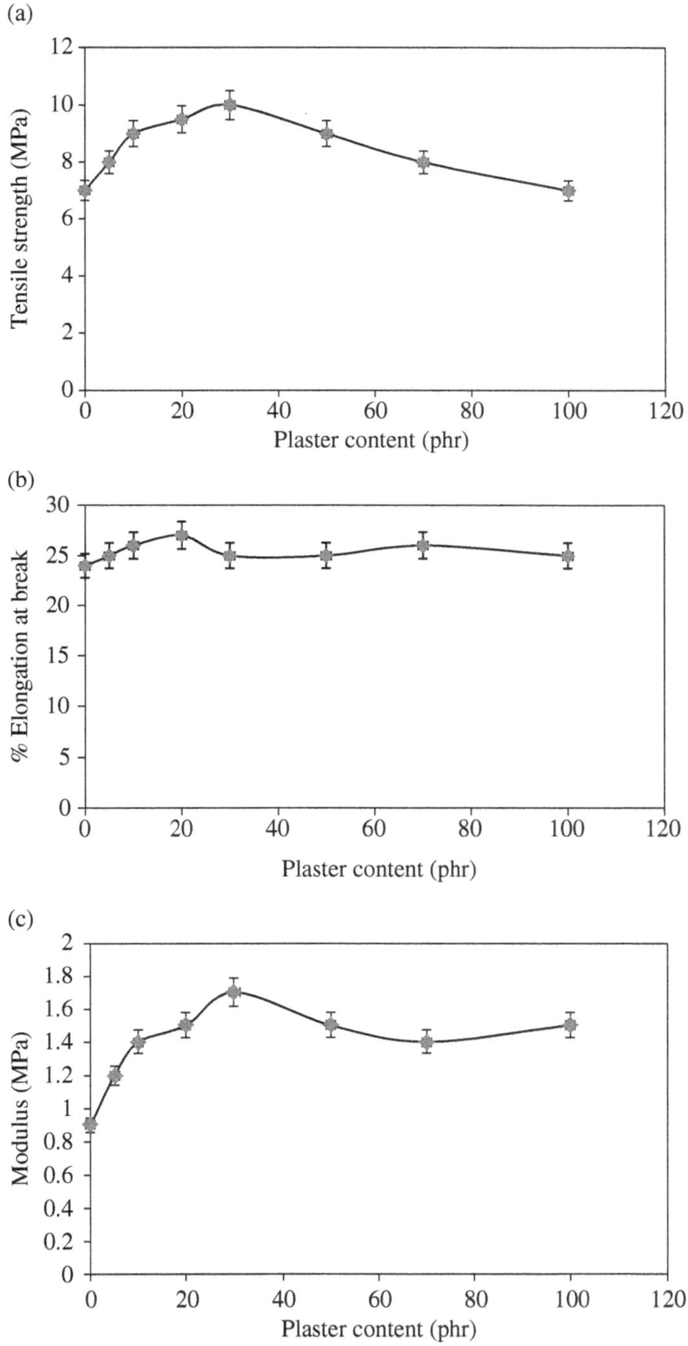

FIGURE 6.31 Influence of plaster content on (a) tensile strength, (b) elongation at break, and (c) modulus (author's experiment)

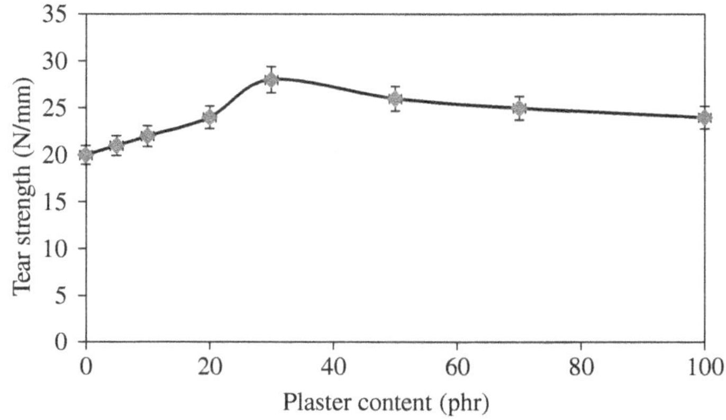

FIGURE 6.32 Influence of plaster content on the tear strength of SPNR/PS blend

content and then decreased as a function of plaster content as given in Figure 6.31a. This result was in good agreement with the impact strength, as mentioned previously. The highest tensile strength of SPNR/PS was found at 30 phr plaster. However, when the amount of plaster was more than 30 phr, the tensile strength of SPNR/PS blend decreased. The lowest tensile strength was found at 100 phr plaster. The elongation at break of SPNR/PS blend containing plaster is presented in Figure 6.31b. Results showed that the elongation of SPNR/PS blend was almost constant after the addition of plaster and their values are 26%. The influence of the plaster content on the modulus of SPNR/PS blend is illustrated in Figure 6.31b. The modulus of SPNR/PS blend was correlated with the tensile strength data. The modulus of SPNR/PS blend increased with increasing plaster content followed by the reduction of modulus as given in Figure 6.31c. The highest modulus was found at 30 phr plaster and its value was 1.3%. The tear strength of SPNR/PS blend with different plaster contents is presented in Figure 6.32. The tear strength results show the same trend with tensile strength and modulus. The tear strength of SPNR/PS blend was 26 N/mm. The tear strength of SPNR/PS blend increased with increasing filler contents and then decreased as a function of plaster.

After the addition of plaster in the SPNR/PS blend, the thermal stability of SPNR/PS blend was enhanced.

Aging of material is a process which takes place at higher temperature through chemical degradation, *leading to the significance of the properties of rubbers*. Oxidation effects on molecular level occurred through crosslinking or breaking molecular chains [35]. Crosslinking makes the rubber harder and the breakage of the chains softens it. Based on the aging tests, it can be

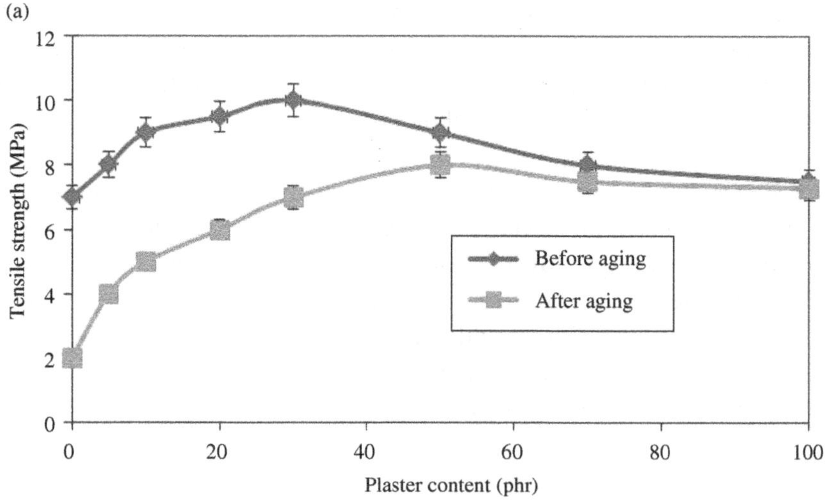

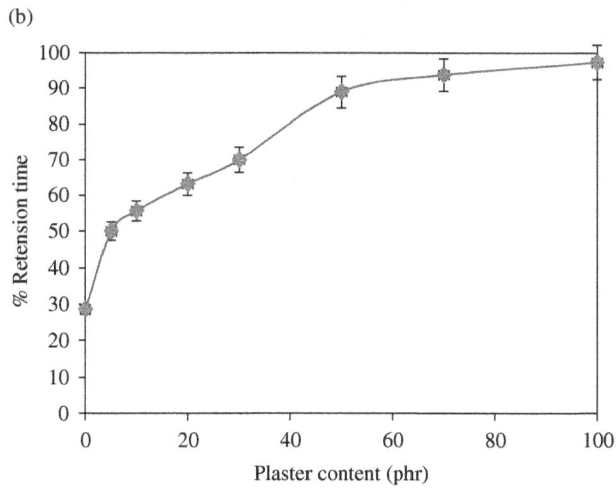

FIGURE 6.33 (a) Tensile strength before and aging at 90°C for 24h and (b) retention in the tensile strength (author's experiment)

concluded that the addition of plaster could provide better aging resistance to SPNR/PS blend by showing higher retention of tensile strength value than that of sample without plaster as given in Figure 6.33a. The retained percentage values of tensile properties were calculated using the following equation:

$$\text{Retention}(\%) = \frac{\text{Tensile strength after ageing}}{\text{Tensile strength before ageing}} \times 100$$

The effect of plaster on the retention is reported in Figure 6.33b. It was clear that when the addition of plaster into the SPNR/PS blend was done, the retention of SPNR/PS blend increased as a function of plaster content in the blend owing to a higher thermal stability of plaster.

6.4.3.5 Creep Properties Creep experiments have a great importance in the design of material products because they reflected the load-bearing capacity of end product [36]. The effect of plaster content on the creep is presented in Figure 6.34. The results show an improvement on the creep behavior of the SPNR/PS blend after the addition of plaster into the polymer blend. The creep values decreased with increasing plaster contents due to the higher content of the hard phase from plaster component.

6.4.3.6 Morphology Figure 6.35 shows the fractured surfaces of plaster filled NR/SPNR composites which was obtained from impact strength experiment. Results showed that plaster was well embedded in the SPNR/PS blend matrix at low plaster content when compared to higher plaster content.

The fractured surface of sample in the presence of plaster content showed more holes and the fractured zone tended to be smooth. This observation explains why the tensile strength of SPNR/PS blend in the presence of plaster

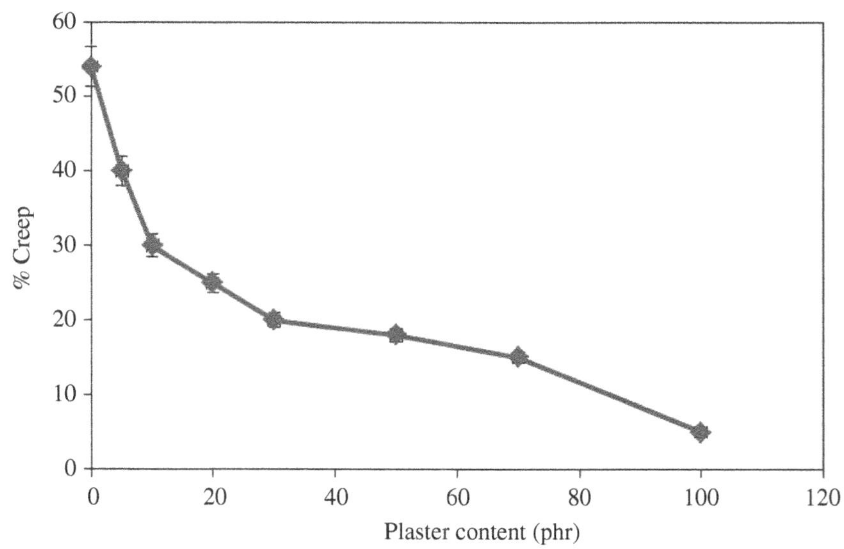

FIGURE 6.34 Influence of plaster content on the creep of SPNR/PS blend (author's experiment)

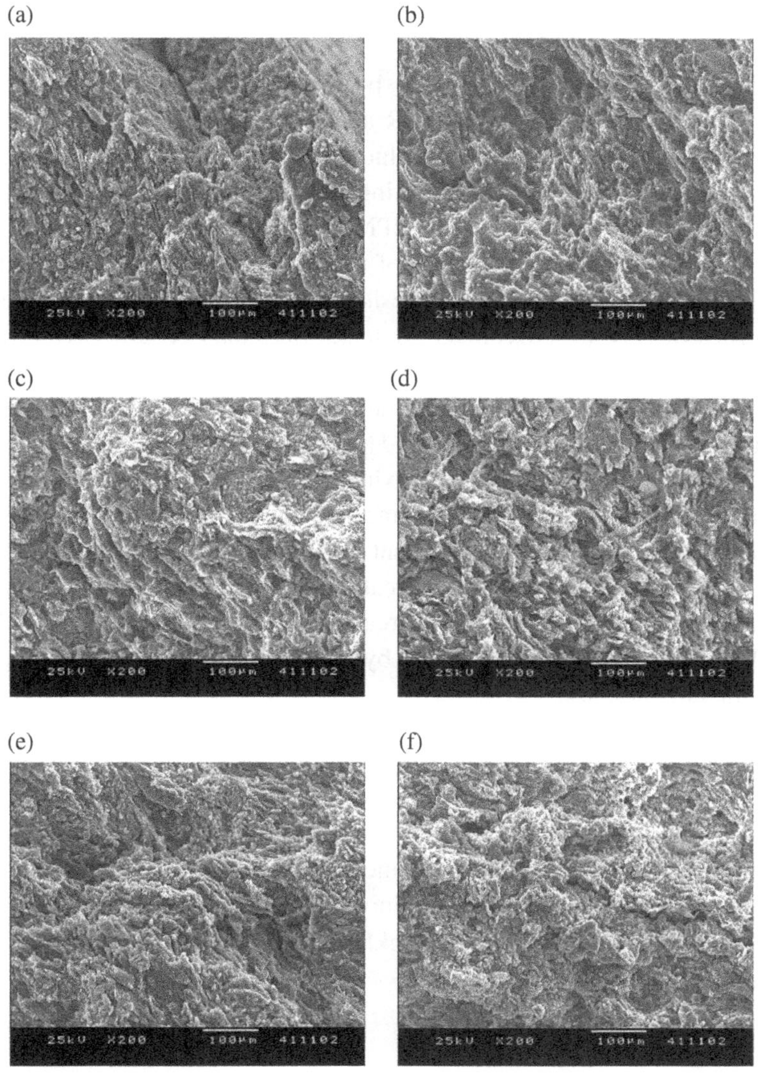

FIGURE 6.35 SEM morphology of SPNR/PS blend with different plaster contents at (a) 0, (b) 10, (c) 20, (d) 30, (e) 50, and (f) 70%wt (author's experiment)

content at 30 phr plaster was higher than that of the sample at higher amount of plaster content. These micrographs support the tensile properties obtained due to efficient stress transfer when an external force was applied to it at low amount of plaster. Therefore, at low amount of plaster content filled SPNR/PS blend composites require more energy to fracture [37].

6.5 CONCLUSIONS

Modification of NR with IPN has been extensively studied over recent decades. IPNs are made from SPNR and organic polymer such as PS and PMMA through two processes including sequential interpenetrating networks and simultaneous interpenetrating networks. Their mechanical strength properties of NR/organic polymer IPN samples are better compared with individual polymer. The properties of the IPN sample show outstanding in tensile strength, creep, and thermal resistance. The parameters controlling the properties of IPN are crosslinker content, crosslinker types, polymer blend ratio, natural polymer, and fillers. But at high level of PS content and high extent of PS crosslinking, the agglomeration of plastic phase was observed. The tensile strength, tear strength, and hardness increased with an increase in PS content. The increase in phase mixing with an increased plastic content is the reason for this enhancement in mechanical performance. The crosslink densities increased with plastic content as the number of crosslinks increased with PS content. The elongation at break values decreased with an increase in plastic content as stretch ability decreased upon the increase of crosslinked PS. This observation was supported by swelling ratio, impact strength, tear strength, and hardness.

ACKNOWLEDGMENTS

The authors are grateful for the fundamental research grant from The Thailand Research Fund/Prince of Songkla University/Thammasat University (Grant No. RTA 5780018) and Prince of Songkla University (Grant No. SCI 540032S) that supported this work.

REFERENCES

[1] Riyajan, S. and Chaiponban, S. (2010) Swelling and physical properties of semi-interpenetrating NR/PVA networks. *Kautschuk Gummi Kunststoffe*, **63**, 70–73.

[2] Vudjung, C., Chaisuwan, U., Pangan, U. *et al.* (2014) Effect of natural rubber contents on biodegradation and water absorption of interpenetrating polymer network (IPN) hydrogel from natural rubber and cassava starch. *Energy Procedia*, **56**, 255–263.

[3] Anon (1993) Study of thermal properties and morphology of interpenetrating polymer networks from natural rubber and polyacrylamide. *Journal of Applied Polymer Science*, **49** (12), 2229–2233.

[4] Amnuaypanich, S., Patthana, J. and Phinyocheep, P. (2009) Mixed matrix membranes prepared from natural rubber/poly(vinyl alcohol) semi-interpenetrating polymer network (NR/PVA semi-IPN) incorporating with zeolite 4A for the pervaporation dehydration of water–ethanol mixtures. *Chemical Engineering Science*, **64** (23), 4908–4918.

[5] Frisch, H.L. and Chen, Z.J. (1994) Electrical conductivity of iodine-doped pseudo interpenetrating polymer networks of poly-(carbonate-urethane) and natural rubber. *Journal of Polymer Science, Part A: Polymer Chemistry*, **32** (7), 1317–1322.

[6] Ismail, H., Rusli, A. and Rashid, A.A. (2005) Maleated natural rubber as a coupling agent for paper sludge filled natural rubber composites. *Polymer Testing*, **24**, 856–862.

[7] Jobish, J., Charoen, N. and Praveen, P. (2012) Dielectric properties and AC conductivity studies of novel NR/PVA full-interpenetrating polymer networks. *Journal of Non-Crystalline Solids*, **358** (8), 1113–1119.

[8] Lu, G., Yu, H., Zeng, Z. and Luo, Y. (2012) Preparation and properties of interpenetrating polymer network films from prevulcanized natural rubber latex/chitosan blends. *Advanced Materials Research*, **396–398**, 400–406.

[9] Murakami, S., Senoo, K., Toki, S. and Kohjiya, S. (2002) Structural development of natural rubber during uniaxial stretching by in situ wide angle X-ray diffraction using a synchrotron radiation. *Polymer*, **43** (7), 2117–2120.

[10] Riyajan, S. and Chaiponban, S. (2009) Activation energy and thermal behaviors of thermoplastic elastomer based on natural rubber and polyvinyl alcohol. *Kautschuk Gummi Kunststoffe*, **63** (4), 28–33.

[11] Riyajan, S., Tanbumrung, K. and Pinyocheep, P. (2010) Physical properties of a 'green' polymer blend based on PVA starch and ENR. *KGK Kautschuk Gummi Kunststoffe*, **63**, 371–376.

[12] Uragami, T. (2010) Selective membranes for purification and separation of organic liquid mixtures comprehensive. *Membrane Science and Engineering*, **2**, 273–324.

[13] Riyajan, S. and Santipanusopon, S. (2010) Influence of ammonia concentration and storage period on properties field NR latex and skim coagulation. *Kautschuk Gummi Kunststoffe*, **63** (6), 240–244.

[14] Mathew, A.P., Packirisamy, S. and Thomas, S. (2001) Studies on the thermal stability of natural rubber/polystyrene interpenetrating polymer networks: thermogravimetric analysis. *Polymer Degradation and Stability*, **72**, 423–439.

[15] Jayasuriya, M.M. and Hourston, D.J. (2009) The effect of composition on the dynamic and mechanical properties of natural rubber-poly(methylmethacrylate) blends. *Journal of Applied Polymer Science*, **112** (6), 3217–3224.

[16] Schneider, M., Pith, T. and Lambla, M. (1996) Preparation and morphological characterization of two- and three-component natural rubber-based latex particles. *Journal of Applied Polymer Science*, **62** (2), 273–290.

[17] Riyajan, S., Intharit, I. and Tangboriboonrat, P. (2012) Physical properties of polymer composite: natural rubber glove waste/polystyrene foam waste/ cellulose. *Industrial Crops and Products*, **36**, 376–382.

[18] Liu, C., Shao, Y. and Jia, D. (2008) Chemically modified starch reinforced natural rubber composites. *Polymer*, **49**, 2176–2181.

[19] Yua, L., Deana, K. and Lib, L. (2006) Polymer blends and composites from renewable resources. *Progress in Polymer Science*, **31**, 576–602.

[20] Bouthegourd, E., Rajisha, K.R., Kalarical, N., Saiter, J.M. and Thomas, S. (2011) Natural rubber latex/potato starch nanocrystal nanocomposites: correlation morphology/electrical properties. *Materials Letters*, **65**, 3615–3617.

[21] Rouilly, A., Rigal, L. and Gilbert, R.G. (2004) Gilbert synthesis and properties of composites of starch and chemically modified natural rubber. *Polymer*, **45**, 7813–7820.

[22] Valodkar, M. and Thakore, S. (2011) Organically modified nanosized starch derivatives as excellent reinforcing agents for bionanocomposites. *Carbohydrate Polymer*, **86**, 1244–1251.

[23] Heidary, S.H., Amraei, I.A. and Payami, A. (2009) Carbon-black-reinforced natural rubber/polystyrene full interpenetrated polymer networks: synthesis and properties. *Journal of Applied Polymer Science*, **113** (4), 2143–2149.

[24] Riyajan, S., Chusri, S. and Voravuthikunchai, S.P. (2012) Mechanical properties and cure test of a natural rubber/poly(vinyl alcohol) blend. *Rubber Chemistry and Technology*, **85**, 147–156.

[25] Kadokawa, J., Kodzuru, K., Kawazoe, S. and Matsuo, T. (2011) Preparation of natural rubber/condensed tannin semi-interpenetrating polymer network composites by hematin-catalyzed cross-linking. *Journal of Polymers and the Environment*, **19** (1), 100–105.

[26] Amnuaypanich, S. and Kongchana, N. (2009) Natural rubber/poly(acrylic acid) semi-interpenetrating polymer network membranes for the pervaporation of water-ethanol mixtures. *Journal of Applied Polymer Science*, **114** (6), 3501–3509.

[27] Riyajan, S., Chaiponban, S. and Tanbumrung, K. (2009) Investigation of the preparation and physical properties of a novel semi-interpenetrating polymer network based on epoxised NR and PVA using maleic acid as the crosslinking agent. *Chemical Engineering Journal*, **153**, 199–205.

[28] Mathew, A.P., Packirisamy, S. and Thomas, S. (2000) Morphology, mechanical properties, and failure topography of semi-interpenetrating polymer networks based on natural rubber and polystyrene. *Journal of Applied Polymer Science*, **78** (13), 2327–2344.

[29] Mathew, A.P., Packirisamy, S., Padmanabhan, A.S. and Thomas, S. (1998) Kinetics of diffusion of styrene monomer containing divinyl benzene through vulcanized natural rubber. *Journal of Polymer Engineering*, **17** (5), 405–429.

[30] Mathew, A.P., Groeninckx, G., Michler, G.H. and Radusch, H.J. (2003) Viscoelastic properties of nanostructured natural rubber/polystyrene interpenetrating polymer networks. *Journal of Polymer Science, Part B: Polymer Physics*, **41** (14), 1680–1696.

[31] Asia, I.O., Akinlabi, A.K. and Egbon, E.E. (2006) Cure characteristics and physico-mechanical properties of blends of epoxidised natural rubber and polystyrene. *Pakistan Journal of Scientific and Industrial Research*, **49** (6), 383–387.

[32] Golub, M.A. and Gargiulo, R.J. (1972) Thermal degradation of 1,4-polyisoprene and 1,4-polybutadidene. *Polymer Letters*, **10**, 41.

[33] Herrero, S., Mayor, P. and Hernández-Olivares, F. (2013) Influence of proportion and particle size gradation of rubber from end-of-life tires on mechanical, thermal and acoustic properties of plaster–rubber mortars. *Materials & Design*, **47**, 633–642.

[34] Mathew, A.P., Packirisamy, S., Radusch, H.J. and Thomas, S. (2001) Effect of initiating system, blend ratio and crosslink density on the mechanical properties and failure topography of nano-structured full-interpenetrating polymer networks from natural rubber and polystyrene. *European Polymer Journal*, **37**, 1921–1934.

[35] Zheng, W., Liu, L., Zhao, X. *et al.* (2015) Effects of lanthanum complex on the thermo-oxidative aging of natural rubber. *Polymer Degradation and Stability*, **120**, 377–383.

[36] Wong, R.H.C., Lin, P., Tang, C.A. and Chau, K.T. (2002) Creeping damage around an opening in rock-like material containing non-persistent joints. *Engineering Fracture Mechanics*, **69** (17), 2015–2027.

[37] Riyajan, S., Intharit, I. and Tangboriboonrat, P. (2013) Physical properties of the maleated sulphur prevulcanized natural rubber latex-g-cellulose fiber. *Journal of Polymer Materials*, **30**, 159–174.

7

THE SYNTHESIS AND THE APPLICATIONS OF IPNs BASED ON SMART POLYMERS

GUILLERMINA BURILLO, EMILIO BUCIO AND LORENA GARCIA-URIOSTEGUI

Departamento de Química de Radiaciones y Radioquímica, Instituto de Ciencias Nucleares, Universidad Nacional Autónoma de México, México City, Mexico

7.1 INTRODUCTION

Hydrogels are networks of hydrophilic polymers that have the capability to absorb large amounts of water without losing their three-dimensional structures. They can absorb water through hydration, which is related to the presence of chemical groups in their structures (such as $-COOH$, $-CONH_2$, $-CONH-$, and $-SO_3H$), the capillary effects, and the osmotic pressure. The swelling behavior of the hydrogels may change in response to the external environment and is dependent on the chemical groups present.

It is known that several hydrophilic polymer gels may undergo a discontinuous volume variation when the solvent composition, temperature, and pH are changed [1]. They are known as "smart," "intelligent," or "stimuli-responsive gels" [2]; there are also other polymeric systems that respond to light, electrical, magnetic, and sonic fields [3]. The stimuli-sensitive hydrogels are readily applied in the medical industry (contact lenses, artificial corneas,

Micro- and Nano-structured Interpenetrating Polymer Networks: From Design to Applications, First Edition. Edited by Sabu Thomas, Daniel Grande, Uroš Cvelbar, K.V.S.N. Raju, Ramanuj Narayan, Selvin P. Thomas and Akhina H.

dressing and coating for sutures, electrode sensors) and as systems for drug delivery and the absorption of liquids [2, 4–8]. Other possible applications include metal extraction, wastewater treatment, soft contact lenses, enzyme supports, wound dressing, water-managing materials, diapers, and agrochemical release among other applications because of their high water content and good biocompatibilities [9–12]. They possess the capability to retain loaded drugs in a collapsed state and releasing drugs in a swollen state because of the volume transition caused by environmental stimuli [13, 14]. The temperature-sensitive polymers comprise one hydrophobic and one hydrophilic group such as poly(*N*-isopropylacrylamide) (PNIPAAm). PNIPAAm is one of the most studied thermosensitive polymers, which exhibits a lower critical solution temperature (LCST) of approximately 32°C. Adjusting the LCST near the body temperature of 37°C is essential for smart drug delivery applications [15, 16]. Generally, the incorporation of hydrophilic monomers leads to an increase in the LCST, while the incorporation of a hydrophobic monomer leads to a decrease in the LCST [17–19]. Recently, many researchers have focused on polymeric systems that respond to more than one type of stimulus [20–22].

Dual stimuli-responsive hydrogels may be prepared by combining PNIPAAm, a widely studied temperature-sensitive polymer, with another stimulus-sensitive polymeric component. Several dual stimuli-responsive hydrogel systems have been investigated.

However, a serious limitation of hydrogels in many applications is the low mechanical strength of such gels when they are highly swollen, as they present significant handling difficulties. To overcome this problem, sensitive hydrogels were prepared with satisfactory mechanical properties as a semi-interpenetrating or an interpenetrating network using flexible polymers. The formation of IPNs instead of synthesizing new types of copolymers constitutes the most useful method of improving the properties of the individual components to satisfy the requirements of specific applications [23].

7.2 STIMULI-RESPONSIVE POLYMERS

Temperature-sensitive polymers exhibit an LCST behavior where phase separation is induced at a certain temperature threshold. They are soluble in water at temperatures below the LCST and insoluble at temperatures above the LCST [24]. Typical thermosensitive polymers include poly(ethylene glycol) [25], ethyl(hydroxyethyl)cellulose, and N-substituted polyamides [26]. Some miscible polymers exhibit the opposite effect, with phase separation on cooling temperature at the upper critical solution temperature (UCST). This UCST behavior was observed in polystyrene/polymethyl methacrylate mixtures by Ougizawa and Walsh [27] and by Yang *et al.* in PS/polystyrene-*co*-acrylonitrile) blends [28] and in the binary grafting of 2-(dimethylamino)ethyl methacrylate/ 4-vinylpyridine (DMAEMA/4VP) onto PP at a pH of 2.2 [29].

Examples of pH-responsive polymers include weak polyelectrolytes that contain acidic or basic functional groups such as carboxylic, phosphoric, or amino functional groups. A change in the pH results in a shift in the chemical equilibrium, therefore changing the ionization degree of the polymer chains [30].

Hydrogels exhibiting a phase transition in response to a change in the external conditions such as pH, ionic strength, temperature, and electric currents are known as "stimuli-responsive" or smart gels that are insoluble in water. These three-dimensional hydrophilic networks can retain a large amount of water, which gives it a good compatibility with blood, while also maintaining a certain degree of structural integrity and elasticity [31].

7.3 INTERPENETRATING POLYMER NETWORKS AND SEMI-INTERPENETRATING NETWORKS

An IPN is a material comprising two or more networks that are fully or partially interlaced on a molecular scale but are not covalently bonded to each other and cannot be separated unless chemical bonds are broken [32]. The most widely investigated IPNs comprise two independent polymers with different chemical compositions. If one polymer is linear and penetrates into another crosslinking network, the IPN is known as a semi-interpenetrated network.

7.4 THE SYNTHESIS AND THE APPLICATIONS OF SEMI-INTERPENETRATING NETWORKS AND INTERPENETRATING NETWORKS

7.4.1 Sequential Semi-Interpenetrating Networks

Sequential networks comprising hydroxypropyl cellulose (HPC) and poly(*N*-isopropylacrylamide) (PNIPAAm) were synthesized by the adsorption of a solution of NIPAAm monomer onto a dried porous network with a thermosensitive volumetric behavior and an LCST that was between the LCST of HPC and the LCST of PNIPAAm, which modulates the thermosensitivity behavior of the SIPNs [33]. A thermosensitive SIPN based on a poly(ethylene glycol)-*co*-poly(epsilon-caprolactone) macromer and sodium alginate was synthesized *in situ* [34]. Its mechanical strength and the cumulative release percentage of bovine serum albumin (BSA) were investigated, with the authors concluding that it was a suitable candidate for being a drug delivery carrier.

7.4.2 The Simultaneous Method for the Synthesis of Semi-Interpenetrating Networks

Thermosensitive poly(*N*-isopropylacrylamide-*co*-vinylpyrrolidone)/chitosan (PNIPAAm-*co*-NVP/CS) semi-interpenetrated hydrogels with improving loading capacity and sustained release for anionic drugs were prepared by free radical polymerization using *N,N*-methylenebisacrylamide (MBA) as a crosslinker simultaneously. The LCST of the hydrogels was adjusted to the vicinity of the human body temperature by introducing hydrophilic NVP. The presence of chitosan (CS) in the SIPNs improves the swelling behavior and provides a high affinity for anionic drugs [35].

Chitosan ((1,4)-2-amino-2-deoxy-β-D-glucan) is a biocompatible, biodegradable, and antigenic pH-sensitive cationic biopolymer and is a promising candidate for employment in the design of controlled delivery systems. Verstiuc and Coll [36] synthesized a series of SIPN networks based on NIPAAm and chitosan in one step using tetraethylene glycol diacrylate as a crosslinker. It was observed that the structure of the network was influenced by the quantities of crosslinker and chitosan that were used. The incorporation of chitosan into the structure induced a pronounced pH sensitivity. The degree of swelling was seen to vary from approximately 100% at basic pH values to more than 210% at acidic pH values. The levels were highly sensitive to temperature variations and had a maximum LCST of 29°C. The system showed little promise in delivering positively charged active pilocarpine, while the effectiveness of the system in delivering neutral or negatively charged therapeutic entities was not investigated yet.

A series of temperature-sensitive hydrogels containing an SIPN comprising a PNIPAAm network and linear poly(vinyl alcohol) (PVA) were prepared simultaneously via radical polymerization with *N,N*-methylenebisacrylamide as a crosslinker. PVA is a water-soluble hydrophilic polymer, which has been extensively explored as a biomaterial and has been used in drug delivery systems, sensors, and surgical repairs because of its excellent mechanical properties. Pure PVA hydrogels are not sensitive to environmental stimuli and conventional PNIPAAm exhibits a relatively low response rate; however,

SIPNs comprising these polymers exhibit an increased sensitivity with responses that were much faster than the responses of conventional PNIPAAm hydrogels [37].

7.4.3 A Comparison of the Properties between Sequential and Simultaneous Semi-Interpenetrating Network Films

Semi-interpenetrated polymer networks of various compositions were synthesized using segmented polyurethane urea (SPUU) by sequential and simultaneous methods, NIPAAm, acrylic acid (AAc), and butyl methacrylate (BMA), resulting in an SIPN film denoted as SPUU/poly(NIPAAm-*co*-AA-*co*-BMA). SIPN films synthesized by the simultaneous and the sequential methods were used for swelling and drug release studies at different pH values and at different temperatures to determine the environmental sensitivity of these gels. The simultaneous SIPNs absorbed more water than the sequential SIPNs, but both had a poor rate of water absorption at a pH of 1.2. A higher thermosensitivity was observed for sequential SIPNs than for simultaneous SIPNs during the drug release tests [38].

The preparation and the properties of a series of polyampholytic (PNIPAAm-*co*-sodium acrylate)/(polyethyleneimine) SIPN hydrogels that were synthesized by the simultaneous method were investigated. Different formulations with polyethyleneimine (PEI) showed a varied swelling behavior, where the hydrogel with 0.5 g showed the maximum swelling capacity [39]. The high swelling capacities and the equilibrium water content values of these hydrogels make them suitable for applications in bioengineering fields.

7.4.4 The SIPNs of Sensitive Star Polymers

It was recently reported that the introduction of poly(amidoamine) dendrimers into SIPN hydrogels could significantly enhance the temperature response rates and the swelling ratios of the resulting structure [40, 41]. Based on the previous reports, it has been found that star polymers have unique conformations, low viscosities at relatively high molecular weights, and nonstructural functionalities in comparison with lineal polymers [42]. Wu *et al.* [43] recently synthesized a series of pH-sensitive SIPN hydrogels comprising PVA and 21-arm star polymers containing poly(2-dimethylamino)ethyl methacrylate at different molecular weights. Riboflavin was used as a model drug to evaluate the loading capacities and the drug release behavior of the polymers. The results suggested that the PVA/star poly(dimethylaminoethyl methacrylate) (PDMAEMA)-50 000 hydrogel exhibited the highest swelling ratio and the highest drug loading capacity. The pH-sensitive SIPN based on the star PDMAEMA polymer could be a promising drug delivery system because of its controllable porous structure.

7.5 INTERPENETRATING POLYMER NETWORKS

7.5.1 Interpenetrating Polymer Networks Synthesized in One Step by the Simultaneous Method

Electrically responsive IPNs comprising polyvinyl alcohol and polyacrylic acid-*co*-acrylamido-2-methyl propyl sulfonic acid were synthesized in one step by Lin *et al.* [44], where the IPN hydrogel showed a significant bending behavior in the presence of an applied electric field.

Lu *et al.* [45] synthesized a new IPN in one step, using poly(3,4-ethylene-dioxythiophene)/poly(styrene sulfonate)/PVA/poly(acrylic acid) for improving the optrode-neural tissue interface in optogenetics. The conducting IPN films exhibited a significantly higher capacitance and lower electrochemical impedance compared to pure conducting polymer films.

7.5.2 Interpenetrating Polymer Networks Synthesized in Two Steps

A pH-sensitive IPN based on chitosan, acrylamide-*g*-polyvinyl alcohol, and hydrolyzed acrylamide-*g*-polyvinyl alcohol that was crosslinked with glutaraldehyde was synthesized in two steps and used in the controlled release of the antibiotic drug cefadroxil [46].

Tanaka *et al.* [47] discovered the deformation of hydrogels in electric fields. Electrically sensitive hydrogels made of polyelectrolytes were synthesized, resulting in the formation of an insoluble, swellable polymer network containing ionic groups [48]. The response of electrically sensitive hydrogels can be exhibited in the form of their swelling, shrinking, and bending behaviors, and the characteristic features of these behaviors are that they can interrupt the change in their volume. The electric-sensitive hydrogels are able to convert free chemical energy directly to mechanical work isothermally. These hydrogels are widely used in artificial muscles, drug delivery transducers, and biomimetic actuators [11, 49, 50]. The equilibrium swelling of polyelectrolyte gels caused by various types of externally applied stimuli arising from possible applications in chemomechanical systems for converting the electrostatic interaction energy working among fixed charges into mechanical energy is of great interest. Kim *et al.* [51] investigated the swelling responses of IPN hydrogels prepared with hyaluronic acid–PVA from electrical and pH stimuli, and Kulkarni and Biswanath [52] investigated the use of electrically responsive hydrogels in drug delivery.

In recent years, the use of polysaccharides to prepare multiunit controlled release dosage forms has attracted the attention of researchers [53–55]. Naturally available polysaccharides are often preferred to synthetic materials because of their nontoxicity, their low cost, their availability, and their biodegradability. However, these natural polysaccharides present some limitations, such as an uncontrolled rate of hydration, microbial contamination, and a decrease in their viscosities upon storage. Many attempts have been made to overcome the limitations of these polysaccharides. Among these attempts, there has been a focus on the development of IPN structures. If the IPN structures contain a pH-sensitive polymer, they could be used for controlled drug release applications. A pH-sensitive IPN containing carboxymethyl cellulose–(polyacrylamide-*g*-alginate) was synthesized by Kulkarni [56] for the controlled release of ketoprofen. The swelling of the beads and drug release was significantly increased when the pH of the medium was changed from an acidic pH to an alkaline pH. Another pH-sensitive drug release experiment used a polyacrylamide-*g*-xanthan–carboxymethyl cellulose IPN that was synthesized in two steps by Kulkarni [57]. This hydrogel was in the form of beads containing —COOH functional groups and remains unionized at the pH of gastric juices, which therefore causes negligible swelling or drug release in the intestines. The hydrogel could reduce the severity of gastric irritation, ulceration, and perforations associated with ketoprofen.

Interpenetrating microgel–hydrogel polymeric networks were prepared by mixing doubly quaternized p(AN-*co*-4VP) core–shell particles with AAm and HEMA before polymerization [58].

Other important applications for the stimuli-sensitive IPNs include ultrafiltration membranes based on PVA and PAM for the pervaporation of ethanol–water mixtures [59], while IPN membranes containing P(acrylic acid-*co*-acrylamide) in a PVA matrix for the pervaporation dehydration of ethylene glycol were synthesized by Kuila [60].

New spherically shaped crosslinked IPN hydrogels based on chitosan and polyvinylpyrrolidone were synthesized in two steps [61]: (i) chitosan beads were crosslinked with ethylene glycol diglycidyl ether and (ii) absorption of a solution of vinylpyrrolidone followed by polymerization and crosslinking inside the CS network. The IPN showed some peculiar properties: the first one was its reversible swelling in water at room temperature at a pH of 7, as neat CS gels swell only in acidic environments. The IPNs swelled quickly even when they were placed in water in a dry state, which could be connected to the thermodynamic interaction between the two polymers.

7.6 INTERPENETRATING POLYMER NETWORKS AND SEMI-INTERPENETRATING NETWORKS SYNTHESIZED BY IONIZING RADIATION

Crosslinking and polymerization by ionizing radiation; by gamma radiation from a ^{60}Co source or by electron beam irradiation have many advantages over other conventional methods in the preparation of IPNs, because it is a fast process and catalysis or additives are not necessary to initiate the reaction [62]. The radiation method is relatively simple, and the degree of crosslinking can be controlled easily by varying the dose rate and the absorbed radiation dose. The radiation can be used in all synthetic routes or in some parts of the reaction pathways (Figure 7.1).

Sequential interpenetrating networks are the most widely obtained networks when ionizing radiation methods are used. The first step of crosslinking is always the use of ionizing radiation with or without the addition of any crosslinking agent. In the second step, either chemical or radiation crosslinking is used, but it was observed that the use of radiation crosslinking without any chemical crosslinking agent could cause the development of a more

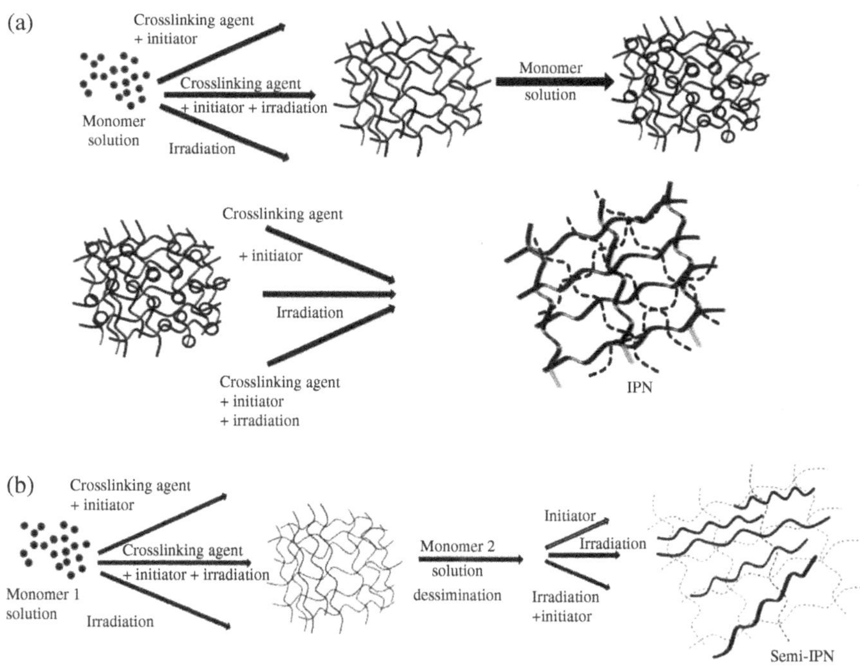

FIGURE 7.1 The sequential method used to obtain IPN (a) and SIPN (b)

homogeneous structure than chemical crosslinking, while the crosslinking density could only be achieved by increasing the radiation dose.

IPNs based on the temperature-sensitive NIPAAm, a well-known thermo-responsive monomer that swells below and deswells above its LCST of 33°C, and the pH-sensitive AAc were prepared in two consecutive steps according to the sequential method. The first hydrogel was synthesized by gamma radiation from a ^{60}Co gamma source or by an electron beam accelerator. A second network was synthesized inside the first AAc hydrogel via polymerization and crosslinked with a redox initiator and a crosslinking agent [63]. The pH of the IPN was conserved and the LCST shifted to a lower temperature with an increase in the radiation dose because there was a higher crosslinking density. These IPNs were characterized by Adem *et al.* [64], who showed that the T_g of the IPN hydrogels increased with the radiation dose because of the increase in the crosslinking density and the formation of a more rigid structure.

A sequential IPN made of PNIPAAm hydrogel and poly(*N*-acryloxysuccinimide), which provides a functional group that is readily displaced by the amino groups of lysine, was studied by Ortega *et al.* [65]. These groups are the anchoring elements of vesicles and also act as agents that can crosslink PNAS.

Lu *et al.* [66] irradiated a sensitive NIPAAm monomer solution containing a crosslinking agent with gamma rays to obtain a PNIPAAm hydrogel that was swollen in a solution of methyl methacrylate (MMA) monomer, hence forming an IPN consisting of a PNIPAAm system containing dispersed MMA after irradiation with γ rays. Guilherme *et al.* [67] reported the development of an IPN based on NIPAAm and acrylamide (PAAm) polymers via UV radiation, where NIPAAm was prepared in a solution containing a crosslinking agent and dispersed through a PAAm hydrogel that was previously prepared by a chemical step, before the NIPAAm monomer was polymerized and crosslinked *in situ* with irradiation from a UV lamp.

An IPN hydrogel where the carboxyl groups on carboxymethyl chitosan (CMCS) were physically crosslinked with Ca^{2+} ions and further crosslinked with PNIPAAm after irradiation was systematically studied [68]. In contrast with chitosan, which is only soluble at low pH in some buffer, CMCS is soluble at almost all pH values, which makes it an attractive option for biomedical applications. The resultant IPN had an LCST that was similar to the pure NIPAAm.

Ahinern *et al.* [69] obtained an IPN hydrogel based on the reaction of polyacrylonitrile (PAN) with poly(*N*-vinyl-2-pyrrolidone) (PVP) in one step. The PVP polymer was dissolved in a solution of acrylonitrile irradiated with gamma rays from a ^{60}Co source. Zhang and Peppas [70] combined pH- and temperature-sensitive materials to synthesize smart IPNs.

A new SIPN based on the preparation of polyacrylic acid/polyhydroxybutyrate (PHB) by gamma radiation was synthesized by Zaldivar *et al.* [71]. PHB is a semicrystalline thermoplastic polymer with mechanical properties that are similar to polypropylene. It is completely biodegradable and has a good biocompatibility, which is suitable for a variety of applications in the field of medicine. The degree of swelling of the IPN network was approximately 600% in water; hence it behaved like a superabsorbent hydrogel.

IPN networks based on the irradiation of *N*-vinylpyrrolidone–AAc copolymers and gelatin were synthesized in one step. Gelatin is a biopolymer and is not found in nature, but it can be produced by the acid or the base hydrolysis of collagen. Gelatin is widely used for wound dressings and drug delivery because of its compatibility and biodegradability. IPN network was prepared by using *N,N'*-methylenebisacrylamide (BIS) and glutaraldehyde as crosslinkers by gamma irradiation. Glutaraldehyde was incorporated after irradiation to crosslink the gelatin chains, while BIS was added to the solutions before irradiation. The thermal stability of the IPN hydrogels increased when the IPNs were formed [72].

PDMAEMA contains aliphatic tertiary amino groups and is a typical stimuli-responsive polymer or hydrogel that exhibits both temperature and pH sensitivity, although its mechanical and swelling properties as a hydrogel are in need of further modification. SIPNs composed of DMAEMA and PEO were synthesized in the presence of crosslinkers by radiation technology. PEO was chosen as the linear polymer in the SIPNs because of its excellent biocompatibility, good swelling, and lower costs [73].

7.7 SEMI-INTERPENETRATING NETWORKS AND INTERPENETRATING NETWORKS IN THE HEAVY IONS IMMOBILIZATION

Hiroki *et al.* [74] synthesized gels based on acryloyl-L-proline methyl ester (A-ProOMe) by a simultaneous radiation-induced polymerization with a crosslinking using an ultraslow thermoresponse for up to 6 months, depending on the type of solvent that was used for irradiation of the A-ProOMe (Figure 7.2). The thermo- and pH-sensitive gel membranes of the polyacrylic acid (PAAc) grafted onto the PA-ProOMe network for the selective permeation of Li, Co, and Ni ions were studied by Hasegawa *et al.* [75]. However, a serious limitation of hydrogels is the low mechanical strength of such gels when they are highly swollen. To overcome this problem, a semi-interpenetrated network of PA-ProOMe penetrating into PAAc was synthesized by gamma irradiation

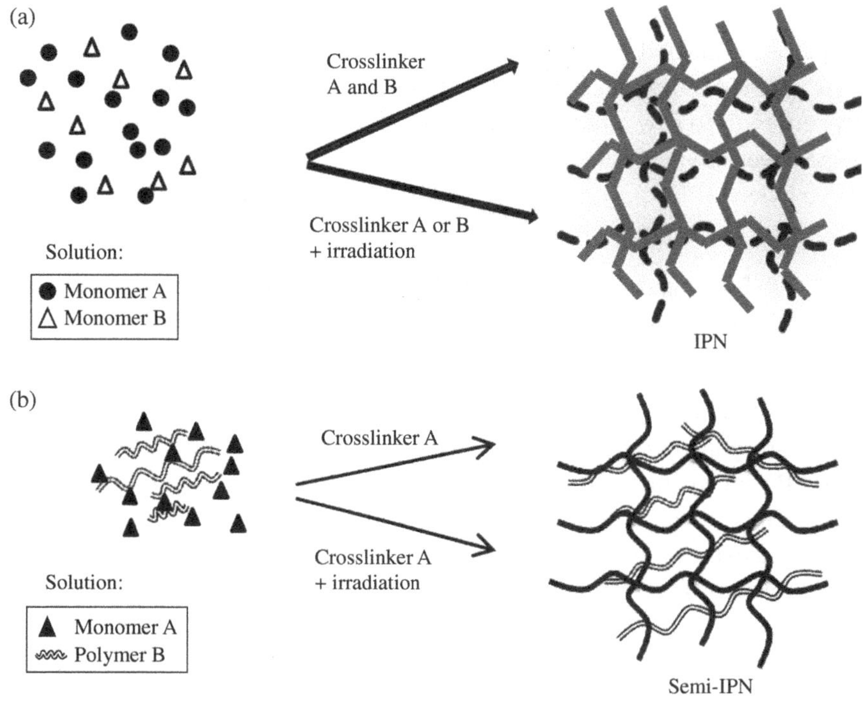

FIGURE 7.2 The simultaneous method used to prepare IPNs (a) and SIPNs (b)

in two steps [76], and the Cu(II) immobilization was studied at different pH values. The response time was faster than the system studied by Hiroki [74] and Hasegawa [75]. An interpenetrating polymer network of PVP and acrylonitrile (AN) was synthesized by simultaneous polymerization and crosslinking via gamma irradiation [69] (Figure 7.2). PVP was dissolved in an AN solution and irradiated by a ^{60}Co source, which converted the AN groups on the PAN backbone into amidoxime groups. Amidoximated PVP-based IPNs are effective in the absorption of various metal ions.

7.8 THE NOVEL ARCHITECTURES OF IPNs DEVELOPED BY IONIZING RADIATION POLYMERIZATION

In a sequential IPN, the first polymer network is synthesized before a monomer and a crosslinker are introduced into the network and polymerized *in situ*, but their sensitivities were often altered or eliminated, and their resultant copolymers had poor mechanical properties [76]. Attempts to solve the problem by creating IPNs with a reduced gel size or by using macroporous structures were successful in preserving the sensitivities of the copolymers, but the mechanical properties remained poor [77]. To improve the mechanical properties, IPNs were grafted onto other polymer matrices with the desired mechanical properties. Two new strategies were used to prepare interpenetrating stimuli-responsive polymer networks grafted onto other polymer matrices such as polyethylene (PE), polypropylene, and poly(tetrafluoroethylene) (PTFE).

7.8.1 Polymer-*g*-Interpenetrating Polymer Networks Synthesized via Irradiation and the Addition of a Chemical Initiator in Three Steps

A new PP-*g*-IPN was synthesized with this strategy by Burillo's research group for the first time [78] (Figure 7.3). The IPNs developed from the thermo- and pH-responsive polymer networks of AAc and PNIPAAm that were grafted onto PP were synthesized in three steps. The first step involved the irradiation of PP films in air with a ^{60}Co source (Gammabeam, 651 PT, Nordion Co.) according to the preirradiation peroxidative method to form peroxy and hydroperoxy groups. The irradiated samples were then placed in ampoules containing aqueous solutions of AAc. The ampoules were saturated with argon, sealed, and heated to 50°C for different periods of reaction time. The ungrafted monomer and homopolymer were washed out with water and the films were dried, where the grafting percentage could then be evaluated based on the

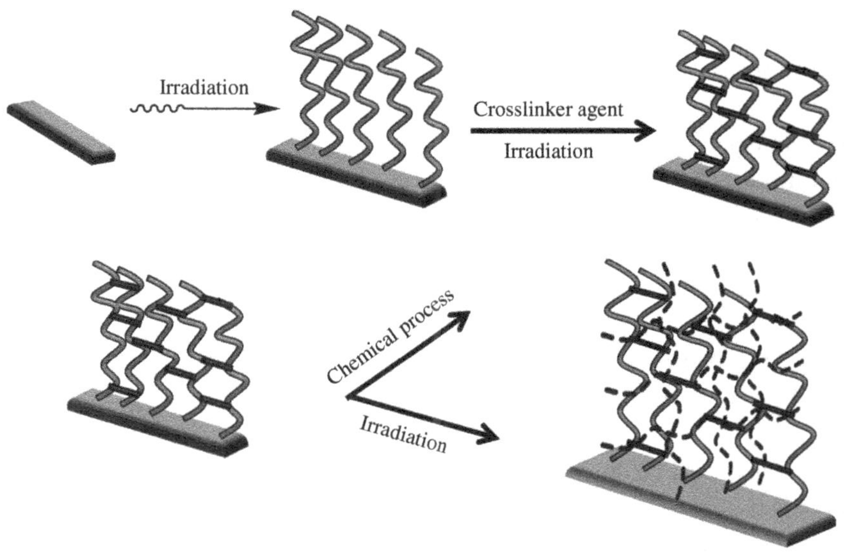

FIGURE 7.3 Polymer-*g*-IPN synthesized by sequential method. Source: Ruiz *et al.* [78]. Reproduced with permission of John Wiley & Sons

increased weight of the film. The grafted films were placed in ampoules with distilled water for the second step, sealed in an argon atmosphere, and crosslinked by gamma radiation to form *net*-(PP-*g*-AAc). As PP required higher irradiation doses than AAc for crosslinking to occur, we end up with a network of AAc grafted onto PP. The second interpenetrated network of PNIPAAm was synthesized within the *net*-(PP-*g*-AAc) by chemical polymerization during the third step and crosslinked with *N,N*-methylenebisacrylamide (NMBAAm), the redox initiator ammonium persulfate (APS), and *N,N,N′,N′*-tetramethylethylenediamine (TEMED). Elastic films with good mechanical properties were obtained, and the velocity of the response (or the swelling equilibrium) occurred rapidly at approximately 5 min, in comparison with the 25 min taken by grafted PP-*g*-AAc films to reach equilibrium. PAAc hydrogels that were crosslinked with gamma radiation were better for forming IPNs because of the larger pore dimensions, which enhance the diffusion of the second monomer into the first network to form a second network within the first network.

New smart surface-modified PP films were prepared with the same strategy with the synthesis of the IPNs of NIPPAm and AAc grafted onto PP by gamma radiation [79] to improve the loading and the sustained delivery of vancomycin and reduce the risk of biofilm formation when it was used as a component of biomedical devices. In this research, NIPAAm was grafted in the first step and then crosslinked. The second network of AAc was synthesized *in situ* by chemical polymerization and crosslinking, as reported by Ruiz *et al.* [78]. The objective of these changes in the monomer order was to increase the NIPAAm content over the AAc content for the immobilization of more vancomycin.

7.8.2 Polymer-*g*-Interpenetrating Polymer Networks Synthesized Only by Radiation in Three Steps

The first step of the process involved the irradiation of PP films in air at a 30 kGy dosage with a ^{60}Co source (Gammabeam, 651 PT, Nordion Co.) according to the preirradiation peroxidative method to form peroxy and hydroperoxy groups. The PP film was irradiated in air and placed in ampoules containing aqueous solutions of NIPAAm. The ampoules were saturated with argon, sealed, and heated to 50°C for different reaction periods. The grafted films were then placed in ampoules containing distilled water in the second step, sealed in an argon atmosphere, and crosslinked by gamma radiation to obtain *net*-(PP-*g*-NIPAAm). As PP requires higher radiation doses than NIPAAm for crosslinking, we have a network of NIPAAm grafted onto PP after irradiation and some residual PP that has not been crosslinked. The crosslinked films were swelled with AAc in DMF in the third step, vacuum sealed, and irradiated with a 2.5 kGy radiation dosage by the direct method, where the irradiation of the monomer and the polymeric film occurs simultaneously. This three-step method of synthesis by gamma radiation is possible only in case when the radiation dosage required for the polymerization and the crosslinking of the second network was low enough to synthesize new surface-modified materials for medical devices and avoid any grafting of the second monomer AAc onto the *net*-(PP-*g*-NIPAAm). This new polymeric film released vancomycin at a suitable rate per unit surface area to prevent bacterial growth [80].

7.9 CONCLUSIONS

Stimuli-responsive systems have broadened their scope of applications by being prepared in different physical forms such as hydrogels. IPNs have shown an improvement in the velocity of the hydrogel response and have enhanced the mechanical properties of the hydrogels. Despite the abundant information about IPNs, few papers report the use of ionizing radiation, which gives a more homogeneous structure and a more controllable porosity. The radiation method used in synthesizing hydrogels has been determined to be very useful in preparing hydrogels for medical applications, where even low amounts of contamination are undesirable, because there is no need for any catalysis or additives to initiate the reaction. The method is relatively simple and the degree of crosslinking can be controlled easily by varying the dose rate and the absorbed radiation dose. The radiation can be used in all synthetic routes or in some parts of reaction pathways.

The surface functionalization of hydrogels with networks that combine sensitivities to two different stimuli has been recently shown to be an interesting approach for the improvement of drug or heavy ions loadings and the release performance of the films. However, stimuli-responsive systems remain as the subject of continuous research academically and industrially, and the large-scale commercial exploitation of smart IPNs still requires some time before it can be fully developed.

ACKNOWLEDGMENTS

The authors would like to thank the DGAPA-UNAM (Grant IN200210) for financial support as well as the CONACyT-CNPq, México (Grant 174378).

REFERENCES

[1] Dusek, K. (1993) *Responsive Gels: Volume Transitions I. Advances in Polymer Science*, 1st edn, Springer-Verlag, New York.

[2] De Rossi, D., Kajiwara, K., Osada, Y. and Yamauchi, A. (1991) *Polymers Gels: Fundamentals and Biomedical Applications*, Plenum Press, New York.

[3] Roy, D., Cambre, J.N. and Sumerlin, B.S. (2010) Future perspectives and recent advances in stimuli-responsive materials. *Progress in Polymer Science*, **35**, 278–301.

[4] Okahata, Y., Noguchi, H. and Seki, T. (1987) Functional capsule membranes. 26. Permeability control of polymer-grafted capsule membranes responding to ambient pH changes. *Macromolecules*, **20**, 15–21.

[5] Cammas, S., Suzuki, K., Sone, C. *et al.* (1997) Thermo-responsive polymer nanoparticles with a core-shell micelle structure as site-specific drug carriers. *Journal of Controlled Release*, **48**, 157–164.

[6] Yoshida, R., Sakai, K., Okano, T. and Sakurai, Y. (1992) Drug release profiles in the shrinking process of thermoresponsive poly (N-isopropylacrylamide-co-alkyl methacrylate) gels. *Industrial and Engineering Chemistry Research*, **31**, 2339–2345.

[7] Kono, K., Henmi, A., Hayashi, H. and Takagishi, H. (1999) Improvement of temperature-sensitivity of poly(N-isopropylacrylamide)-modified liposomes. *Journal of Controlled Release*, **59**, 63–75.

[8] Ichikawa, H. and Fukumori, Y. (2000) A novel positively thermosensitive controlled-release microcapsule with membrane of nano-sized poly(N-isopropylacrylamide) gel dispersed in ethylcellulose matrix. *Journal of Controlled Release*, **63**, 107–119.

[9] Schild, H.G. (1992) Poly(N-isopropylacrylamide): experiment, theory and application. *Progress in Polymer Science*, **17**, 163–249.

[10] Tanaka, Y., Gong, J.P. and Osada, Y. (2005) Novel hydrogels with excellent mechanical performance. *Progress in Polymer Science*, **30**, 1–9.

[11] Gil, E.S. and Hudson, S.M. (2004) Stimuli-responsive polymers and their bioconjugates. *Progress in Polymer Science*, **29**, 1173–1222.

[12] Osada, Y., Gong, J.P. and Tanaka, Y. (2004) Polymer gels. *Journal of Macromolecular Science—Polymer Reviews*, **44**, 87–112.

[13] Chaterji, S., Kwon, I.K. and Park, K. (2007) Smart polymeric gels: redefining the limits of biomedical devices. *Progress in Polymer Science*, **32**, 1083–122.

[14] Liu, T.Y. and Lin, Y.L. (2010) Novel pH-sensitive chitosan-based hydrogel for encapsulating poorly water-soluble drugs. *Acta Biomaterialia*, **6**, 1423–1429.

[15] Lin, C.C. and Metters, A.T. (2006) Hydrogels in controlled release formulations: network design and mathematical modeling. *Advanced Drug Delivery Reviews*, **58**, 1379–1408.

[16] Iyer, G., Tillekeratne, L.M.V., Coleman, M.R. and Nadarajah, A. (2008) Equilibrium swelling behavior of thermally responsive metal affinity hydrogels: I. Compositional effects. *Polymer*, **49**, 3737–3743.

[17] Wei, H., Zhang, X.-Z., Cheng, H. *et al.* (2006) Self-assembled thermo- and pH responsive micelles of poly(10-undecenoic acid-b-N-isopropylacrylamide) for drug delivery. *Journal of Controlled Release*, **116**, 266–274.

[18] Liu, W., Zhang, B., Lu, W.W. *et al.* (2004) A rapid temperature-responsive sol–gel reversible poly(*N*-isopropylacrylamide)-*g*-methylcellulose copolymer hydrogel. *Biomaterials*, **25**, 3005–3012.

[19] Hu, X., Xiong, L., Wang, T. *et al.* (2009) Synthesis and dual response of ionic nanocomposite hydrogels with ultrahigh tensibility and transparence. *Polymer*, **50**, 1933–1938.

[20] He, C., Zhao, C., Chen, X. *et al.* (2008) Novel pH- and temperature-responsive block copolymers with tunable pH-responsive range. *Macromolecular Rapid Communications*, **29**, 490–497.

[21] Stavrouli, N., Katsampas, I., Aggelopoulos, S. and Tsitsilianis, C. (2008) pH/thermosensitive hydrogels formed at low pH by a PMMA-PAA-P2VP-PAA-PMMA pentablock terpolymer. *Macromolecular Rapid Communications*, **29**, 130–135.

[22] Dayananda, K., He, C., Park, D.K. *et al.* (2008) pH- and temperature-sensitive multiblock copolymer hydrogels composed of poly(ethylene glycol) and poly(amino urethane). *Polymer*, **49**, 4968–4973.

[23] Hoffman, A.S. (1987) Applications of thermally reversible polymers and hydrogels in therapeutics and diagnostics. *Journal of Controlled Release*, **6**, 297–305.

[24] Burillo, G. and Bucio, E. (2009) *Gamma Radiation Effects on Polymer Materials and Applications*, Research Singpost, Trivandrum, pp. 45–62.

[25] Saeki, S., Kuwahara, N., Nakata, M. and Kaneko, M. (1976) Upper and lower critical solution temperatures in poly (ethylene glycol) solutions. *Polymer*, **17**, 685–689.

[26] Percot, A., Zhu, X.X. and Lafleur, M. (2000) Immobilization of lipid vesicles on polymer support via an amphiphilic peptidic anchor: application to a membrane enzyme. *Bioconjugate Chemistry*, **11**, 674–678.

[27] Ougizawa, T. and Walsh, D.J. (1993) Upper critical solution temperature behavior in polystyrene/poly(methyl methacrylate) mixture. *Polymer Journal*, **25**, 1315–1318.

[28] Yang, Q., Mao, Y., Li, G. *et al.* (2004) Study on the UCST behavior of polystyrene/poly(styrene-co-acrylonitrile) blend. *Materials Letters*, **58**, 3939–3944.

[29] Burillo, G., Bucio, E., Arenas, E. and Lopez, G.P. (2007) Temperature and pH-sensitive swelling behavior of binary DMAEMA/4VP grafts on poly(propylene) films. *Macromolecular Materials and Engineering*, **292**, 214–219.

[30] Contreras-García, A., Burillo, G. and Bucio, E. (2012) Polymeric Nano-, Micellar and Core-shell Materials, in *Intelligent Nanomaterials: Processes, Properties,*

and Applications (eds A. Tiwari, A.K. Mishra, H. Kobayashi and A.P.F. Turner), John Wiley & Sons, Inc., Hoboken, NJ, pp. 319–346.

[31] Bajpai, A.K., Shukla, S.K., Bhanu, S. and Kankane, S. (2008) Responsive polymers in controlled drug delivery. *Progress in Polymer Science*, **33**, 1088–1118.

[32] Nič, M. (1997) *IUPAC Compendium of Chemical Terminology. Volume 2, IUPAC Compendium of Chemical Terminology Gold Book*, 2nd edn, Blackwell Scientific Publications, Oxford.

[33] Marsano, E., Bianchi, E. and Viscardi, A. (2004) Stimuli responsive gels based on interpenetrating network of hydroxy propylcellulose and poly (N-isopropylacrylamide). *Polymer*, **45**, 157–163.

[34] Zhao, S., Cao, M., Li, H. *et al.* (2010) Synthesis and characterization of thermo-sensitive semi-IPN hydrogels based on poly(ethylene glycol)-co-poly(ε-caprolactone) macromer, N-isopropylacrylamide, and sodium alginate. *Carbohydrate Research*, **345**, 425–431.

[35] Li, G., Guo, L., Chang, X. and Yang, M. (2012) Thermo-sensitive chitosan based semi-IPN hydrogels for high loading and sustained release of anionic drugs. *International Journal of Biological Macromolecules*, **50**, 899–904.

[36] Verestiuc, L., Ivanov, C., Barbu, E. and Tsibouklis, J. (2004) Dual-stimuli-responsive hydrogels based on poly(N-isopropylacrylamide)/chitosan semi-interpenetrating networks. *International Journal of Pharmaceutics*, **269**, 185–194.

[37] Zhang, J.-T., Bhat, R. and Jandt, K.D. (2009) Temperature-sensitive PVA/PNIPAAm semi-IPN hydrogels with enhanced responsive properties. *Acta Biomaterialia*, **5**, 488–497.

[38] Thimma Reddy, T. and Takahara, A. (2009) Simultaneous and sequential microporous semi-interpenetrating polymer network hydrogel films for drug delivery and wound dressing applications. *Polymer*, **50**, 3537–3546.

[39] Mohan, Y.M. and Geckeler, K.E. (2007) Polyampholytic hydrogels: poly(N-isopropylacrylamide)-based stimuli-responsive networks with poly(ethyleneimine). *Reactive and Functional Polymers*, **67**, 144–155.

[40] Wu, X.Y., Huang, S.H., Zhang, J.Y. and Zhuo, R.X. (2004) Preparation and characterization of novel physically cross-linked hydrogels composed of poly(vinyl alcohol) and amine-terminated polyamidoamine dendrimer. *Macromolecular Bioscience*, **4**, 71–75.

[41] Zhang, J.T., Huang, S.W. and Zhuo, R.X. (2004) Temperature-sensitive polyamidoamine dendrimer/poly(N-isopropylacrylamide) hydrogels with improved responsive properties. *Macromolecular Bioscience*, **4**, 575–578.

[42] Connal, L.A., Li, Q., Quinn, J.F. *et al.* (2008) pH-Responsive poly(acrylic acid) core cross-linked star polymers: morphology transitions in solution and multilayer thin films. *Macromolecules*, **41**, 2620–2626.

[43] Wu, W., Liu, J., Cao, S. *et al.* (2011) Drug release behaviors of a pH sensitive semi-interpenetrating polymer network hydrogel composed of poly(vinyl

alcohol) and star poly[2-(dimethylamino) ethyl methacrylate]. *International Journal of Pharmaceutics*, **416**, 104–109.

[44] Lin, S.-B., Yuan, C.-H., Ke, A.-R. and Quan, Z.-L. (2008) Electrical response characterization of PVA-P(AA/AMPS) IPN hydrogels in aqueous Na2SO4 solution. *Sensors and Actuators, B: Chemical*, **134**, 281–286.

[45] Lu, Y., Li, Y., Pan, J. *et al.* (2012) Poly(3,4-ethylenedioxythiophene)/poly(styrenesulfonate)-poly(vinylalcohol)/poly(acrylic acid) interpenetrating polymer networks for improving optrode-neural tissue interface in optogenetics. *Biomaterials*, **33**, 378–394.

[46] Krishna Rao, K.S.V., Vijaya Kumar Naidu, B., Subha, M.C.S. *et al.* (2006) Novel chitosan-based pH-sensitive interpenetrating network microgels for the controlled release of cefadroxil. *Carbohydrate Polymers*, **66**, 333–344.

[47] Tanaka, T., Nishio, I., Sun, S. and Ueno-Nishio, S. (1982) Collapse of gels in an electric field. *Science*, **218**, 467–469.

[48] Kaewpirom, S. and Boonsang, S. (2006) Electrical response characterisation of poly(ethylene glycol) macromer (PEGM)/chitosan hydrogels in NaCl solution. *European Polymer Journal*, **42**, 1609–1616.

[49] Galaev, I.Y. and Mattiasson, B. (1999) “Smart” polymers and what they could do in biotechnology and medicine. *Trends in Biotechnology*, **17**, 335–340.

[50] Jeong, B. and Gutowska, A. (2002) Lessons from nature: stimuli-responsive polymers and their biomedical applications. *Trends in Biotechnology*, **20**, 305–311.

[51] Kim, S.J., Lee, C.K., Lee, Y.M. *et al.* (2003) Electrical/pH-sensitive swelling behavior of polyelectrolyte hydrogels prepared with hyaluronic acid-poly(vinyl alcohol) interpenetrating polymer networks. *Reactive and Functional Polymers*, **55**, 291–298.

[52] Kulkarni, R.V. and Biswanath, S. (2007) Electrically responsive smart hydrogels in drug delivery: a review. *Journal of Applied Biomaterials & Biomechanics*, **5**, 125–139.

[53] Davis, T.P. and Huglin, M.B. (1990) Effect of composition on properties of copolymeric N-vinyl-2-pyrrolidone methyl methacrylate hydrogels and organogels. *Polymer*, **31**, 513–519.

[54] Desai, N.P. and Hubbel, J.A. (1992) Surface physical interpenetrating networks of poly(ethylene terephthalate) and poly(ethylene oxide) with biomedical applications. *Macromolecules*, **25**, 226–232.

[55] Khare, A.R. and Peppas, N.A. (1993) Investigation of hydrogel water in polyelectrolyte gels using differential scanning calorimetry. *Polymer*, **34**, 4736–4739.

[56] Kulkarni, R.V. and Sa, B. (2008) Novel pH-sensitive interpenetrating network hydrogel beads of carboxymethylcellulose-(polyacrylamide-grafted-alginate) for controlled release of ketoprofen: preparation and characterization. *Current Drug Delivery*, **5**, 256–264.

[57] Kulkarni, R.V. and Sa, B. (2008) Evaluation of pH-sensitivity and drug release characteristics of (polyacrylamide-grafted-xanthan)-carboxymethyl cellulose-based pH-sensitive interpenetrating network hydrogel beads. *Drug Development and Industrial Pharmacy*, **34**, 1406–1414.

[58] Silan, C., Akcali, A., Otkun, M.T. *et al.* (2012) Novel hydrogel particles and their IPN films as drug delivery systems with antibacterial properties. *Colloids and Surfaces B: Biointerfaces*, **89**, 248–253.

[59] Ruckenstein, E. and Liang, L. (1996) Pervaporation of ethanol-water mixtures through polyvinyl alcohol-polyacrylamide interpenetrating polymer network membranes unsupported and supported on polyethersulfone ultrafiltration membranes: a comparison. *Journal of Membrane Science*, **110**, 99–107.

[60] Kuila, S.B., Ray, S.K., Das, P. and Singha, N.R. (2011) Synthesis of full interpenetrating network membranes of poly(acrylic acid-co-acrylamide) in the matrix of polyvinyl alcohol for dehydration of ethylene glycol by pervaporation. *Chemical Engineering and Processing*, **50**, 391–403.

[61] Marsano, E., Bianchi, E., Vicini, S. *et al.* (2005) Stimuli responsive gels based on interpenetrating network of chitosan and poly(vinylpyrrolidone). *Polymer*, **46**, 1595–1600.

[62] Bhattacharya, A. (2000) Radiation and industrial polymers. *Progress in Polymer Science*, **25**, 371–401.

[63] Burillo, G., Briones, M. and Adem, E. (2007) IPN's of acrylic acid and N-isopropylacrylamide by gamma and electron beam irradiation. *Nuclear Instruments and Methods in Physics Research Section B*, **265**, 104–108.

[64] Adem, E., Burillo, G., Bucio, E. *et al.* (2009) Characterization of interpenetrating networks of acrylic acid (AAc) and N-isopropylacrylamide (NIPAAm) synthesized by ionizing radiation. *Radiation Physics and Chemistry*, **78**, 549–552.

[65] Ortega, A., Bucio, E. and Burillo, G. (2008) New interpenetrating networks of N-isopropylacrylamide/N-acryloxysuccinimide: Synthesis and characterization. *Polymer Bulletin*, **60**, 515–524.

[66] Xuequan, L., Maolin, Z., Jiuqiang, L. and Hongfei, H. (2000) Radiation preparation and thermo-response swelling of interpenetrating polymer network hydrogel composed of PNIPAAm and PMMA. *Radiation Physics and Chemistry*, **57**, 477–480.

[67] Guilherme, M.R., Campese, G.M., Radovanovic, E. *et al.* (2006) Thermo-responsive sandwiched-like membranes of IPN-PNIPAAm/PAAm hydrogels. *Journal of Membrane Science*, **275**, 187–194.

[68] Chen, J., Sun, J., Yang, L. *et al.* (2007) Preparation and characterization of a novel IPN hydrogel membrane of poly(N-isopropylacrylamide)/carboxymethyl chitosan (PNIPAAm/CMCS). *Radiation Physics and Chemistry*, **76**, 1425–1429.

[69] Şahiner, N., Pekel, N. and Güven, O. (1999) Radiation synthesis, characterization and amidoximation of N-vinyl-2-pyrrolidone/acrylonitrile interpenetrating polymer networks. *Reactive and Functional Polymers*, **39**, 139–146.

[70] Zhang, J. and Peppas, N.A. (2000) Synthesis and characterization of pH- and temperature-sensitive poly(methacrylic acid)/poly(N-isopropyl acrylamide) interpenetrating polymeric networks. *Macromolecules*, **33**, 102–107.

[71] Zaldivar, M.P., Fernández, N.G., Peña, C.G. *et al.* (2011) Synthesis and characterization of a new semi-interpenetrating polymer network hydrogel obtained by gamma radiations. *Journal of Thermal Analysis and Calorimetry*, **106**, 725–730.

[72] Singh, D., Choudhary, V. and Koul, V. (2007) Radiation synthesis of interpenetrating polymer networks based on N-vinyl pyrrolidone—acrylic acid copolymer and gelatin. I. Swelling, morphology, and thermal characterization for biomedical applications. *Journal of Applied Polymer Science*, **104**, 1456–1463.

[73] Liu, Z., Yi, M., Zhai, M. *et al.* (2004) Radiation synthesis and environmental responsiveness of semi-interpenetrating polymer networks composed of poly(dimethyl–aminoethyl methacrylate) and poly(ethylene oxide). *Journal of Applied Polymer Science*, **92**, 2995–3001.

[74] Hiroki, A., Maekawa, Y., Katakai, R. *et al.* (2006) Ultraslow thermoresponse in hydrogels synthesized with radiation technique. *Macromolecules*, **39**, 4132–4137.

[75] Hasegawa, S., Ohashi, H., Maekawa, Y. *et al.* (2005) Thermo and pH sensitive gel membranes based on poly(acryloyl-L-proline methyl ester) graft- poly(acrylic acid) for selective permeation of metal ions. *Radiation Physics and Chemistry*, **72**, 595–600.

[76] González, G. and Burillo, G. (2010) Synthesis of comb type and semi-interpenetrating networks of acryloyl-L-proline methyl ester and poly (acrylic acid) for cu (II) immobilization. *Radiation Physics and Chemistry*, **79**, 870–875.

[77] Zhao, Y., Kang, J. and Tan, T. (2006) Salt-, pH- and temperature-responsive semi-interpenetrating polymer network hydrogel based on poly(aspartic acid) and poly(acrylic acid). *Polymer*, **47**, 7702–7710.

[78] Ruiz, J.-C., Burillo, G. and Bucio, E. (2007) Interpenetrating thermo and pH stimuli-responsive polymer networks of PAAc/PNIPAAm grafted onto PP. *Macromolecular Materials and Engineering*, **292**, 1176–1188.

[79] Ruiz, J.-C., Alvarez-Lorenzo, C., Taboada, P. *et al.* (2008) Polypropylene grafted with smart polymers (PNIPAAm/PAAc) for loading and controlled release of vancomycin. *European Journal of Pharmaceutics and Biopharmaceutics*, **70**, 467–477.

[80] Muñoz-Muñoz, F., Ruiz, J.-C., Alvarez-Lorenzo, C. *et al.* (2012) Temperature- and pH-sensitive interpenetrating polymer networks grafted on PP: cross-linking irradiation dose as a critical variable for the performance as vancomycin-eluting systems. *Radiation Physics and Chemistry*, **81**, 531–540.

8

MICROSCOPY OF IPNs

Rameshwar Adhikari
Central Department of Chemistry, Tribhuvan University, Kathmandu, Nepal

8.1 INTRODUCTION AND OVERVIEW

It is not necessary to give a detailed introduction to polymer networks, their fabrication routes, and applications as the whole book is devoted to the characterization of interpenetrating polymer networks (IPNs). The readers have already noticed that every chapter in this book has direct or indirect reference to different microscopic techniques for elucidation of the structure–property correlations of IPNs. This illustrates the significance of microscopic tool in the characterization of multicomponent polymers including polymer blends, alloys, and composites. Additionally, there are several references available in the literatures which elucidate different aspects of materials science and engineering of IPNs [1–18].

It can be noticed that the IPNs are characterized by rich variety of structural and functional diversities. Their end-use properties are determined by their morphology on different length scales, the latter being controlled by molecular structure and chain architecture of the constituent molecules as well as their synthetic history. Further, the construction of micro- and nanostructured polymers with tailored mechanical and functional properties is an important facet of contemporary materials science and engineering. There is a growing

Micro- and Nano-structured Interpenetrating Polymer Networks: From Design to Applications, First Edition. Edited by Sabu Thomas, Daniel Grande, Uroš Cvelbar, K.V.S.N. Raju, Ramanuj Narayan, Selvin P. Thomas and Akhina H.

interest in the field of nanostructured polymeric materials due to their attractive functional properties. These materials have indeed stimulated a great deal of interactions among physicists, chemists, biologists, and engineers. As a result, new techniques for the characterization of polymer network, hybrids, and composites have evolved and got advancement. The latter has also triggered the advancement of new tools in electron microscopy and scanning probe techniques.

The objective of this chapter is to present an overview of the application of electron microscopy, in general, for the study of IPNs with emphasis on sample preparation. The chapter is organized in the following ways. First, a brief overview of the sample preparation will be given. Then, after a brief account of the overview of microscopy of IPNs, typical examples on characterization of various classes of full IPNs, semi-IPNs, gels, and IPN-like systems will be presented.

8.2 SAMPLE PREPARATION TECHNIQUES FOR MICROSCOPIC ANALYSIS

Generally, the quality of results delivered by microscopic techniques depends much on the sample preparation steps. Only with proper care during specimen preparation, reasonable conclusions on IPN morphology can be drawn. Thin films, cut surface, fractured surface, and etched surfaces are the principal sample preparation methods used for the study of polymer morphology. For the detailed account of sample preparation for microscopy, the readers are referred to specific literatures [17–19].

8.2.1 Microtomy and Ultramicrotomy

Microtomy and ultramicrotomy (sometimes equipped with cryo devices) are employed for the preparation of thin sections of polymeric materials. Generally, freshly prepared glass knives or diamond knives are used for this purpose [18].

As most of the polymeric materials are soft at room temperature, the specimens should be hardened before sectioning in order to avoid local mechanical deformation of one or more of the components present in the polymer samples. Treatment of sample with high-energy radiations (such as γ rays and electron beam) or with chemical staining agents (such as osmium tetroxide or ruthenium tetroxide) can help in hardening the sample so that it can be sectioned at room temperature [17, 18]. The samples are sectioned at cryogenic temperatures using liquid nitrogen as coolant. The sectioning temperature is generally below or close to the glass transition temperatures (T_g) of all the components of the polymer alloy. The thickness of the films required for the TEM investigations should be in the range of 50–70 nm. Film surfaces of any thicknesses can be studied by scanning electron microscopy (SEM) and atomic force microscopy (AFM).

8.2.2 Staining of Thin Sections

Electron microscopy of materials is based on the interaction of electron beam with the atoms of the materials under study. The contrast generally arises from the production of secondary or back scattered electrons (BSEs) (in SEM) or scattering of electrons by the specimen (in TEM). Thus, it is necessary to have sufficient intrinsic atomic mass contrast and appropriate surface topography for TEM and SEM investigations, respectively.

Polymers comprise mainly of low molar mass elements such as carbon, hydrogen, nitrogen, etc. Thus, the atomic mass contrast is very low. Therefore, in conventional TEM of heterophase polymers, selective staining of one or more of the phases using heavy metal compounds (such as osmium tetroxide, ruthenium tetroxide, uranyl acetate, etc.) is required [20–23]. The staining can be performed either on the bulk material prior to sectioning or on the thin films. As mentioned earlier, an advantage of staining the bulk sample is that the sample is hardened and can be sectioned at room temperature.

8.2.3 Etching of Surfaces

The sample surfaces are studied by SEM and AFM. Smooth surfaces are prepared by ultramicrotomy, polishing, and etching. The free specimen surface can be etched by reactive ions and radiations or by using chemicals such as in permanganic etching [24–30].

8.2.4 Fracture Surface Preparation

It is a common practice in materials science of polymers to investigate their cryofracture surfaces produced using liquid nitrogen by SEM. In addition to the study of the phase morphology, this way permits the investigation of deformation and fracture behavior of the polymers. Generally, the fracture surfaces are generally sputter coated with thin film of noble metals such as gold, silver, etc., to render the fracture surface conducting, to avoid surface charging, and to minimize the damage by electron beam irradiation [18, 19].

8.3 MICROSCOPY OF INTERPENETRATING POLYMER NETWORKS: AN OVERVIEW

The whole text in the present book concerns with the IPNs, their structure and morphologies, their properties, and the ways of manipulating them by designing the structures relevant to those properties. Ultimately, we are interested in the morphology of the materials which are best characterized by electron microscopy in combination with several other analytical techniques. In this section, we present a review of literature works which have a direct reference to electron microscopy and associated techniques for morphological characterization of the IPNs.

The general structure–property correlations in the IPNs synthesized *via* a number of routes and using a wide range of starting materials have been established in several monographs [1–6, 9, 12], review papers [6–8, 13], and encyclopedia [11]. Several of the early works on IPNs were concentrated on the nature of their phase morphologies. One of the key issues sufficiently debated in the literature was about the extent and dimension of phase segregation and phase continuity of the IPNs. In the early 1980s the issue was resolved with the help of electron microscopy [10]. It was fairly established that the chains constituting the polymer networks are well segregated and form interconnected three-dimensional (3D) network structures also in the mesoscale.

The objective of this chapter is to introduce the application of electron microscopy and related techniques for the morphological characterization of polymer network, in general, and semi- and fully IPNs in particular. It is not the scope of the chapter to discuss the classification of those polymers on the basis of synthetic and processing methodologies. Wherever appropriate, the sample preparation and treatment techniques will be highlighted.

The chemical structure of the semi- and fully interpenetrating polymers are best illustrated by the simplified sketches outlined in Figure 8.1a, which clearly deliver the notion of segregation of polymer A and B provided that the polymers are inherently incompatible [6]. The extent of phase segregation and the dimension of the phase-separated structures, of course, depend on the degree of incompatibility, A/B composition ratio, and physicochemical properties of the polymers as well as on the ways and sequences of how the polymer alloys are fabricated.

Figure 8.1b shows an AFM height image of an IPN based on vinyl ester (VE) and epoxy (EP) resin containing a small amount of mineral filler [31]. The filler, however, cannot be ascertained in the AFM micrograph. The alternating and co-continuously existing network of two different polymers can be recognized in Figure 8.1b, an ideal picture of the interpenetrating network morphology.

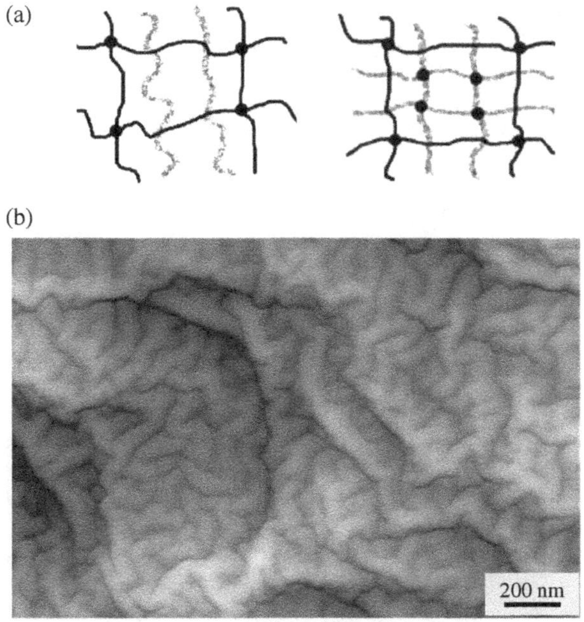

FIGURE 8.1 General picture of structure of IPNs: schemes showing the structures of semi- and fully IPN (a) (Source: Lohani *et al.* [6], http://www.hindawi.com/journals/jdd/2014/583612/. Used under CC-BY-3.0 http://creativecommons.org/licenses/by/3.0/); bright line—polymer A, faint line—polymer B, and filled circles—crosslinks; and AFM image of a polished and physically etched surface of VE/EP (b) (Source: Kocsis *et al.* [31]. Reproduced with permission of Elsevier)

This typical IPN morphology has been made visible by polishing and etching of the surface prior to the analysis with the aid of the AFM [31].

The early works establishing structure–property correlation with the aid of electron microscopy dealt with semi- and fully IPNs and ionomers. Some noteworthy examples include the systems based on polycarbonates (PC) [32], polystyrene (PS) [33, 34], polyacrylates, vinyl acetates [35–39], and others [10, 40, 41]. Here, we present two examples from early works on IPNs studied by TEM and SEM.

Transmission electron micrographs of semi-IPNs comprising polyurethane (PU) and polyvinyl acetate (PVAc) are presented in Figure 8.2 [36]. The thin sections of the samples were treated and hardened by osmium tetroxide (OsO_4) which preferentially stained the PU phase. Thus, the PU phase appears darker in the electron micrograph. It should be kept in mind that the choice of the staining agent and staining procedure is the issue that should be handled cautiously while studying the multiphase polymeric materials by electron microscopy.

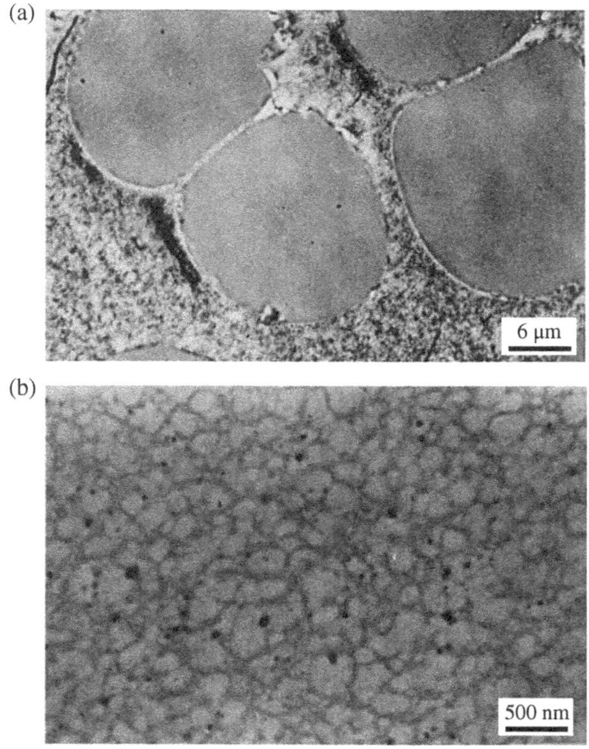

FIGURE 8.2 (a) TEM image of thin section of semi-IPN comprising polyurethane (PU) and polyvinyl acetate (PVAc) (Source: Hourston and Zia [36]. Reproduced with permission of John Wiley & Sons); (b) fully IPN of the same composition as (a) (Source: Hourston *et al.* [42]. Reproduced with permission of John Wiley & Sons)

Figure 8.2a illustrates the TEM image of the semi-IPN sample containing 20 wt.% of the crosslinked PU phase, which apparently forms the matrix and appears to possess its own internal morphology. The crosslinked PU phase matrix encloses large domains of PVAc phase, some 20 μm in diameter. The formation of large domains of one of the constituents is facilitated by ease of segregation of the other components (the non-crosslinked phase). The morphology of the network polymers, as for other multicomponent systems, changes with the crosslink density, phase composition, and processing or synthetic procedures. The morphology of the system changes dramatically when fully interpenetrating network structure is designed [42] from the same composition as for the sample presented in Figure 8.2a. The result is presented in Figure 8.2b where the continuity of the phases and domain size presents an entirely different picture than for the corresponding semi-IPN leading to different physical properties. These are the consequences of the domains turning into the nanoscale. The fully interpenetrating network structure of the macromolecules prevents the macroscopic segregation of the domains, hence contributing toward the formation of nanostructured morphology.

With the development of new synthetic routes and characterization tools, the application areas of IPNs have widened significantly. Remarkable developments have been achieved in the field of polymer networks based on biodegradable and bioresorbable polymers, biobased materials, and those used as biomaterials and for applications in tissue engineering. Chemically modified soybean oil [43–45], collagen [46–48], soy protein [49], gelatin [50, 51], castor, and vernonia oils [52–54] have been used to prepare various forms of polymer networks, whereby the electron microscopy has offered the most important direct phase information on the structure and properties of those materials.

The network polymer materials having potential application as biomaterials and drug delivery systems so far developed include antibacterial semi-IPN based on PHEMA and polyvinyl pyrrolidone (PVP) [52], the biodegradable endotracheal tube materials [53, 54], and drug delivery vehicles [55]. Similarly IPNs gels and porous materials useful in filter application as well as superadsorbents have been developed [56–58].

The electron microscopy has played significant role in structural elucidation of elastomers and latex-based INPs and semi-INPs. Notable works in this regard include the natural rubber-based IPNs [59], latex IPNs based on polychloroprene rubber [60, 61], and others such as elastomeric block copolymers [40, 62].

Various works correlated the morphological features of micro- and nanostructured IPNs and nanostructured hybrid derivatives obtained by electron microscopy with their physical properties including thermal, thermomechanical, and mechanical properties [32, 63–66]. Other systems developed in the recent years are those having potential applications in energy and charge storage transfer devices such as fuel cell materials, ion exchange resins, electrochemical devices, etc. [33, 34, 67–71]. Several materials development strategies include doping [34, 67, 70] and also involve the use of intrinsically conducting polymers such as polyaniline, polythiophene, and polypyrrole [68, 70, 71].

The new developments in IPNs with new functionalities have laid foundation to the advancement of IPN-like systems based on different polymers such as polyethylene (PE) and vinyl polymers [72–78], PU [37], epoxy resin and polyacrylates [79], and nanostructured thermoplastics and thermosets [31, 80, 81]. The electron microscopy does not only provide information about the local structure of the polymers but also delivers important clues on deformation and fracture behavior of the polymers. These techniques have been utilized for the study of deformation behavior of various IPNs [45, 82–86].

It should be mentioned that electron microscopy alone has not contributed to the morphological characterization of the IPNs, but there are several other complementary techniques. Very frequently used tools in this respect, in particular for phase morphology elucidation of the network polymers, are dynamic mechanical analysis (DMA) and differential scanning calorimetry (DSC). Very important roles in the structural clarification of the IPNs and composites have been played by AFM which provides the direct evidence of phase structure and continuity with nanometer resolution without the need of any chemical modification of the samples. Some typical examples on literature works utilizing AFM as morphology analytical tool are contained in the references [31, 43, 45, 60, 77, 83, 84, 87]. In the SEM, secondary electron (SE) mode imaging is the most popular one, but there are references utilizing BSE mode imaging [43], using field emission cathodes [60, 67], and humidity chambers [88] in scanning electron microscopes. Further, optical microcopy including confocal and fluorescence techniques [43, 47, 49, 53], along with various spectroscopic and scattering techniques [31, 46, 71, 88], has been used in the morphological inspection of the IPNs, in particular that of the materials used for tissue engineering.

By definition, IPNs comprise the macromolecules organized in such a way that they are intermingled on molecular scale and still forming the phase-separated structures. Thus, the structures hence formed span a length scale ranging from molecular to nano-, micro-, and mesolevel. The microscopic techniques characterize the structures on nano- and microscale. In the following paragraphs, we present the examples of application of electron microscopy for the morphological characterization of micro- and nanostructured IPNs.

8.4 MORPHOLOGICAL CHARACTERIZATION OF POLYMER NETWORKS

8.4.1 Biomaterials and Biomedical Materials

As introduced earlier, fully and semi-IPNs have attracted considerable research interest in designing materials for drug delivery and tissue engineering applications. In a recent study, the IPNs based on poly(dimethyl siloxane) (PDMS) and poly(2-hydroxyethyl methacrylate) (PHEMA) were synthesized via sequential method. The IPN films were further modified by using collagen (type I) protein. The biocompatibility of the resulting PDMS film was analyzed by culture of the fibroblast cells [46].

The behavior of the biomaterial in contact with the living cells is an important parameter for the biocompatibility of the material. Hence, the *in vitro* tests of different samples were conducted using L929 fibroblast cells under suitable conditions [46].

The optical micrographs of the cell culture using different kinds of films are presented in Figure 8.3 [46]. It can be seen that the L929 fibroblast cells were attached well to the IPN surface (Figure 8.3b) and grew to significant extent compared to the neat PDMS film surface (Figure 8.3a). Much higher degree of cell spreading and adhesion was observed on the surface modified by the collagen protein (Figure 8.3c) implying that collagen offered optimum bioderived substrate for the growth of the cells.

The optical microscopy does not provide structural information of the samples on nanoscale but offers an overview of the morphological details with an opportunity to carry out the tests under native or designed physiological conditions, which the electron microscopy does not offer. It should be mentioned that generally, there is a need of sample treatments with some staining agent for optical imaging for the contrast enhancement. For instance, the samples were stained with 5% Giemsa solution for the investigations presented in Figure 8.3 [46].

As already mentioned, for the detailed survey of structures of IPNs, the application of electron microscopy becomes inevitable as the resolution of optical techniques is limited. As an example of the IPN system useful for tissue engineering application investigated by electron microscopy, we present the SEM micrograph of highly porous hydrogel based on collagen and hyaluronic acid [47], as shown in Figure 8.4. The material has similar chemistry as for the one depicted in Figure 8.3. The sample surfaces for the SEM analysis were prepared by freeze-drying the hydrogel followed by sputter coating with gold film and SEM investigation in the SE imaging mode.

The fibrils in the porous framework appearing highly interconnected do not exactly give the dimension of the network as the sample structures can be significantly altered during the drying procedure (dehydration). However, the electron micrograph of the material presented in Figure 8.4 attests the highly enhanced surface area of the fibrillar framework of the IPN that allows a rapid cell adhesion and proliferation. These structures also allow straightforward micropatterning using a physical template enabling the formation of complex tissue engineering devices [47].

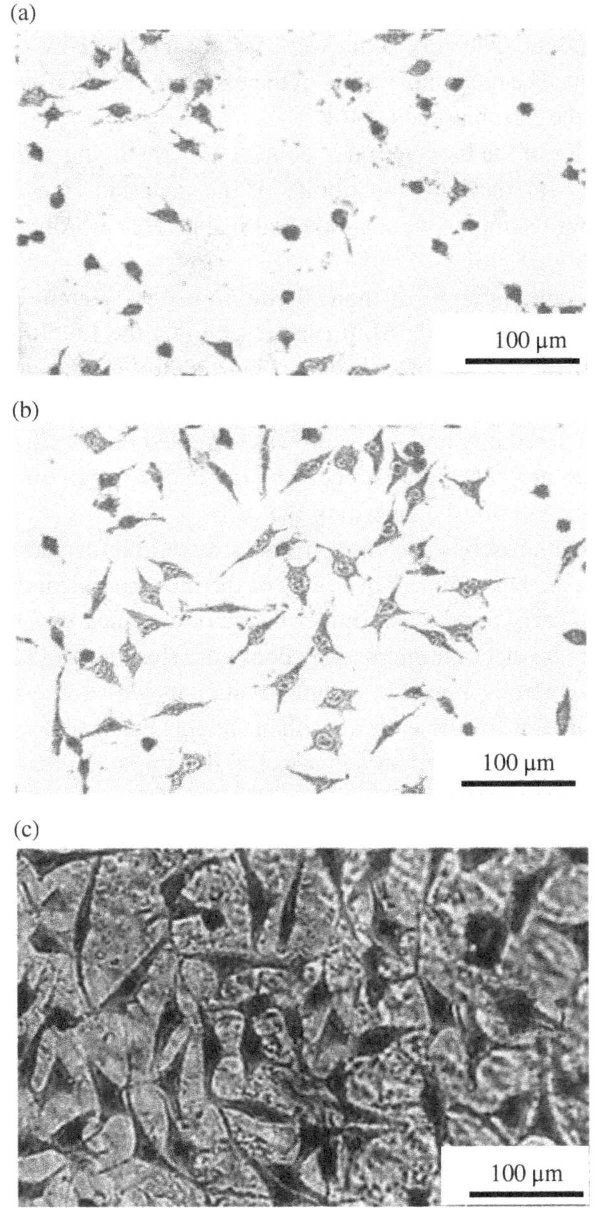

FIGURE 8.3 Optical micrographs of cell culture on (a) unmodified PDMS, (b) IPN comprising PDMS/PHEMA, and (c) collagen-grafted PDMS/PHEMA IPN. Source: Rezaei and Mohd Ishak [46], https://doaj.org/article/9bf941f09d4c4013b1740cc36770c4b4. Used under CC-BY-SA-4.0 https://creativecommons.org/licenses/by-sa/4.0/

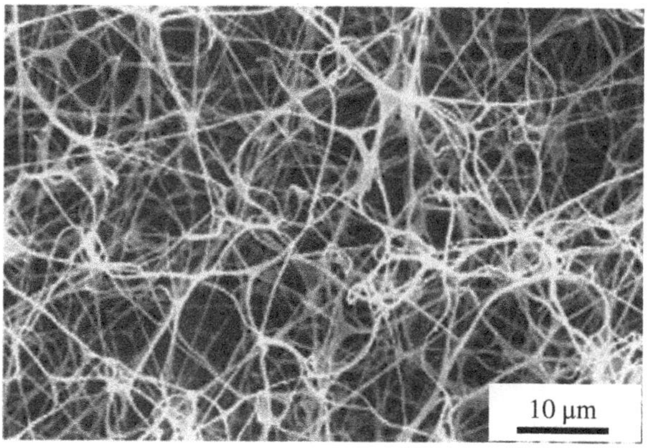

FIGURE 8.4 Secondary electron (SE) mode SEM micrograph of representative IPN structure based on PDMS and collagen. Source: Suri and Schmidt [47]. Reproduced with permission of Elsevier

It should be noted that the SEM results presented in Figure 8.4 represent the surface of porous material. The porous nature of the material is related basically to processing pathways utilized during fabrication process. For the comprehensive study of the microstructure of the IPNs and modified IPNs, SEM using field emission gun (FEG) was used. The results obtained on two different samples (PDMS/PHEMA IPN and collagen-grafted version) are presented in Figure 8.5.

The SEM micrograph of the cross section of the IPN sample (Figure 8.5a) clearly confirms the presence of two-phase morphology as expected for the IPN. It can be seen that the PHEMA droplets (brighter areas in Figure 8.5a) are homogeneously distributed throughout the PDMS matrix. This unique homogeneous distribution has been attributed to the increase in hydrophilicity of the samples permitting a sufficient cell adhesion and proliferation (compare with Figure 8.3b). It can be further observed that the surface morphology of collagen-grafted IPN (Figure 8.5b) depicts enhanced roughness allowing still higher hydrophilicity and much larger areas for cell adhesion and growth (compare with Figure 8.3c). The results were consistently explained and correlated with contact angle measurements.

As a second example for microscopic investigation of the IPN system comprising bioderived material, let us examine the morphology of a semi-IPN system based on PDMS and silylated soybean oil [43], in which only the results obtained from SEM of two different compositions are presented (Figure 8.6).

In the first glance, one can observe a bright contrast between the constituent phases in the SEM images in Figure 8.6 unlike that in Figure 8.5. The reason

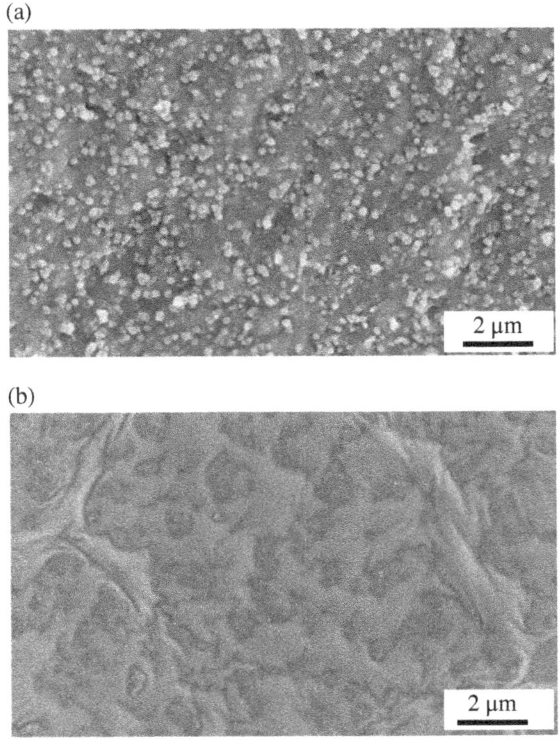

FIGURE 8.5 SEM images of (a) IPN comprising PDMS/PHEMA and (b) collagen-grafted PDMS/PHEMA IPN; the sample fracture surface was sputter coated with a thin film of gold prior to SEM imaging. Source: Rezaei and Mohd Ishak [46], https://doaj.org/article/9bf941f09d4c4013b1740cc36770c4b4. Used under CC-BY-SA-4.0 https://creativecommons.org/licenses/by-sa/4.0/

lies in the technique used in the SEM. In the present case, the cryofracture surfaces were prepared by using liquid nitrogen followed by treatment of the fresh fracture surface with OsO_4 vapor. The latter renders the areas comprising unsaturated oil stained thus making the soybean oil phase producing more BSEs and appearing brighter in the SEM images.

Thus, the IPN with 20 wt.% silylated soybean (Figure 8.6a) shows the morphology comprising PDMS as a major component (darker areas in the SEM images). As higher concentration of soybean oil component, the PDMS phase becomes dispersed and gives rise to tiny microscopic domains.

The fact that the morphology changes dramatically with the composition of the IPN, the mechanical properties of these materials can be tailored over a wide range by changing the composition of the IPN constituents.

(a)

(b)

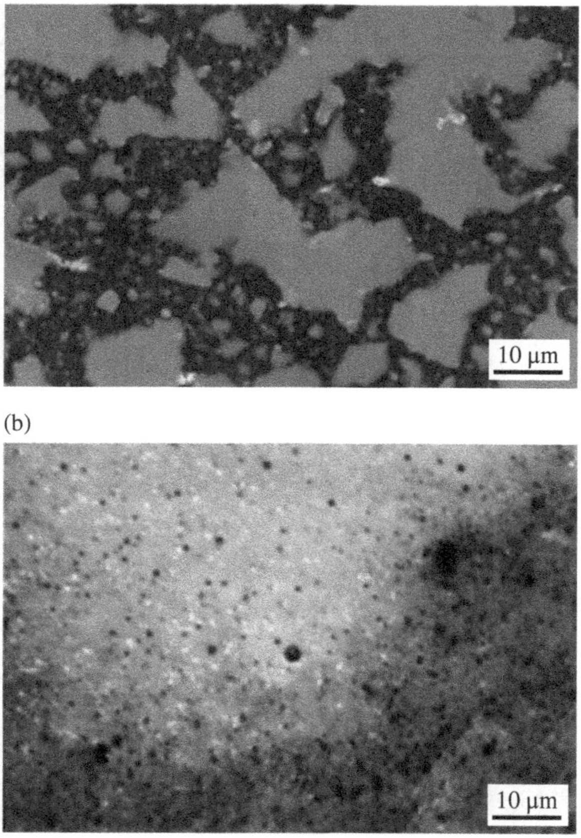

FIGURE 8.6 Back scattered electron (BSE) mode SEM images of IPNs composed of different silylated soybean oil/PDMS compositions: (a) 20:80 and (b) 80:20. Source: Dewasthale *et al.* [43]. Reproduced with permission of John Wiley & Sons

8.4.2 Porous Networks

Porous polymeric materials have many useful applications including thermal insulation by rigid foamed PU. The foamed materials fabrication *via* IPN route provides environmentally friendly alternative for the fabrication of these technically interesting materials. The synthesis of the IPNs in the presence of epoxide (EP) resins enables the formation of such foamed materials [56]. An example of such material is depicted by Figure 8.7a. The SE mode SEM image shows the highly porous structure of the PU/EP IPN system. By changing the curing time of the resin, it was possible to vary the pore density of the network polymer. For instance, by increasing the curing time, the pore size was decreased [56].

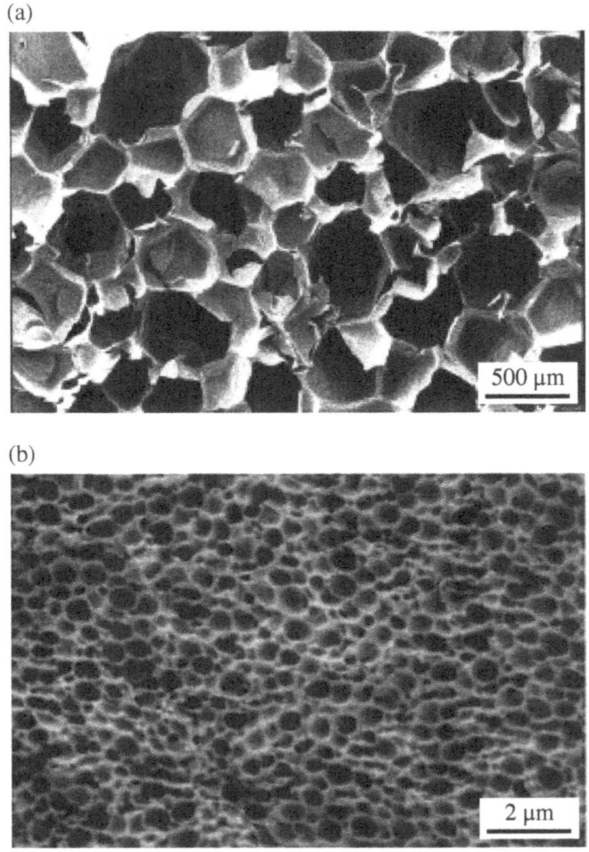

FIGURE 8.7 SEM micrographs of (a) foamed PU/EP network polymer cured for 60 min [56] (Source: Zhang *et al.* [56]. Reproduced with permission of John Wiley & Sons) and (b) porous surface structure of the IPN comprising PE/MMA copolymer (Source: Pionteck *et al.* [57]. Reproduced with permission of John Wiley & Sons)

Alternatively, the foamed materials can be prepared by selectively degrading one of the components of the IPN system, such as the removal of polymethacrylate (PMA) from the interpenetrating network comprising PE and PMA copolymers [57] *via* electron beam irradiation. By suitably selecting the pendant groups in the PMA copolymers, the pore size of the resulting porous material can also be tailored. An example of the foamed material produced by this method is presented in Figure 8.7b.

The porous materials are interesting also for the application as biomaterials and in regenerative medicine [53]. Confocal laser-scanning microscopy (CLSM) is a technique of choice in such applications to study how the biomaterials respond when placed into a host, for instance. In a recent study, the

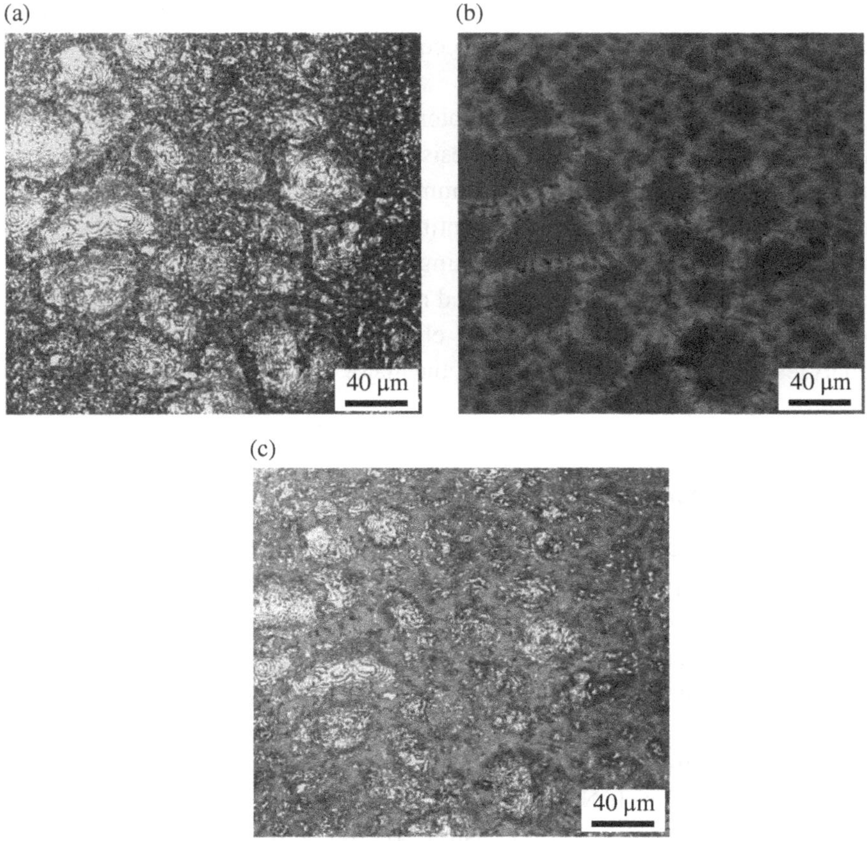

FIGURE 8.8 Confocal laser-scanning micrographs showing a model semi-IPN based on p(HEMA-*co*-NIPAA) and 10 wt.% PCL: (a) stained with acridine orange, (b) stained with rhodamine B, (c) simultaneously stained with both acridine orange and rhodamine B. Source: Jones *et al.* [53]. Reproduced with permission of Elsevier

development of thermoresponsive semi-IPN system based on copolymer of *N*-isopropylacrylamide (NIPA) and 2-hydroxyethyl methacrylate (HEMA) and polycaprolactone (PCL) has been reported. The target was to develop a new efficient drug delivery vehicle with optimized mechanical properties.

As an example of the application of the CLSM in IPN research, the CLSM images of the model semi-IPN material are presented in Figure 8.8. In those micrographs, the hydrophilic part is stained green (Figure 8.8a), and hydrophobic regions are stained red (Figure 8.8b). The two dyes used for staining fluoresced independent of each other.

In Figure 8.8c, the hydrophilic and hydrophobic domains were stained simultaneously, illustrating the biphasic distribution of the semi-IPN. In other words, the hydrophobic PCL phase was uniformly dispersed within the IPN hydrogel matrix as confirmed by the confocal microscopy and supported by calorimetric measurements.

It has been proposed that these materials could be used as bioactive endotracheal tubes to offer enhanced resistance to ventilator-associated pneumonia. These have been further recommended as a suitable candidate for the effective release of the hydrophilic antimicrobial drug [53]. This study is one of the few investigations demonstrating the application of confocal microscopy for imaging the phase-segregated morphology of multicomponent polymers. This method enables, unlike electron microscopy, the microscopic analyses of the sample in native state, the major limitation, however, being the poor resolution power.

8.4.3 Elastomer and Latex-Based Networks

So far we have addressed the morphological studies on IPNs by means of optical microscopy, confocal microscopy, and SEM. Let us introduce some systems studied with transmission electron microscopy (TEM) using latex-based IPNs. Preparation of IPN systems on the basis of latex particles has been an active area of research since long, and it still continues to grow [40, 59–61]. One of the early works on such systems was carried out by Silverstein and Narkis investigating the structural details of the IPNs by TEM [40].

Elastomeric latex IPNs were prepared using polyacrylates (PA) and PS as monomers *via* two-stage emulsion polymerization followed by crosslinking. The multiphase structures were first ascertained by the presence of two distinct glass transition temperatures [40]. The authors were able to produce excellent micrographs of the submicron-sized elastomer particles reinforced with glassy domains of PS [40]. A typical example is presented in Figure 8.9.

TEM image presented in Figure 8.9 depicts the morphology of the latex IPN comprising polybutyl acrylate (PA) and PS synthesized by polymerizing PS onto the PA seed latex. The TEM image of latex IPN shows the spherical latex particles with characteristic nanoscale inclusions which might have resulted from freezing-induced voids and local degradation of the matrix due to electron beam irradiation. It was suggested that the lightly crosslinked PS chains possibly formed microscopic domains with its chains contained as an interwoven continuous network in the particle together with the relatively strongly crosslinked PA chains.

It is worth to mention the preparation of the samples of such emulsions. The latex particles were squeezed between the microscope grids coated

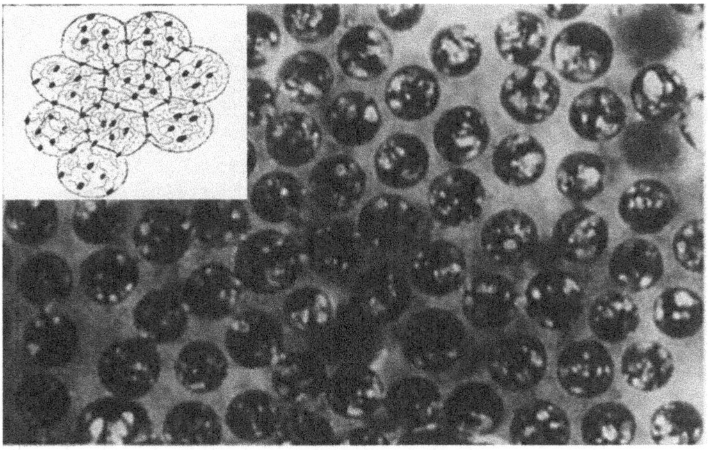

FIGURE 8.9 TEM image of an IPN comprising PS and PA containing 25 wt.% of PS with 20 000× magnification; in the insert: scheme showing the cross section of molded latex IPN showing PS microdomains and an interwoven network structure: • PS microdomains, (-) PS network, (- - -) PA network. Source: Silverstein and Narkis [40]. Reproduced with permission of John Wiley & Sons

with polyimide, dried in liquid nitrogen, and transferred directly to the TEM using a cooling holder [40]. The contrast observed in the electron micrograph results primarily from the local thickness of the latex particles that is originated from the interaction of the electron beam with the polymers involved.

A decade later the studies were extended to the latex IPN systems based on styrene–butadiene rubber (SBR) and chloroprene rubber [60, 61]. The osmium tetroxide (OsO_4) treatment was used to selectively stain double bond-containing phase, that is, the rubbery phase. Such a procedure enables the unambiguous identification of the constituent phases in the latex IPNs. Two typical examples are discussed in Figure 8.10.

8.4.4 Micro- and Nanostructured Materials and Hybrids

In the TEM images of PS-based latex IPNs presented in Figure 8.10a, the nanostructured domains of the IPN can be easily identified, the shell of the latex particles being made up of the PS phase (appearing brighter) enclosing the SBR phase (appearing gray). On careful examination of Figure 8.10a, the tiny particles of PS are also visible as internal morphology of the spherical SBR domains [60]. Similarly, for the chloroprene/PS network polymer, the PS shell (appearing bright in the TEM image; Figure 8.10b) around the rubbery

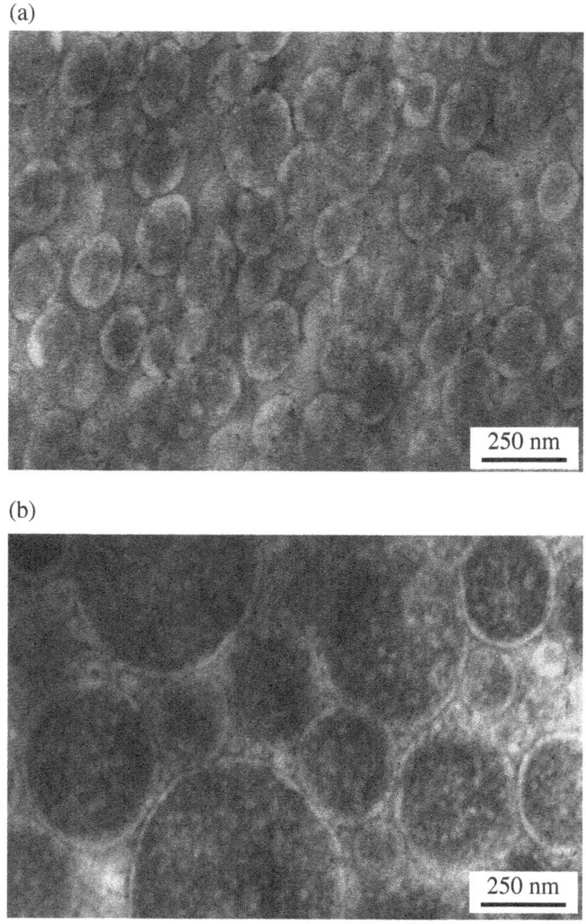

FIGURE 8.10 TEM micrographs of (a) SBR/PS (60 : 40) latex IPN (Source: Vo and Burford [60]. Reproduced with permission of John Wiley & Sons) and (b) chloroprene/PS (60 : 40) latex IPN; the rubbery phase appears dark due to staining by OsO_4 (Source: Burford and Vo [61]. Reproduced with permission of John Wiley & Sons)

particles can be observed. The particles have distinctly visible droplets of PS within them [61]. Thus, both the systems presented in Figure 8.10 represent the nanostructured ones.

Nowadays, the nanomaterials and nanoanalytical techniques have attracted considerable interest not only of the researchers but also of entrepreneurs, economists, and policy makers as the nanotechnology is invading the society as a new industrial revolution worldwide. Further, nanomaterials have special position in polymer science as the nanostructured polymers are playing

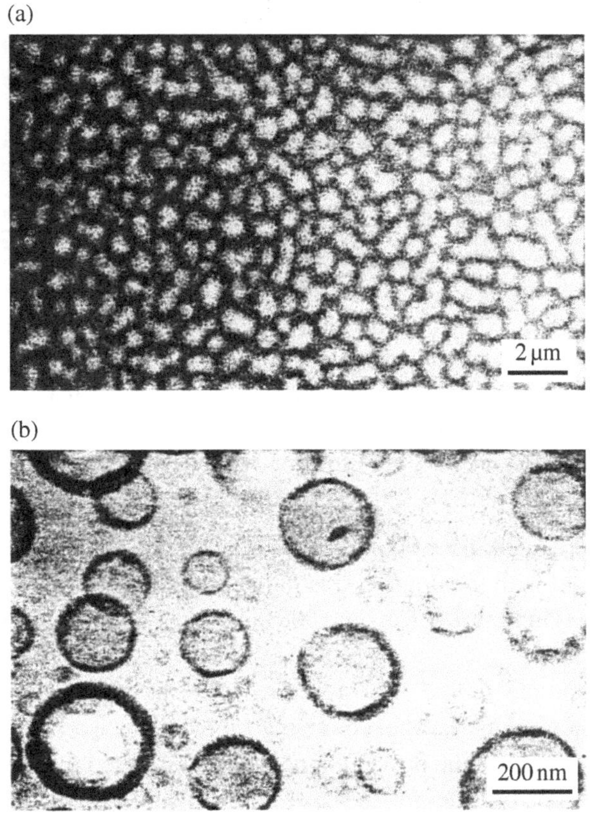

FIGURE 8.11 TEM images of (a) linear polymer blend with PU phase domain of thickness 300 nm and (b) semi-IPN with diameter of the PU phase domain between 50 and 150 nm. Source: Shyu and Chen [32]. Reproduced with permission of John Wiley & Sons

leading role in the new technological breakthroughs and so is the importance of the nanostructured IPNs.

One of the early reports on the nanostructured IPNs was published by Shyu and Chen [32]. They prepared semi-IPN based on PC and PU and hence studied the microphase-separated morphology by using TEM.

Figure 8.11 shows TEM images of two different kinds of blends of PC and PU: the linear blend (Figure 8.11a) and semi-interpenetrating network (Figure 8.11b). The specimens were treated with OsO_4 for the selective staining of the PU phase. However, it has been reported that due to short staining time, the PU phase is not stained for the linear blends, thus making PU phase appear bright in Figure 8.11a [32]. The adequate time of staining for the semi-IPN renders the PU phase appear dark in Figure 8.11b.

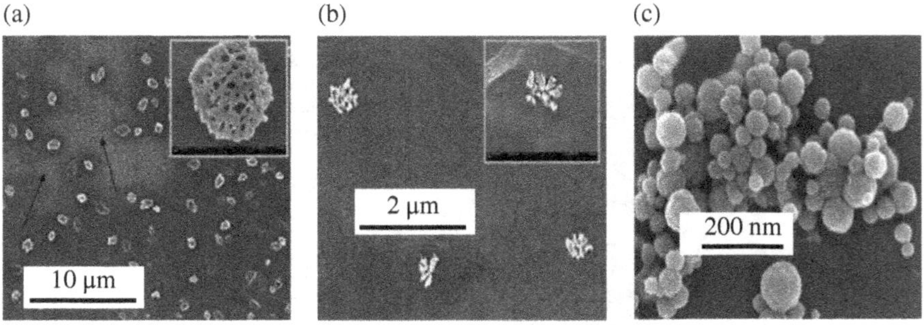

FIGURE 8.12 Morphology of semi-IPN and IPN nanogels after different treatment times: (a) SEM images of semi-IPN after 7 days of dialysis; porous structure evidence the leaching of free PAA chains; (b) broken semi-IPN nanogels after extensive dialysis (11 days); and (c) intact spherical IPN nanogels even after 11 days of dialysis. Source: Koul *et al.* [66]. Reproduced with permission of Elsevier

It can be observed that in both the kinds of blends, the PU phase is microphase separated with the domain size in the range of 50–150 nm in the semi-IPN (dark areas in Figure 8.11b) and 300 nm in the linear blend (brighter areas in Figure 8.11a).

Further example of the nanostructured IPN based on polyacrylic acid (PAA) and gelatin has been taken from a more recent work [66] and is presented in Figure 8.12. The novel IPN nanogels composed of PAA and gelatin were synthesized by one-pot inverse miniemulsion (IME) technique. The nanogel structures were prepared which have been recommended for targeted drug delivery systems for the treatment of solid tumors. Some of the samples were treated in hot water which made the PAA phase to leach out of the samples. Scanning electron microcopy was used to evaluate the structural changes caused by various treatments in the nanogels.

When the semi-IPN was subjected to extensive dialysis (for 7 days), a highly porous gel resulted (Figure 8.12a). Further dialysis for extended period of 11 days led to the breaking of the semi-IPN (Figure 8.12b). The broken part of the semi-IPN samples can be observed as white powderlike substance in the background of Figure 8.12a (shown by arrows). In contrast to semi-IPN, the IPN nanogels retain their structure even after 11 days of dialysis (Figure 8.12c).

8.4.5 Interpenetrating Polymer Networks-Like Systems

There are multicomponent polymers having similar nature in terms of morphology and properties as IPNs but do not strictly fall under such classical categories as sequential or simultaneous IPNs, semi-IPNs, gradient IPNs, etc.

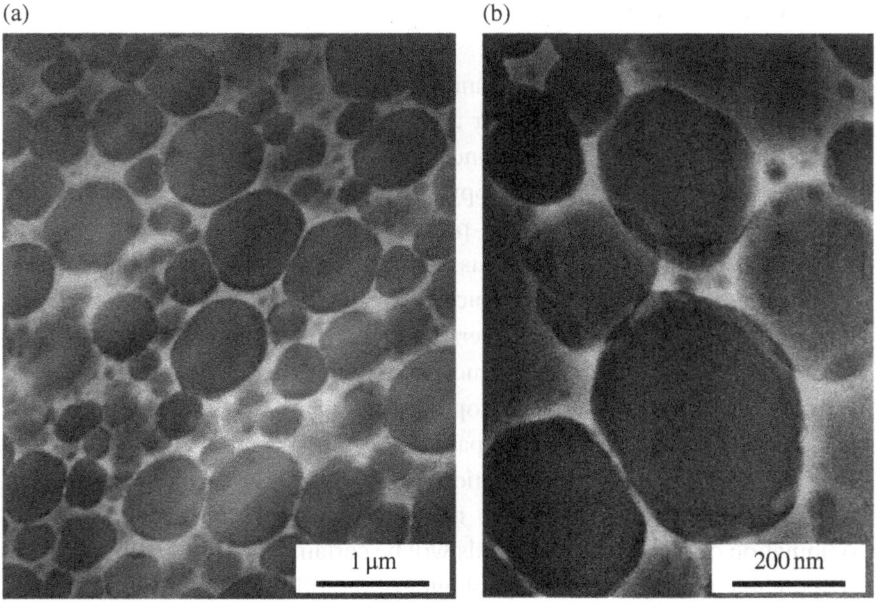

FIGURE 8.13 TEM micrographs of lower (a) and higher (b) magnifications of IPN sample with PE/PS copolymer (70:30 by weight), crosslinker is 1.5 mol.%. Source: Borsig *et al.* [77]. Reproduced with permission of John Wiley & Sons

For instance, PE/vinyl copolymer-based IPN-like systems were synthesized by copolymerizing and subsequent crosslinking of diverse monomer pairs dissolved in solution of low-density PE [31, 37, 72–82]. Similar IPN-like systems can be prepared by embedding physically crosslinkable block copolymer networks within thermosetting resins [31, 37, 82]. These may be termed as IPN-like systems. Electron microscopy has been extensively used to study such IPN-like systems which have a number of potential technical applications.

The IPN-like systems consisting of PE and butyl methacrylate (BMA)–styrene (S) copolymer PE/(BMA-*co*-S) were studied with emphasis of the effect of the crosslinking agent on the morphology of IPN. The vinyl portions of the samples were stained with ruthenium tetroxide (RuO_4) prior to the TEM imaging. The microscopic techniques used were TEM and AFM. The results are presented in Figure 8.13.

The copolymer domains (the dark-appearing domains in Figure 8.13) having diameters ranging from a few tens of nanometers to few hundred nanometers can be identified. It has been found that the IPN-like system represents a two-phase morphology. In spite of the crosslinking of the PE phase, the observation of individual lamellae was made possible with the help of tapping mode AFM [77].

8.5 CONCLUDING NOTES

IPNs have a large number of technical applications in various fields. They have a wide diversity of molecular and supramolecular assemblies and thus possess complex phase behavior and morphologies. Microscopic technique, particularly the electron microscopy, has made significant contribution in unraveling the complex structure–property correlations in these important classes of polymeric materials. Thus, electron microscopy, complemented by other microscopic and spectroscopic tools, has offered most straightforward solution to the comprehensive understanding of morphology–property correlation of the IPNs. In this chapter, morphology of different classes of polymer networks investigated by microscopic techniques has been discussed with special attention to specimen preparation. It has been highlighted that the selection of the specimen preparation method and the specific microscopic technique depends largely on the nature of the problem at hand, but the combination of two or more methods will be certainly helpful in obtaining the most reliable picture of the morphologies of the IPNs.

ACKNOWLEDGMENTS

The author acknowledges Alexander von Humboldt (AvH) Foundation for supporting his several research visits to Germany, which made the compilation of this chapter possible. He further thanks Prof. Goerg H. Michler, Prof. Wolfgang Grellmann, and Prof. Hans-Joachim Radusch and their research groups at Martin Luther University Halle-Wittenberg, Germany; Prof. Sabu Thomas, Mahatma Gandhi University, Kottayam, Kerala, India and Prof. Jean Marc Saiter, Rouen University, Rouen, France, for their kind supports.

REFERENCES

[1] Klempner, D. and Frisch, K.C. (1983) *Polymer Alloys III: Blends, Blocks, Grafts, and Interpenetrating Networks*, Plenum Press, New York.

[2] Dusek, K. and Kuchanov, S.I. (eds.) (1992) Proceedings of the International Conference "Polymer Networks '91," April 21–26, 1991, Moscow, VSP, Utrecht, the Netherlands.

[3] Karger-Kocsis, J. and Fakirov, S. (eds) (2009) *Nano- and Micromechanics of Polymer Blends*, Carl Hanser Verlag, Munich.

[4] Sperling, L.H. (1981) *Interpenetrating Polymer Networks and Related Material*, Plenum Press, New York.

[5] Klempner, D., Sperling, L.H. and Utracki, L.A. (1994) *Interpenetrating Polymer Networks*, American Chemical Society, Washington, DC.

[6] Lohani, A., Singh, G., Bhattacharya, S.S., and Verma, A. (2014) Interpenetrating polymer networks as innovative drug delivery systems. *Journal of Drug Delivery*, Article ID 583612.

[7] Dragan, E.S. (2014) Design and applications of interpenetrating polymer network hydrogels: a review. *Chemical Engineering Journal*, **243**, 572–590.

[8] Seal, B.L., Otero, T.C. and Panitch, A. (2001) Polymeric biomaterials for tissue and organ regeneration. *Materials Science and Engineering R: Reviews*, **34**, 147–230.

[9] Harrats, C. (2009) *Multiphase Polymer-Based Materials: An Atlas of Phase Morphology at the Nano and Micro Scale*, CRC Press, Taylor & Francis Group, Boca Raton, FL.

[10] Sperling, L.H., Widmaier, J.M., Yeo, J.K. and Michel, J. (1983) Phase domain size and continuity in sequential IPN's: a review. *Polymer Science and Technology*, **20**, 191–218.

[11] L.H. Sperling, *Interpenetrating Polymer Networks*. In: Encyclopedia of Polymer Science and Technology, Mark, H.F. (eds). John Wiley & Sons, New York, 2004.

[12] Thomas, S., Rajisha, K.R., Maria, H.J. *et al.* (eds) (2014) *Natural Rubber Materials, Volume 1: Blends and IPNs*, The Royal Society of Chemistry, Cambridge, UK.

[13] Shivashankar, M. and Mandal, B.K. (2012) A review on interpenetrating polymer network. *International Journal of Pharmacy and Pharmaceutical Sciences*, **4**, 1–7.

[14] Woodward, A.E. (1988) *Atlas of Polymer Morphology*, Hanser Publication, Munich.

[15] Michler, G.H. and Baltá Calleja, F.J. (eds) (2005) *Mechanical Properties of Polymers Based on Nanostructure and Morphology*, CRC Press, Boca Raton, FL.

[16] Michler, G.H. and Baltá-Calleja, F.J. (2012) *Nano- and Micromechanics of Polymers: Structure Modification and Improvement of Properties*, Hanser Verlag, Munich.

[17] Michler, G.H. (ed) (2008) *Electron Microscopy of Polymers*, Springer Verlag, Heidelberg.

[18] Michler, G.H. and Lebek, W. (2004) *Ultramicrotomie in der Materialforschung*, Carl Hanser Verlag, Munich.

[19] Sawyer, L.C. and Sawyer, L.C. (1996) *Polymer Microscopy*, Chapmann and Hall, London.

[20] Trent, J.S., Scheinbeim, J.I. and Couchman, P.R. (1983) Ruthenium tetraoxide of polymers for electron microscopy. *Macromolecules*, **16**, 589–598.

[21] Froehling, P.E. and Pijpers, A.J. (1987) Ruthenium tetroxide as a staining agent for transmission electron microscopy of polyamide-polyether blends and block copolymers. *Journal of Polymer Science Part B: Polymer Physics*, **25**, 947–952.

[22] Kato, K. (1966) Osmium tetroxide fixation of rubber lattices. *Journal of Polymer Science, Part B: Polymer Letters*, **4**, 35–38.

[23] Vitali, R. and Montani, E. (1980) Ruthenium tetroxide as a staining agent for unsaturated and saturated polymers. *Polymer*, **21**, 1220–1222.

[24] Collins, S., Hamley, I.W. and Mykhaylyk, T. (2003) An atomic force microscopy study of ozone etching of a polystyrene/polyisoprene block copolymer. *Polymer*, **44**, 2403–2410.

[25] Lammertink, R.G.H., Hempenius, M.A., van den Enk, J.E. *et al.* (2000) Nanostructured thin films of organic–organometallic block copolymers: one-step lithography with poly(ferrocenylsilanes) by reactive ion etching. *Advanced Materials*, **12**, 98–103.

[26] Park, M., Harrison, C., Chaikin, P.M. *et al.* (1997) Block copolymer lithography: periodic arrays of ~10^{11} holes in 1 square centimeter. *Science*, **276**, 1401–1404.

[27] Adhikari, R., Godehardt, R., Lebek, W. *et al.* (2005) Morphological and micromechanical study of ethylene/octene copolymers and their blends with high density polyethylene. *Polymers for Advanced Technologies*, **16**, 156–166.

[28] Olley, R.H. and Bassett, D.C. (1982) A permanganic etchant for polyolefines. *Polymer*, **23**, 1707–1710.

[29] Olley, R.H., Bassett, D.C., Hine, P.J. and Ward, I.M. (1993) Morphology of compacted polyethylene fibres. *Journal of Materials Science*, **28**, 1107–1112.

[30] Henning, S., Adhikari, R., Michler, G.H. *et al.* (2004) Micromechanical mechanisms for toughness enhancement in β-modified polypropylene. *Macromolecular Symposia*, **214**, 157–171.

[31] Karger-Kocsis, J., Gryshchuk, O., Froehlich, J. and Muellhaupt, R. (2003) Interpenetrating vinylester-epoxy resins modified with organophilic layered silicates. *Composites Science and Technology*, **63**, 2045–2054.

[32] Shyu, S.S. and Chen, D.S. (1987) Polycarbonate-polyurethane semi-interpenetrating polymer networks: T_g behavior and morphology. *Journal of Applied Polymer Science*, **34**, 2151–2162.

[33] Hsieh, K.H. and Chou, L.M. (1989) Interpenetrating polymer networks of polyurethane and polystyrene ionomers. *Journal of Applied Polymer Science*, **38**, 645–654.

[34] Hourston, D.J., Schaefer, F.-U., Walter, N.J. and Gradwell, M.H.S. (1998) Mechanical and morphological study of polyurethane/polystyrene interpenetrating polymer networks containing ionic groups. *Journal of Applied Polymer Science*, **67**, 1973–1985.

[35] Hourston, D.J. and Zia, Y. (1983) Semi- and fully interpenetrating polymer networks based on polyurethane-polyacrylate systems. III Polyurethane-poly(methyl acrylate) semi-2-interpenetrating polymer networks. *Journal of Applied Polymer Science*, **28**, 3849–3857.

[36] Hourston, D.J. and Zia, Y. (1984) Semi- and fully interpenetrating polymer networks based on polyurethane-polyacrylate systems. V. An isomerically related semi-l- interpenetrating polymer network system. *Journal of Applied Polymer Science*, **29**, 2951–2961.

[37] Frisch, H.L., Xue, Y.-P., Maaref, S. *et al.* (1996) Pseudo-interpenetrating polymer networks and interpenetrating polymer networks of zeolite-13X and polystyrene and poly(ethyl acrylate). *Macromolecular Symposia*, **106**, 147–166.

[38] Chen, Q., Ge, H., Chen, D. *et al.* (1994) Investigation on damping behavior and morphology of polyurethane/polymethacrylates and polyacrylates interpenetrating polymer networks. *Journal of Applied Polymer Science*, **54**, 1191–1197.

[39] Hourston, D.J., Huson, M.G. and McCluskey, J.A. (1986) Semi- and fully interpenetrating polymer networks based on polyurethane-polyacrylate systems. X. Polyurethane-poly(ethyl acrylate) interpenetrating polymer networks. *Journal of Applied Polymer Science*, **32**, 3881–3888.

[40] Silverstein, M.S. and Narkis, M. (1987) Elastomeric domain-type interpenetrating polymer networks. *Journal of Applied Polymer Science*, **33**, 2529–2547.

[41] Mai, K. and Zeng, H. (1994) High-performance semi-interpenetrating polymer networks based on acetylene-terminated sulfone: phase separation and morphology. *Journal of Applied Polymer Science*, **53**, 1653–1661.

[42] Hourston, D.J., Huson, M.G. and McCluskey, J.A. (1986) Semi- and fully interpenetrating polymer networks based on polyurethane-polyacrylate systems. IX. Properties of an isomerically related interpenetrating network. *Journal of Applied Polymer Science*, **31**, 709–716.

[43] Dewasthale, S., Shi, X., Hablot, E. *et al.* (2013) Interpenetrating polymer networks derived from silylated soybean oil and polydimethylsiloxane. *Journal of Applied Polymer Science*, **130**, 2479–2486.

[44] Sahiner, N., Godbey, W.T., McPherson, G.L. and John, V.T. (2006) Microgel, nanogel and hydrogel–hydrogel semi-IPN composites for biomedical applications: synthesis and characterization. *Colloid & Polymer Science*, **284**, 1121–1129.

[45] Grishchuk, S. and Karger-Kocsis, J. (2011) Hybrid thermosets from vinyl ester resin and acrylated epoxidized soybean oil (AESO). *Express Polymer Letters*, **5**, 2–11.

[46] Rezaei, S.M. and Mohd Ishak, Z.A. (2014) Grafting of collagen onto interpenetrating polymer networks of poly(2-hydroxyethyl methacrylate) and poly (dimethyl siloxane) polymer films for biomedical applications. *Express Polymer Letters*, **8**, 39–49.

[47] Suri, S. and Schmidt, C.E. (2009) Photopatterned collagen–hyaluronic acid interpenetrating polymer network hydrogels. *Acta Biomaterialia*, **5**, 2385–2397.

[48] Liu, W., Deng, C., McLaughlin, C.R. *et al.* (2009) Collagen-phosphoryl-choline interpenetrating network hydrogels as corneal substitutes. *Biomaterials*, **30**, 1551–1559.

[49] Liu, Y. and Cui, Y. (2011) Thermosensitive soy protein/poly(N-isopropylacrylamide) interpenetrating polymer network hydrogels for drug controlled release. *Journal of Applied Polymer Science*, **120**, 3613–3620.

[50] Burugapalli, K., Bhatia, D., Koul, V. and Choudhary, V. (2001) Interpenetrating polymer networks based on poly(acrylic acid) and gelatin. I: swelling and thermal behavior. *Journal of Applied Polymer Science*, **82**, 217–227.

[51] Singh, D., Choudhary, V. and Koul, V. (2007) Radiation synthesis of interpenetrating polymer networks based on N-vinyl pyrrolidone—acrylic acid copolymer and gelatin. I. Swelling, morphology, and thermal characterization for biomedical applications. *Journal of Applied Polymer Science*, **104**, 1456–1463.

[52] Krezovic, B.D., Dimitrijevic, S.I., Filipovic, J.M. *et al.* (2013) Antimicrobial P(HEMA/IA)/PVP semi-interpenetrating network hydrogels. *Polymer Bulletin*, **70**, 809–819.

[53] Jones, D.S., Andrews, G.P., Caldwell, D.L. *et al.* (2012) Novel semi-interpenetrating hydrogel networks with enhanced mechanical properties and thermoresponsive engineered drug delivery, designed as bioactive endotracheal tube biomaterials. *European Journal of Pharmaceutics and Biopharmaceutics*, **82**, 563–571.

[54] Saha, K., Irwin, E.F., Kozhukh, J. *et al.* (2007) Biomimetic interfacial interpenetrating polymer networks control neural stem cell behavior. *Journal of Biomedical Materials Research*, **81A**, 240–249.

[55] Kaity, S., Isaac, J. and Ghosh, A. (2013) Interpenetrating polymer network of locust bean gum-poly (vinyl alcohol) for controlled release drug delivery. *Carbohydrate Polymers*, **94**, 456–467.

[56] Zhang, Y., Heath, R.J. and Hourston, J. (2000) Morphology, mechanical properties, and thermal stability of polyurethane-epoxide resin interpenetrating polymer network rigid foams. *Journal of Applied Polymer Science*, **75**, 406–416.

[57] Pionteck, J., Hu, J. and Schulze, U. (2003) Membrane properties of microporous structures prepared from polyethylene/polymethacrylate IPN. *Journal of Applied Polymer Science*, **89**, 1976–1982.

[58] Chavda, H.V., Patel, R.D., Modhia, I.P. and Patel, C.N. (2012) Preparation and characterization of superporous hydrogel based on different polymers. *International Journal of Pharmaceutical Investigation*, **2**, 134–139.

[59] Mathew, A.P., Packirisamy, S. and Thomas, S. (2000) Morphology, mechanical properties, and failure topography of semi-interpenetrating polymer networks based on natural rubber and polystyrene. *Journal of Applied Polymer Science*, **78**, 2327–2344.

[60] Vo, C.D. and Burford, R.P. (1999) Latex interpenetrating polymer networks based on high styrene resin. *Journal of Applied Polymer Science*, **74**, 622–628.

[61] Burford, R.P. and Vo, C.D. (1999) Morphology development in polychloroprene-polystyrene latex interpenetrating polymer networks. *Journal of Applied Polymer Science*, **74**, 629–638.

[62] Burford, R.P., Markotsis, M.G. and Knott, R.B. (2003) Small angle neutron scattering and transmission electron microscopy studies of interpenetrating polymer networks from thermoplastic elastomers. *Nuclear Instruments and Methods in Physics Research B*, **208**, 58–65.

[63] John, J., Suriyakala, R., Thomas, S. *et al.* (2010) Morphology, mechanical and thermal properties of nano-structured full IPNs based on polyisoprene and PMMA. *Journal of Materials Science*, **45**, 2892–2901.

[64] Zu, L., Li, R., Jin, L. *et al.* (2014) Preparation and characterization of polypropylene/silica composite particle with interpenetrating network *via* hot emulsion sol–gel approach. *Progress in Natural Science: Materials International*, **24**, 42–49.

[65] Tan, J.H., Wang, X.P., Tai, J.J. *et al.* (2012) Novel blends of acrylonitrile butadiene rubber and polyurethane-silica hybrid networks. *Express Polymer Letters*, **6**, 588–600.

[66] Koul, V., Mohamed, R., Kuckling, D. *et al.* (2011) Interpenetrating polymer network (IPN) nanogels based on gelatin and poly(acrylic acid) by inverse miniemulsion technique: synthesis and characterization. *Colloids and Surfaces B: Biointerfaces*, **83**, 204–213.

[67] Tang, Q., Huang, K., Qian, G. and Benicewicz, B.C. (2013) Phosphoric acid-imbibed three-dimensional polyacrylamide/poly(vinyl alcohol) hydrogel as a new class of high-temperature proton exchange membrane. *Journal of Power Sources*, **229**, 36–41.

[68] Nagaraju, D.H., Rebis, T., Gabrielsson, R. *et al.* (2014) Charge storage capacity of renewable biopolymer/conjugated polymer interpenetrating networks enhanced by electroactive dopants. *Advanced Energy Materials*, **4**, 1300443.

[69] Choi, Y.-J., Kang, M.-S. and Moon, S.-H. (2003) Characterization of semi-interpenetrating polymer network polystyrene cation-exchange membranes. *Journal of Applied Polymer Science*, **88**, 1488–1496.

[70] Liao, D.C., Hsieh, K.H., Chern, Y.C. and Ho, K.S. (1997) Interpenetrating polymer networks of polyaniline and maleimide-terminated polyurethanes. *Synthetic Metals*, **87**, 61–67.

[71] Yin, W., Liu, C., Li, J. *et al.* (1997) High-quality conductive polypyrrole within a secondary crosslinked interpenetrating polymer network. *Journal of Applied Polymer Science*, **65**, 1–4.

[72] Borsig, E., Fiedlerova, A., Häusler, K.G. *et al.* (1994) Structure and properties of polystyrene-polyethylene interpenetrating polymer network (IPN)-like systems. *Macromolecular Symposia*, **83**, 147–156.

[73] Borsig, E., Fiedlerová, A., Haüsler, K.G. *et al.* (1996) Structure and properties of an interpenetrating polymer network-like system consisting of polystyrene-polyethylene: 2. Electron microscopic investigation. *Polymer*, **37**, 3959–3963.

[74] Greco, R., Iavarone, M., Fiedlerovà, A. and Borsig, E. (2002) Optical properties of polyethylene/styrene-co-methacrylate copolymers IPN-like networks: effect

of different methacrylate styrene comonomers on properties. *Journal of Materials Science*, **37**, 3389–3395.

[75] Greco, R., Iavarone, M., Fiedlerova, A. and Borsig, E. (2001) Optical properties of IPN-like networks polyethylene/poly (butylmethacrylate-co-styrene) copolymer systems. III. Influence of copolymer cross-linkers. *Polymer*, **42**, 5089–5095.

[76] Wang, Y., Liang, D., Ying, Q. and Chu, B. (2005) Quasi-interpenetrating network formed by polyacrylamide and poly(N,N-dimethylacrylamide) used in high-performance DNA sequencing analysis by capillary electrophoresis. *Electrophoresis*, **26**, 126–36.

[77] Borsig, E., Thomann, R., Fiedlerova, A. and Muelhaupt, R. (2001) Morphology of the transparent IPN-like system PE: (BMA-co-S). *Journal of Applied Polymer Science*, **81**, 2615–2620.

[78] Borsig, E., Fiedlerova, A., Greco, R. *et al.* (1999) Morphology of IPN-like systems consisting of polyethylene and vinyl polymer. *Journal of Macromolecular Science, Part B*, **38**, 541–547.

[79] Chakrabarty, D. and Das, B. (1996) Epoxy-poly(buty1 methacrylate) interpenetrating polymer networks: morphology and various physical, mechanical, and thermal properties. *Journal of Applied Polymer Science*, **60**, 2125–2132.

[80] Pandit, R., Michler, G.H., Lach, R. *et al.* (2014) Epoxidation of styrene/butadiene star block copolymer by different methods and characterization of the blends with epoxy resin. *Macromolecular Symposia*, **341**, 67–74.

[81] Harismendy, I., Delrio, M., Marieta, C. *et al.* (2001) Dicyanate ester-polyetherimide semi-interpenetrating polymer networks. II. Effects of morphology on the fracture toughness and mechanical properties. *Journal of Applied Polymer Science*, **80**, 2759–2767.

[82] Chakrabarty, D., Das, B. and Roy, S. (1998) Epoxy resin–poly(ethyl methacrylate) interpenetrating polymer networks: morphology, mechanical, and thermal properties. *Journal of Applied Polymer Science*, **67**, 1051–1059.

[83] Karger-Kocsis, J. (2006) Simultaneous interpenetrating network structured vinylester/epoxy hybrids and their use in composites, in *Micro- and Nanostructured Multiphase Polymer Blend Systems Phase Morphology and Interfaces* (eds C. Harrats, S. Thomas and G. Groeninckx), CRC Press, Taylor & Francis Group, Boca Raton, FL.

[84] Irwin, E.F., Saha, K., Rosenbluth, M. *et al.* (2008) Modulus-dependent macrophage adhesion and behavior. *Journal of Biomaterials Science. Polymer Edition*, **19**, 1363–1382.

[85] Li, J.K. (2010) Investigation of nano-scale, self-assembled, polymeric systems by atomic force microscopy. Ph.D. thesis. Department of Chemistry, University of Toronto.

[86] Ould MBareck, C., Metayer, M., Nguyen, Q.T. *et al.* (2003) Microporous anion-exchange membranes made of semi-interpenetrating polymer network: I. Elaboration and characterization. *Journal of Membrane Science*, **221**, 53–68.

[87] Xuay, M.-X., Liu, W.-G., Li, H.- P. *et al.* (1997) Visualization of the phase-separated surface of an unsaturated polyester/polyurethane interpenetrating polymer network by atomic force microscopy. *Materials Chemistry and Physics*, **47**, 9–12.

[88] Mishra, S., Bajpai, R., Katare, R. and Bajpai, A.K. (2007) Radiation induced crosslinking effect on semi-interpenetrating polymer networks of poly(vinyl alcohol). *Express Polymer Letters*, **1**, 407–415.

9

VISCOELASTIC PROPERTIES OF INTERPENETRATING POLYMER NETWORKS

Sudipta Goswami

Chemical Engineering and Technology, Birla Institute of Technology, Mesra, Ranchi, Jharkhand, India

9.1 INTRODUCTION

Interpenetrating polymer networks (IPNs) represent a unique type of heterogeneous polymer blends in which microphase separation due to thermodynamic incompatibility of the components takes place during their formation [1–6]. Development of such system is governed by the superposition of two processes—the chemical kinetics of formation of IPNs and the physical kinetics of phase separation [7]. The initial stages of the latter process obey the spinodal decomposition mechanism. An increase in the molecular mass of linear components and the development of the three-dimensional chemical network facilitate fixation of microphase separation at a certain level. As a consequence, the thermodynamically nonequilibrium system with incomplete phase separation is formed, resulting in an interfacial region (IFR) coexisting with the two formed phases. Earlier it is shown that the level of phase separation in an IPN depends on the chemical nature of the components and their ratio in the system,

Micro- and Nano-structured Interpenetrating Polymer Networks: From Design to Applications, First Edition. Edited by Sabu Thomas, Daniel Grande, Uroš Cvelbar, K.V.S.N. Raju, Ramanuj Narayan, Selvin P. Thomas and Akhina H.

cure temperature, and catalyst, initiator, and crosslinker concentrations [8–11]. On the whole, the nonequilibrium microphase structure of IPNs may be regarded as microheterogeneous, in which there is no molecular mixing of the components over the entire volume, but, within each of the segregated phases, the state of molecular mixing is frozen [12–14]. Compatibility can be substantially enhanced by IPN modification with compatibilizing agents, additives that can act on the basis of not only physical but also chemical interaction at the interface between two polymers [15, 16]. Physical and viscoelastic properties of IPNs are controlled by the two formed phases and by the volume fraction of the IFR. Sequential IPN synthesis leads to an enhancement of the compatibility of the components in the system, preferably in contrast to the simultaneous curing technique where it is possible to vary the chemical composition of the IPN and its component ratio over wide ranges [1]. During sequential IPN synthesis, a finished network polymer (matrix) is swollen in monomer of another variety, followed by its polymerization and the formation of the second network of another chemical nature. Thus the polymerization of the second monomer occurs in the space confined to the network of the primary polymer matrix and results in the compatibilization of the components due to the formation of a large number of topological entanglements between polymer-matrix chains and the second polymer chains. With simultaneous IPNs, it is found [17] that the network which is first formed represents the continuous phase. As the synthesis of both the networks is initiated at the same time and proceeds to completion more or less simultaneously, bigger domains may form, due to lack of topological restriction from an already present structure [18].

Dynamic mechanical analysis (DMA) can be simply described as applying an oscillating force to a sample and analyzing the material response to that force. From this, one may calculate the quantitative values of properties like the tendency to flow (called viscosity) from the phase lag and the stiffness (modulus) from the sample recovery. These properties are often described as the ability to lose energy as heat (damping) and the ability to recover from deformation (elasticity). One way to describe what we are studying is the relaxation of the polymer chains. Another way would be to discuss the changes in the free volume of the polymer that occur [19]. As the free volume of the chain segment increases, its ability to move in various directions also increases. This increased mobility in either side chains or small groups of adjacent backbone atoms results in a greater compliance (lower modulus) of the molecule:

$$E^* = E' + iE'' \tag{9.1}$$

The modulus measured in DMA is, however, not exactly the same as Young's modulus of the classical stress–strain curve. Young's modulus is the slope of a stress–strain curve in the initial linear region. In DMA, a complex modulus (E^*), expressed in terms of elastic (storage) modulus (E') and imaginary (loss) modulus (E''), is calculated from the material response to the sine wave (Eq. 9.1).

These different moduli allow better characterization of the material, because we can now examine the ability of the material to return or store energy (E'), its ability to lose energy (E''), and the ratio of these effects ($\tan\delta$), which is called damping. As the material becomes elastic, the phase angle, δ, becomes smaller, that is, damping is less. In Figure 9.1, F_d and F_s represent the damping and storage stress, respectively.

When subjected to a stress (σ), a material will exhibit a deformation or strain, γ. As the temperature is increased from very low value, where the molecule is tightly compressed, it passes through the solid-state transitions. Firstly, the free volume increases so that localized bond movements (bending and stretching) and side-chain movements can occur. This is termed as gamma transition, $T\gamma$ [20]. As the temperature and the free volume continue to increase, the whole side chains and localized groups of four to eight backbone atoms begin to have enough space to move, and the material starts to develop some toughness. This transition, called the beta transition $T\beta$, is not clearly defined everywhere. Often it is described as the T_g or glass transition of a secondary component in a blend or of a specific block in a block copolymer. A correlation of this temperature with toughness is seen empirically [21]. As heating continues, T_g or glass transition is reached, where the chains in the amorphous regions begin to coordinate large-scale motions. This transition is also called the alpha transition, $T\alpha$ (not shown in the figure). One classical description of this region is that the amorphous regions have begun to melt.

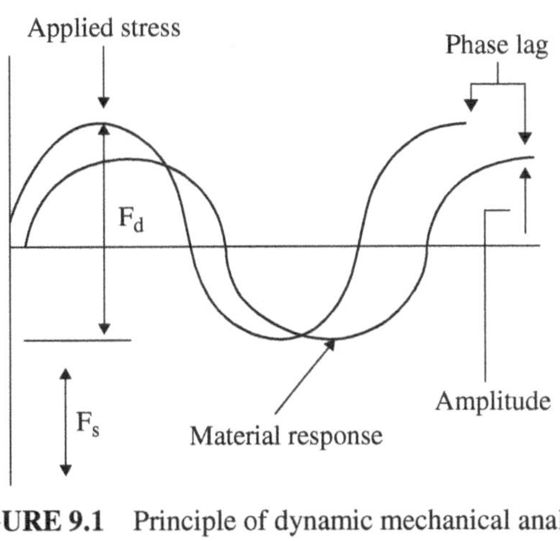

FIGURE 9.1 Principle of dynamic mechanical analysis

Since the T_g only occurs in amorphous material, in a 100% crystalline material, T_g is not seen. Continuous heating causes slippage of the crystallites past each other in crystalline or semi-crystalline polymer. At the similar stage for segments in the amorphous phase, coordinated movement occurs, contributing to reduced viscosity. However, these two transitions are not accepted by everyone, and their existence is still a matter of some disagreement. Finally, we reach the melting temperature, T_m (Figure 9.2), at which large-scale chain slippage starts and the material flows. For a cured thermoset, nothing happens after the T_g until the sample begins to burn and degrade because the crosslinks prevent the chains from slipping past each other [22].

Viscoelastic property analysis is useful in determining crosslink density(ν_e) of polymer matrix within polymer blends and/or IPNs (Eq. 9.2) which in turn may help in correlating the basis of the effect of variables, for example, cross-linker concentration and its type, blend ratio of the components, order of blending the components, etc:

$$\nu_e = \frac{E'}{3RT} \tag{9.2}$$

where E' is the storage modulus in the rubbery plateau region, R is the gas constant, and T is the absolute temperature [23]. IPNs and in situ blends or linear polymers are characterized by numerous transition states, which control

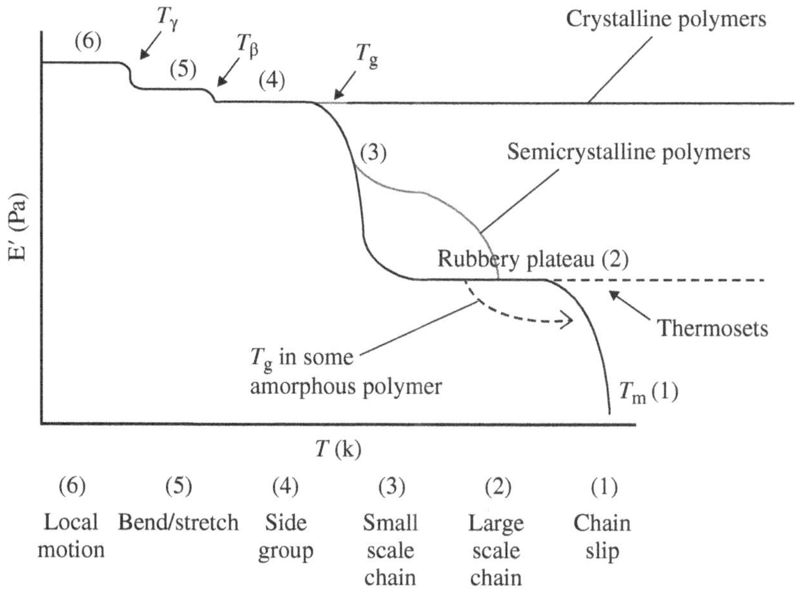

FIGURE 9.2 Ideal loss modulus versus temperature curve of polymer

the thermodynamic equilibrium of the system. The transition layers are formed during the synthesis of such systems and during the physical mixing and non-equilibrium phase separation in the course of the chemical reactions, and their development depends on the processing conditions. It is established that the physical aging of IPNs has a specific character related to the microphase separation.

Viscoelastic properties of semi-interpenetrating polymer networks (semi-IPNs) based on crosslinked PU and linear polystyrene (PS) were studied [24] after physical aging was performed in three stages: (1) for 15 days at 60 and 110°C (samples 1 and 2), (2) conditioned at thermal chamber for 1 month under same temperature, (3) after stage B, samples 1 and 2 that were kept for 4 months at 25°C. In this case, the range of temperature is chosen in such a way that all possible changes in both PU and PS phases due to molecular mobility in semi-IPNs could be accounted for. Before aging the semi-IPNs (50 : 50 wt.%) were two-phase systems with well-defined IFR. Upon thermal aging acceleration of the relaxation processes occurred to this heterogeneous system, resulting in structural rearrangements (after stage A and B). Even when the system came back to the initial conditions and all relaxation processes slowed down (after stage C), these processes went on. Therefore, a new nonequilibrium state, with completely different arrangement of molecules in the IPNs, was formed. In this system an increased fraction of IFR was interpreted as an apparent increase in compatibility though it was not really giving rise to increase in thermodynamic stability of the system.

Viscoelastic properties of IPNs may vary depending on the following factors:

- Mode of synthesis—simultaneous, sequential, semi-I IPN or semi-II IPN, full-IPN, and gradient IPN.
- Chemical nature of the constituents—that is, substituent groups and chemical links present in the main chain, and crosslink density
- Condition of testing—for example, frequency and amplitude of oscillation, temperature range, ramp rate, etc.
- Thermal properties of constituent polymers—thermoplastic or thermosetting resin
- Flow property of the constituents—viscosity, melt flow index, and elasticity
- Intermolecular as well as intramolecular interaction involving the polymers, for example, crystalline/amorphous, polar/nonpolar, hydrophilic/hydrophobic, and other combinations of polymers which control the morphology of the IPNs subject to miscibility and hence response to applied stress

9.2 VISCOELASTIC PROPERTIES OF SIMULTANEOUS IPNs

Thermal and viscoelastic properties of IPNs, formed from radically initiated unsaturated polyester resin (UPR) (maleic type) and acid anhydride-cured epoxy polyester (EP) as well as the neat UPR and EP, had been studied by nonisothermal DSC, DMA, and TG analyses [23]. The IPNs consisting of 10, 30, 50, 70, and 90 wt.% of EP component were prepared. Below T_g the storage modulus decreased with EP content up to 30 wt.% in the blends due to plasticization effect of EP. Higher E′ of the blends with EP > 30 wt.% in the rubbery region (above T_g) was due to the formation of stiffer network through diester linkages during reaction of epoxy and anhydride groups. It is seen from Table 9.1 that the crosslink density decreased regularly with EP content in the blends when benzoyl peroxide (BPO)/tetrahydrophthalic anhydride (THPA) curing agent was used. Heterogeneity of the system increased with decrease in EP content in the IPNs as was evidenced from two distinct peaks in tanδ curve. Among the two distinct transition temperatures in the tanδ curve, the first peak corresponding to secondary β-relaxation ($\tan\delta_{max1}$) was responsible for polyester end group mobility or motions of the segments formed between two crosslinks in the system, and the second one was corresponding to α-relaxation ($\tan\delta_{max2}$) related to overall molecular motion. The width of the tanδ curves was connected with the degree of structural heterogeneity of polymer networks. The broader the peak, the more heterogeneous is the polymer network, with a wide distribution of relaxation times. More homogeneous network was observed with higher EP-containing IPNs (Figure 9.3). Blending had not impaired the curing ability of individual component.

Use of maleic anhydride (MA) instead of THPA while keeping all other variables unchanged caused considerable increase of T_{max2} values in tanδ

TABLE 9.1 Data from DMA and DSC Curves of the BPO/THPA-Cured IPNs, BPO-Cured UPR, and THPA-Cured EP

Composition	Curing Agent	FWHM/°C	Crosslink Density ($mol/cm^3 \times 10^{-3}$)	T_g (K)
100UPR	BPO	81	8.58	323.3
90UPR:10EP	BPO/THPA	68	3.88	318.8
70UPR:30EP	BPO/THPA	63	2.96	335.3
50UPR:50EP	BPO/THPA	62	2.7	343
30UPR:70EP	BPO/THPA	54	2.12	357.7
10UPR:90EP	BPO/THPA	44	1.72	372.3
100EP	THPA	44	1.2	374.8

Source: Worzakowska [23]. Reproduced with permission from Springer.

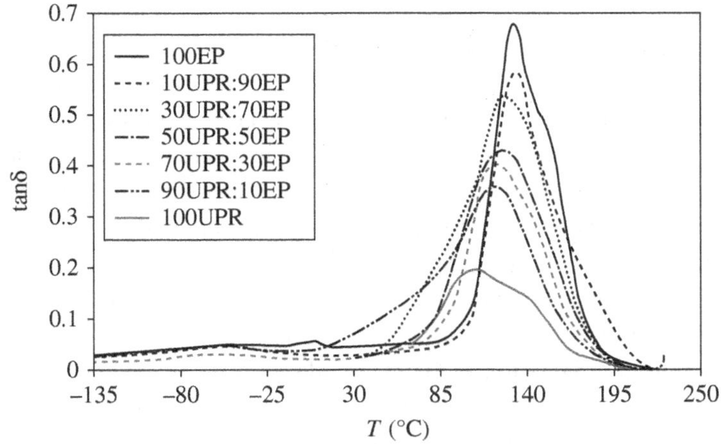

FIGURE 9.3 Tanδ curves of the BPO/THPA-cured UPR/EP IPNs, BPO-cured UPR, and THPA-cured EP Source: Worzakowska [23]. Reproduced with permission from Springer

TABLE 9.2 Data Obtained from DMA and DSC Curves for IPNs Containing 50 : 50 wt.% UPR: EP Cured with Different Hardener System, CHP-Cured and MA-Cured EP

Composition	Curing Agent	FWHM/°C	Crosslink Density (mol/cm^3 × 10^{-3})	T_g (K)
100UPR	CHP	78	7.98	325.4
UPR:EP(50 : 50)	CHP/THPA	63	3.57	336
	BPO/THPA	62	2.7	343
	CHP/MA	75	4.87	351.9
	BPO/MA	72	4.51	353.5
100EP	MA	50	2.84	401

Source: Song *et al.* [25]. Reproduced with permission from Elsevier.

curves probably due to the chemical connection between two components of the blends and epoxy hardener. The higher stiffness of these CHP/MA or BPO/MA cured IPNs compared to that of neat resins was understood from their crosslink density, glass transition temperature and FWHM values (Table 9.2). Acceleration in the rate of cure of UPR component causing a significant change in the viscosity of the medium was caused by the interactions of double bonds of MA and styrene, UPR, and radical initiator. Such increase in viscosity altered the rate of curing of the second component (EP). The 50 : 50 wt.% UPR: EP IPN cured with CHP was characterized by higher crosslink density and lower molecular mobility compared to those cured with BPO. Thermal properties determined by DSC showed that the glass transition temperatures (T_g) increased with EP content in the blends following the rule of additivity (Table 9.1). On the other hand, plasticization effect, as determined by the peak height of tanδ, increased with increase in EP content and was explained in terms of lowering the degree of crosslinking (Figure 9.3).

The free radical copolymerization involving unsaturated polyester and styrene, additional copolymerization of MA carbon–carbon double bonds and styrene or unsaturated polyester, and polyaddition reaction of MA groups to epoxy groups in EP were suggested to explain the higher crosslink density in these systems compared to the THPA-cured one.

Hourston and Schäfer [25] investigated the rate of network formation of IPNs between polyurethane (PU) and PS by means of FTIR spectroscopy. Their study confirmed that the PU network was formed first. Simultaneous PU/PS IPNs were synthesized by the one-shot route. By the variation of the crosslink level in both polymer networks, the controlled introduction of internetwork grafting, and the incorporation of compatibilizing segments into the PS network, the compatibility of two networks was improved.

The dC_p'/dT signal for the IPNs, from modulated-temperature differential scanning calorimetry (M-TDSC), showed a nonlinear change over a broad temperature range between the glass transitions of PU and PS. This implied that both the networks were heterogeneously distributed in space, but with an interfacial zone containing a mixture of the two networks. The two components formed several kinds of separated domains, which had different segment concentration distributions at the microstructural level. Moreover the interface in these IPNs was considered as a phase having average concentration of the components as was evidenced from the quantitative analysis of the dC_p'/dT signal from M-TDSC when IPNs were compared with the corresponding physical blends. The effect of the variation of the crosslink level in PU phase was assessed by dynamic mechanical thermal analysis (DMTA) for the IPNs having PU of variable triol to diol ratios. With increasing PU crosslinking, the PU transition broadened (a<b<c Figure 9.4), and the PS transition remained at the same location. This was in agreement with the M-TDSC results.

Phase separation that occurred [26] during formation of semi-IPNs of various compositions based on crosslinked poly(urethane) and linear poly(methyl methacrylate) (PMMA) containing 1 wt.% iron, copper, and chromium chelates was studied by Y.S. Lipatov *et al.* SIPNs of compositions 70 : 30 and 50 : 50 (by wt.) showed two phase transitions in DMA curves with weak shoulder like transition on tanδ curve corresponding to PU. The addition of

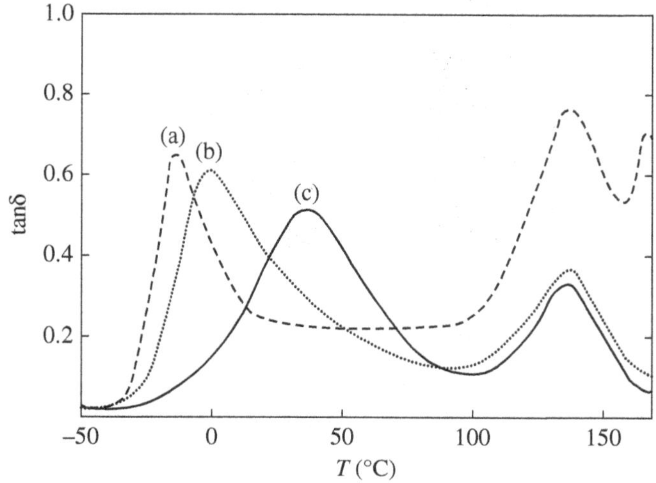

FIGURE 9.4 DMTA data for different PU network crosslink levels: (a) IPN1 (7 : 1 diol/triol), (b) IPN2 (3 : 1 diol/triol), and (c) IPN3 (1 : 1 diol/triol) with 5 mol% DVB in each Source: Kosyanchuk *et al.* [26]. Reproduced with permission from Springer

metal chelates strongly affected the relaxation behavior of SIPNs depending on the β-diketonate type.

The relaxation behavior was also found to be function of the blend ratio. A rise in $\tan\delta_{max}$ in the region of relaxation transition of the PU-enriched phase relative to the corresponding value of the metal-free SIPNs and with increase in the PU content in SIPNs containing chromium β-diketonate was evidenced. Moreover, in the case of chromium-containing SIPNs, the values of $\tan\delta_{max}$ for the PMMA-enriched phase decreased, for example, PU/PMMA (70 : 30 wt.%) system showed lower values relative to 50 : 50 composition. The surface area under the loss maximum was determined by the net contribution of various types of structural units to the relaxation process. DMA was used here as a method of estimating the share of the IFR from the surface area under the intermediate maximum S_m. In contrast to chromium chelates, the incorporation of iron and copper β-diketones into the SIPNs (50 : 50 wt.%) caused retardation of phase separation owing to high rates of poly(urethane) and PMMA formation and the appearance of coordination complexes involving both the components in the blend. A decrease in the share of IFR in these systems proved that the complexation processes dominated over the kinetics of formation of the above blends. More complete phase separation was found with 70 : 30 SIPNs containing metal chelates in which chemical kinetics and complexation processes were opposing each other. The latter

TABLE 9.3 Relaxation Transition Parameters for Metal-Free SIPNs and SIPNs with Metal Chelates

		$Tan\delta_{max}$		
PU : PMMA (wt.%)	Metal Chelate	PU	PMMA	S_m
100:0	—	1.0	—	—
0:100	—	—	1.23	—
30:70	—	—	0.75	—
50:50	—	0.23	0.73	32
	Iron	0.21	1.09	43
	Copper	0.22	1.13	76
	Chromium	0.31	0.05	34
70:30	—	0.36	0.59	22
	Iron	0.25	0.90	7
	Copper	0.26	0.90	10
	Chromium	0.59	0.34	11

Source: Kosyanchuk *et al.* [26]. Reproduced with permission from Springer.

processes prevail, and the influence of the type of metal ion that predominantly interacted with PU was decreased. The S_m values of metal-free SIPNs with 50:50 blend ratio were lower than that of corresponding SIPNs with iron and copper chelating compounds though opposite trend was found with 70:30 SIPNs (Table 9.3).

The effects of compatibilizing additive (oligourethane dimethacrylate (OUDM)) on the kinetics of IPN formation based on crosslinked polyurethane (CPU) and linear PS and its influence on the microphase separation and viscoelastic and thermophysical properties were investigated by T.T. Alekseeva and coworkers [27]. It was established that the introduction of 10 and 20 mass% of ODUM into the initial reaction system prevented microphase separation and led to the formation of compatible IPNs. To estimate the compatibilizing action of OUDM on the viscoelastic properties of IPNs, they selected the semi-IPN based on crosslinked PU and uncrosslinked PS in the weight ratio of 70/30. Semi-IPNs without compatibilizer were characterized by two sharp maxima corresponding to two phases enriched in any one of the components. For IPNs with 5 and 10 mass% of OUDM, these two maxima were preserved, although the shape of the relaxation curves changed. The T_g of PS-enriched phase under the same conditions diminished from 408 K up to 375 K. Decreasing of the heights of relaxation maxima of PU and PS constituents and their narrowing and broadening as well as increasing amount of two phases due to transition of some part of PU and PS into the IFR was the result of interaction with OUDM. The segregation degree (α_{segr}) of the components in IPN was determined from the parameters of the maxima of mechanical loss using the following relation [28]:

$$\alpha_{segr} = \frac{h_1 + h_2 - (h_1 \cdot l_1 + h_2 \cdot l_2) \cdot L^{-1}}{h_1^0 + h_2^0} \quad (9.3)$$

where h_1^0 and h_2^0 are the values of tanδ of the initial homopolymers (TPU and CPU), h_1 and h_2 are tanδ values corresponding to the suitable components of IPNs in the mechanical spectra, l_1 and l_2 are shifts of the tanδ peaks of the IPN components along the temperature scale, and L^{-1} is the difference of the glass transition temperatures (T_g) of the initial homopolymers (TPU and CPU). For complete mixing $\alpha_{segr}=0$. The degree of segregation value for the PU/PS simultaneous IPN without compatibilizer diminished from 0.31 to 0.10 upon use of 10 mass% OUDM, indicating that the introduction of compatibilizing additive during the IPN formation prevented microphase separation. Furthermore, 20 mass% of OUDM introduction in the IPNs caused only one broad maximum at 385 K, whereas a maximum corresponding to PU phase degenerated and revealed only as a shoulder.

Wen-yen Chiang and Dong-Meau Chang [29] prepared polyurethane/allyl novolac resin simultaneous interpenetrating network and carried out DMA to see the compatibility of the components. PU was prepared by reacting 4,4′-diphenylmethane diisocyanate (MDI) with poly(tetramethylene oxide) (PTMO) whose molecular weight falls in 600–700(UT1), 900–1050(UT2), and 1900–2100(UT3). Allyl novolac was synthesized by substituting the hydroxyl group of phenolic resin with allyl group. PU and allyl novolac were mixed by the ratio of 100/0, 75/25, 50/50, and 25/75 separately and were dissolved in DMF at 75–80°C. The resultant solution was added with chain extender of PU, 1,4-butanediol, and crosslinker trimethylolpropane. Initiator of allyl novolac resin, BPO, was also added simultaneously to this mixture and heated for 24 h at 60°C in air oven, followed by 12 h heating at 70°C in vacuum oven. It was found that the lower peak temperature in tanδ versus temperature curve for the IPNs, corresponding to the glass transition temperature of the soft segment of PU, moved toward higher temperature and magnitude of $\tan\delta_{max}$ decreased as the component allyl novolac resin increased. Similarly the peak magnitude at a higher temperature, corresponding to the glass transition temperature of allyl novolac resins, increased with the proportion of allyl novolac resins in the IPNs. Extensive interpenetration of the components was confirmed from the peak broadening for the IPNs in contrast to that of the pure PU.

A series of polyurethane/vinyl ester resin (PU/VER) simultaneous and gradient IPNs were synthesized by introducing acrylic esters as VER's comonomers and cured at room temperature [30]. The effects of thermodynamic factor, kinetic factor, component ratio, and gradient techniques on their damping properties were studied by DMA. It was found that the damping properties of simultaneous PU/VER IPN were improved by introducing acrylic esters as VER's comonomer instead of styrene due to its less rigid structure. Also the damping properties of PU/VER (ethyl acrylate) IPN were better than those of PU/VER (methyl methacrylate) IPN for a broad temperature range. PU/VER (butyl methacrylate) IPN showed the best damping properties due to its broader and higher damping platform compared with two separate peaks and low tanδ values between peaks of the others. The results were attributed primarily to the compatibility of the comonomers on the basis of their solubility parameter differences. However the compatibility of comonomers in PU/VER (butyl methacrylate) was further improved due to hydrogen bonding between polar carboxy in PBMA and amino of PU. The catalyst dosage for PU and initiators for VER and the component ratios also affected the damping properties of IPN. When the amounts of BPO initiator and the catalyst stannous octoate were, respectively, 0.67 wt.% of VER and 0.5 wt.% of PU, 80:20, 70:30, and 60:40 PU/VER (BMA) IPNs had tanδ values higher than 0.3 in the temperature range of at least near 80°C, and these were excellent damping materials (Figure 9.5). The damping properties of IPNs with component ratios from 50:50 to 30:70 were poor due to two damping peaks and low tanδ

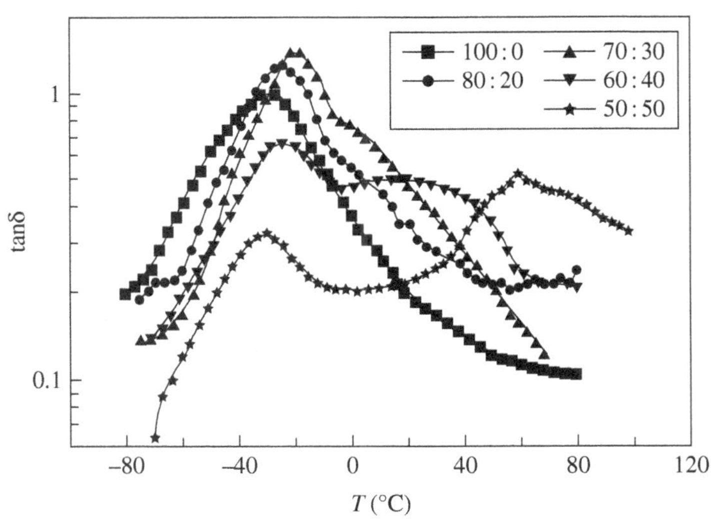

FIGURE 9.5 DMA curves of PU/VER (BMA) IPNs with component ratios from 100:0 to 50:50 Source: Qin *et al.* [30]. Reproduced with permission from Elsevier

values between them. The damping properties were further broadened by controlling the techniques of gradient IPN.

Poly(urethane–imide) (PUI) membranes were prepared by blending polyurethane prepolymer with different proportion of polyamide acid (PAmA) [31]. PU prepolymer were synthesized by two-stage process with hydroxyl-terminated polybutadiene (HTPB), 4,4′-dicyclohexylmethane diisocyanate, and 1,4-butanediol. Also 4,4′-diaminodiphenyl ether (ODA) and 3,3′-4,4′-benzophenonetetracarboxylic dianhydride (BTDA) were reacted to form the polyimide (PI) oligomer solution of PAmA. IPN structures were formed through reaction between PU prepolymer and PAmA and unsaturated C—C double bond between different HTPB segments of PU. DMA result of the IPNs of different compositions such as 0.9 : 0.1, 0.8 : 0.2, 0.5 : 0.5, and 0.2 : 0.8 PU: PI (w/w) was reported. With increase in PU content in the IPNs the storage modulus decreased to lower than 500 MPa, that is, lower than that of pure PI (1901 MPa), while the storage modulus of PUI with high PI content, for example, 0.2-PUI, was 2019 MPa. Similarly tanδ peaks shifted to higher value for the compositions with higher PI content. Storage moduli values are largely decreased at temperature higher than T_g, and the tanδ curves presented disrupt decrease of damping loss peak. This indicated that the damping loss peak of PUI was lower than that of pure PI membrane. The compatibility between PU and PI was found to be best in IPNs with PU content lower than 50 wt.%.

9.3 VISCOELASTIC PROPERTIES OF SEQUENTIAL IPNs

The viscoelastic behavior of sequential semi-IPNs prepared via swelling of network polyurethane in different monomers (butyl methacrylate, styrene) followed by their polymerization in the polyurethane matrix was investigated by means of DMA [32, 33]. The relaxation behavior of the sequential semi-interpenetrating PU/PBMA and PU/PS networks showed a degree of phase separation in the system that was altered by varying the network inner space in the PU matrix through a change in the molecular mass of the polyether block. At a low crosslink density of the PU matrix, its intranetwork volume was sufficient for the phase separation to occur in the system during monomer polymerization at a certain degree of segregation of the components. Such a polymer system is characterized by two relaxation transitions as was found with various simultaneous IPNs in the previous section.

However, in contrast to simultaneous IPNs, enhancement of the forced compatibility of the components was evidenced for this sequential IPN owing to the confinement of the first network inner space available for polymerization of second monomer, leading to a decrease in the degree of segregation and an increase in the number of topological entanglements between growing chains of the polymerizing monomer and between these chains and polymer-matrix macrochains. Such entangled network altered the second monomer polymerization kinetics and, presumably, impeded microphase separation. The network inner space might be limited to such an extent that the growing monomer chains could penetrate through the polymer matrix, forming either a homogeneous polymer system or a system with a very small size of phase domains. Thus the sequential semi-IPN was characterized by the appearance of one relaxation transition.

Brovko *et al.* [34] studied the dynamic mechanical properties of polyurethane–polyurethane semi-IPNs filled with γFe_2O_3.Thermoplastic polyurethane (TPU),basedonpoly(1,4-butylene)glycoladipateand4,4′-methylenebis(phenyl isocyanate),was used as the linear component of semi-IPNs. CPU was prepared by the homogeneous mixing of a solution of vinyl chloride/vinyl acetate/vinyl alcohol, a terpolymer (TCP), in cyclohexane with the adduct of 1,1,1′-trimethylolpropane and 2,4-toluene diisocyanate (TMPTDI) at a ratio of NCO/OH = 0.55 for crosslinking. Semi-IPNs were prepared in solvent mixtures by mixing the initial components of CPU with TPU. A ratio of TPU/CPU = 28.5/71.5% (w/w) was selected. Filled compositions were prepared using a ball mill. All compositions were dispersed at room temperature for 50 h, and thin films (15–20 μm) were prepared by casting a 13% solution onto a Mylar film. The iron oxide volume fraction in the filled semi-IPNs was varied from 0.3 to 21.5%. The temperature dependence of viscoelastic functions, E′, E″ and tanδ for TCP, TPU, CPU, blends of TCP and TPU and semi-IPNs (filled and unfilled) were studied by DMA. The position and shift of the dynamic glass transitions (T_gs) corresponding to the two polymers in the filled IPNs (filler content >5 wt.%) evidenced for the phase separation in contrast to the unfilled IPNs or blends that exhibited a single T_g (Figure 9.6). Crosslinking of TCP with TMPTDI resulted in a shift in the T_g of CPU to a higher temperature in comparison with that of TCP. Introduction of γFe_2O_3 into TPU did not affect the glass transition value but rather led to a decrease in the intensity of tanδ. Also at a temperature below T_g the storage modulus (E′) of filled TPU increased considerably in comparison to that of unfilled one which was attributed to the change in the intermolecular interaction, caused by adsorption of TPU onto the filler surface due to hydrogen bonding of OH— groups of the filler with NH— and C—O groups of the TPU.

Introduction of filler into CPU affected both the system of physical bonds and the crosslink density. In this case, interaction of NCO groups of TMPTDI

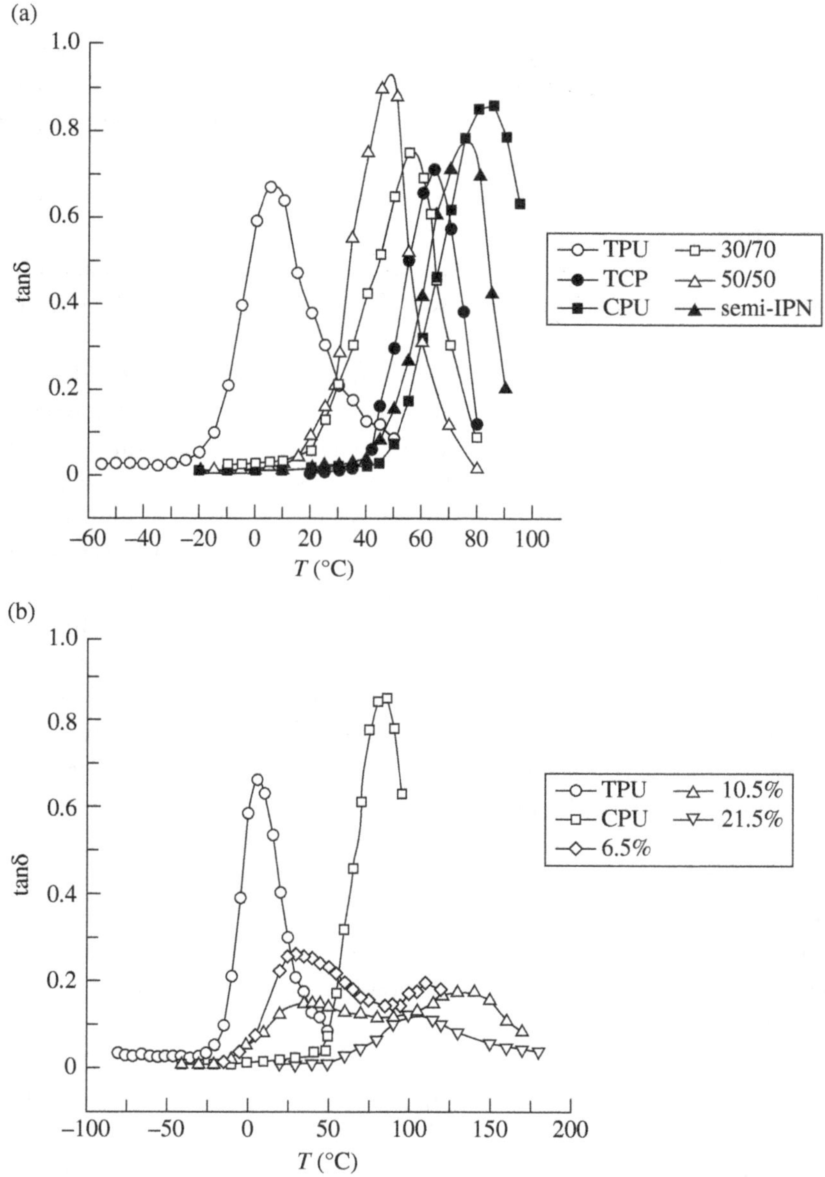

FIGURE 9.6 Viscoelastic properties of (a) unfilled pure TPU, TCP, CPU, and TPU/TCP blends (30/70 and 50/50 wt.%) and semi-IPNs (28.5/71.5 wt.%) and (b) comparison of mechanical properties of TPU, CPU, and filled semi-IPNs with a high loading of γ-Fe_2O_3 Source: Brovko *et al.* [34]. Reproduced with permission from Elsevier

with OH— groups on the filler surface due to partial chemisorptions was the reason of the decrease in glass transition for CPU and compatibility between the two polymer systems at lower concentration (0.3–1.4%) of the filler. Dual-phase continuity was observed in the semi-IPNs with filler content of 6.5 and 10.5% volume fraction. However with increase in filler concentration to 21.5%, increase in T_g (single) was postulated to be due to the cumulative effect of adsorption and steric effect. For semi-IPNs containing 6.5 or 10.5 vol.% of the filler, phase segregation coefficient (α_{segr}) was equal to approximately 0.23 and 0.26, respectively. With increase in filler content in IPNs, the degree of phase separation was negligibly increased.

Conducting IPNs for actuator applications were described by C. Plesse and coworkers [35]. The IPNs were synthesized from poly(ethylene oxide) (PEO) and polybutadiene (PB) networks in which the conducting polymer, poly(3,4-ethylenedioxythiophene) (PEDOT), was gradually dispersed, that is, its content decreased from the outside toward the center of the film. The typical viscoelastic behavior of PEO/PB (80/20) IPN was studied by DMA. Only one relaxation was observed in the tanδ curve at −45°C. DMA of the PEO/PB IPN sample swollen in ionic liquid (RTIL), 1-ethyl-3-methylimidazolium bis(trifluoromethylsulfonyl)imide (EMITFSI) at room temperature, was carried out, and the α-relaxation was found to be 15°C lower than that of the unswollen IPN, demonstrating plasticization effect of the solvent. The influence of PEDOT content in the IPNs on their viscoelastic properties was determined by DMA. As the weight proportion of PEDOT increased, the rubber plateau value increased and conversely the damping properties decreased. Thus at room temperature the stiffness of samples increased (lower tanδ) though glass transition point, T_g, corresponding to the tanδ peak hardly shifted with increase in the PEDOT. Therefore the actuator material was suggested to be operated in open air at low temperature (−30°C)—EMITFSI being the electrolyte in that case.

In a later work D. Teyssié and coworkers studied the PEO/HTPB-based IPNs prepared by in situ process [36]. A series of IPNs containing 80, 60, 50, 40, and 20 wt.% PEO were synthesized by mixing all components together and allowing the networks to be formed independently but sequentially. The PEO network was obtained by free radical copolymerization and crosslinking by poly(ethylene glycol) methacrylate and dimethacrylate in presence of peroxy initiator. Addition processes between HTPB and a pluri-NCO crosslinker (Desmodur® N3300) in the presence of dibutyltin dilaurate (DBTDL) led to the HTPB network. The effects of interpenetrating PEO and HTPB networks with various compositions were derived from the tanδ versus temperature curves (Figure 9.7). Two distinct α-transitions corresponding to the HTPB

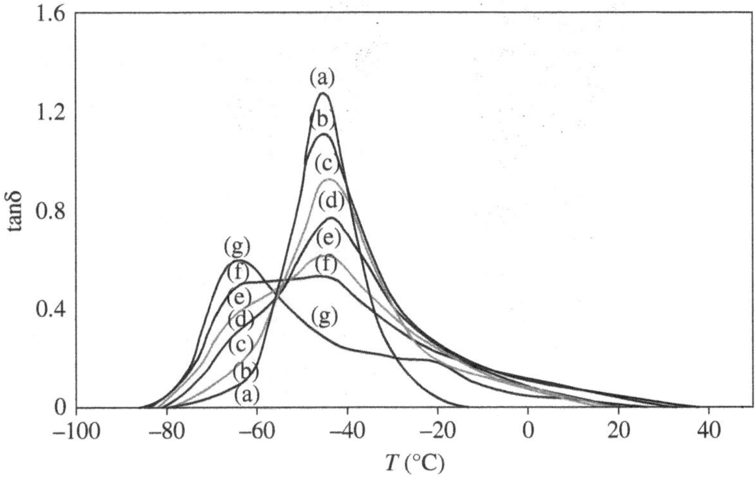

FIGURE 9.7 Tanδ as a function of the temperature for (a) pure PEO network, (g) for pure HTPB network, and for PEO/HTPB (X/100-X) IPNs. X = (b) 80 wt.%, (c) 60 wt.%, (d) 50 wt.%, (e) 40 wt.%, (f) 20 wt.%. DCPD = 2.2 wt.% with respect to PEGM/PEGDM; [DBTDL]/[OH] = 0.06 Source: Plesse *et al.* [36]. Reproduced with permission from Elsevier

and PEO single networks, that is, at −61 and at −44°C, respectively, were observed for compositions where the PEO weight proportion increased from 20 to 50 wt.% in the IPNs.

These results indicated the existence of dual-phase morphology in this composition range of IPNs and suggested that the components were not interpenetrating on a molecular scale rather on a larger scale. Also IPNs with greater proportion of PEO showed a very similar appearance to that of the single PEO network. However, the close approach of α-relaxation temperature of the two networks and the small intensity and broadness of the HTPB α-relaxation was clearly depicted in Figure 9.7. The overlapping tendency of α-relaxation temperature of the two networks outweighed the possibility of intermediate transition corresponding to the interpenetrating phase, if any. On the basis of DMA and TEM results, it was concluded that the HTPB phase was continuous in space for all compositions and that macrophase separation occurred. The PEO phase appeared as nodules in the matrix of IPNs with 80 wt.% PEO content where its domains were in close contact and the material displayed dual-phase continuity (Figure 9.8).

Kinetics of sequential semi-IPN formation based on CPU with different crosslinking densities and linear PS and polybutyl methacrylate had been studied [37, 38]. Polyurethane networks were synthesized, differing in

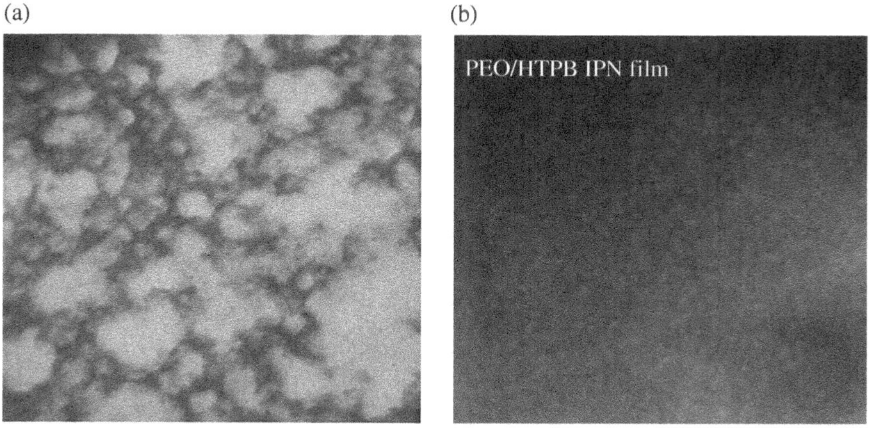

FIGURE 9.8 TEM pictures of PEO/HTPB(X/100-X) IPNs. (a) X = 80 wt.%; (b) X = 50 wt.% at ×50 000 magnification Source: Plesse *et al.* [36]. Reproduced with permission from Elsevier

molecular mass M_c of the chains between crosslink points and the size of cells in the intranetwork space. Styrene and butyl methacrylate monomers were introduced into these networks by swelling them in monomers up to equilibrium. The experimental data showed the dependence of the kinetic parameters of polymerization on M_c and of the chemical nature of the second monomer. Sharp discrepancy in molecular mass distribution of polymers formed in various matrices had been observed. The differences in dependencies of reaction kinetics and molecular mass distribution were proposed to be due to the difference in the chain growth and termination of various monomers on the density of the network, that is, on the confinements imposed by the intranetwork space. DMA of the samples showed increase in glass transition temperature and the elastic modulus in the rubbery state with decreasing molecular mass of polyoxypropylene glycol (POPG) of PU networks. Also it was found that the increase in elastic modulus was in good agreement with the decrease in Flory–Rehner crosslink density. Semi-IPNs of compositions, PU-2000/PBMA 35/65 mass% and PU-1000/PBMA 59/41 mass%, showed two phases, as were indicated by the existence of two relaxation maxima. The level of heterogeneity, as estimated from the degree of segregation (α_{segr}) in the two-phase systems from the parameters of the relaxation maxima, decreased for PU-1000/PBMA 59/41 mass% (α_{segr} = 0.04) compared with the corresponding value for PU-2000/PBMA 35/65 mass% (α_{segr} = 0.22). This showed an enhancement in the compatibility in these sequential PU/PBMA semi-IPNs with increase in crosslink density of PU network. Similar results were found with the PU/PS IPNs. However, a decrease in α value for PU-1000/PS52/48 mass% (α_{segr}=0.2) compared with the corresponding value for PU-2000/PS28/72 mass% (α_{segr}=0.34) was interpreted in terms of the enhancement of the compatibility of components in the system. Also, with decreasing value of M_c of PU network (PU 500), only single broad relaxation maximum was found for semi-IPN PU-500/PS76/24 mass%. Semi-IPNs with PU-500 showed that a single relaxation maximum irrespective of the chemical nature of the second component may be due to the formation of a thermodynamically compatible single-phase system or to a change in its morphology as a result of increased component compatibility and decreased dimensions of the resultant domains.

IPNs of polydimethylsiloxane (PDMS) and poly(2-hydroxyethyl methacrylate) (PHEMA) were prepared by sequential method by F. Abbasi *et al.* [39] to see the property variation in comparison to that of physical blends of the same two polymers. The tensile properties and damping factor (tanδ) were assessed by stress–strain measurement and DMTA, respectively. In comparison with physical blends, the T_g of the IPNs having about 10, 20, 30, and 40 wt.% hydrogel contents were shifted from those for the pure components (T_{gPDMS} = −35°C and T_{gPHEMA} = 136°C) because of an incomplete phase separation of the components and formation of two varying composition regions. IPNs with small hydrogel content showed low and wide single maximum, but IPN with higher hydrogel content showed pronounced maximum in the region of α-transition of PHEMA. The tanδ curve of the IPN with 40 wt.% PHEMA hydrogel showed a second maximum at approximately 50°C (shown in Figure 9.9) due to multiphase structure formation.

An *in situ* sequential IPN system was developed using diglycidyl ether of bisphenol A (DGEBA) cured with 4,4′-methylenebiscyclohexanamine as the crosslinked phase and poly(methacrylated phenyl glycidyl ether) (MPGE) as the linear polymer phase [40]. The percentage of MPGE was varied from 0 to 40 wt.%. The glass transition point was determined via DMA. Storage and loss modulus curves showed single T_g with values ranging from 165 to 119°C and decreased steadily with increasing MPGE content. The T_g of pure MPGE was 58°C. With the exception of the 40 wt.% MPGE formulation, the DMA plots showed sharp transition, indicating homogeneity. The variation of T_g values was well characterized by the Fox equation (r^2 = 0.990). However the broadening of the loss modulus peak for the 40 wt% MPGE formulation suggested that microphase separation might have occurred in this case. Glassy storage modulus increased with MPGE content, suggesting that MPGE had a stiffening effect on the DGEBA–PACM system. Since the repeat unit of MPGE was structurally similar to DGEBA, MPGE might be stacking alongside DGEBA within the crosslinked network.

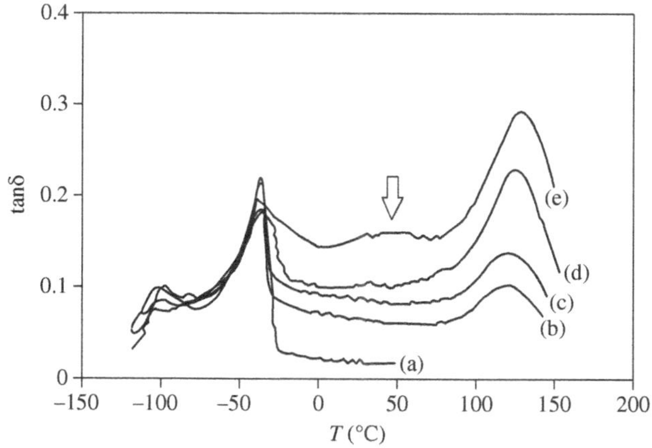

FIGURE 9.9 Bending damping factor, tanδ, versus temperature for PDMS/PHEMA sequential IPNs having PHEMA contents of (a) 0 wt.%, (b) 10.8 wt.%, (c) 19.5 wt.%, (d) 30.4 wt.%, and (e) 38.8 wt.% Source: Abbasi *et al.* [39]. Reproduced with permission from John Wiley & Sons

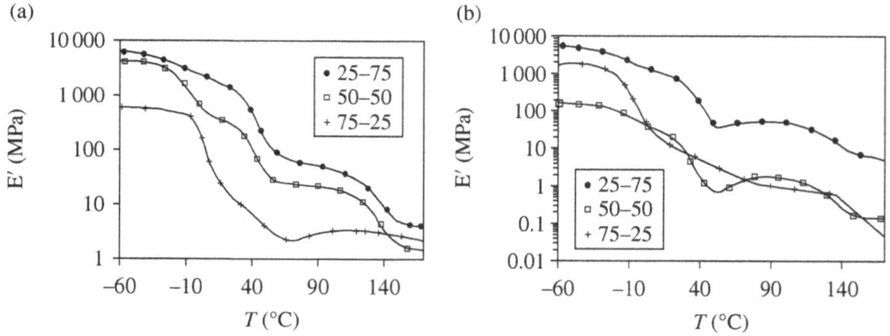

FIGURE 9.10 Storage modulus versus temperature curve for PU/PAA sequential IPN (a) with 3 wt.% CA in PU and (b) with 3 wt.% HEC in PU. Source: Hourston *et al.* [41]. Reproduced with permission from Elsevier

DMA was used to study the viscoelastic behavior of sequential IPNs prepared from castor oil, 2,4-toluene diisocyanate (2,4-TDI), cellulose acetate (CA) and 2-hydroxyethyl cellulose (HEC)-based polyurethane (PU), and polyacrylic acid (PAA) [18]. A series of three PU/PAA ratios (75/25, 50/50, and 25/75) were prepared with two varieties of PU. The storage modulus curve for PU/PAA with 3 wt.% CA showed two-step decreases with increase in temperature indicating the occurrence of two-phase system. With increase in PAA content in the IPN, the rubbery plateau between T_{gPU} and T_{gPAA} increased to a higher modulus. IPNs with CA had higher storage modulus than IPNs with HEC (Figure 9.10) as CA having less bulkier acetyl groups compared to that in HEC possessed a better rearrangement within the PU structures and provided better viscoelastic behavior. The storage modulus decreased because of PU in the amorphous phase, which has a diminishing capacity for energy absorption when is heated. Tanδ curves showed two main peaks, suggesting phase separation in all compositions. A γ relaxation appeared at around 60°C for 50/50 and 25/75 IPN ratios which was attributed to the fact that lateral chains of PU were relaxed and the segmental motions were typical of the glass–rubber transition, and also these chains could be associated to CA structure present in the PU network [41].

DMTA was performed on PVC/PMMA IPNs and blends of corresponding compositions (Goswami, J Macromol Sci, #3090, Communicated, 2013). The storage modulus of blends was found lower than that of IPNs. With increase in PMMA content in the physical blends, the modulus value decreased gradually for them. IPNs showed opposite trend, that is, with increase in PMMA content storage modulus increased. Inclusion of *in situ* formed linear PMMA in the intermolecular space of PVC matrix has caused plasticization effect in the case of blends, whereas in IPNs elastic nature and restricted mobility of crosslinked PMMA entrapped in the PVC matrix have resulted in higher modulus. Damping of IPNs was found to be lower though the glass transition temperature was higher than that of the corresponding blends. Lowering in $T_{\tan}\delta_{max}$ with increase in the PMMA content is evidenced particularly with MMA 73(70 : 30 blends ratio of PVC/PMMA by weight) and MMA64. A single T_g was found in the tanδ curves for all the samples though the Cole–Cole plot [42] of the IPN samples with higher PMMA concentration (>40%) has shown slight deviation from miscibility (Figure 9.11).

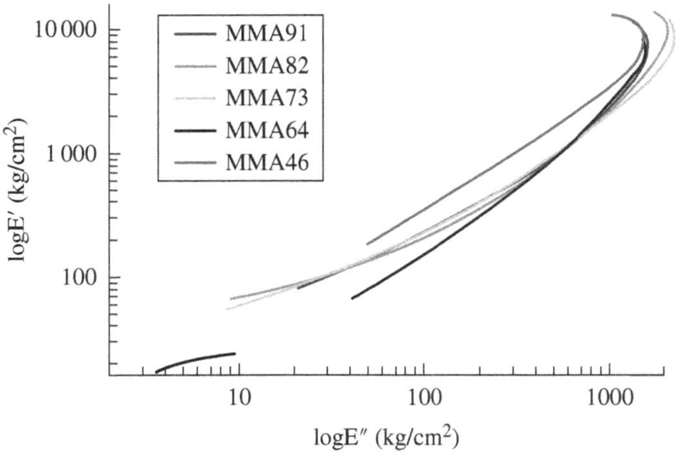

FIGURE 9.11 Cole–Cole plot of different PVC/PMMA IPNs at frequency of 1 Hz

Theoretical analysis of storage modulus is performed considering various composite models such as the parallel model, series model, Davies model, and Coran–Patel model which have been applied to predict the viscoelastic behavior of the polymer mixtures [43, 44]. The upper-bound parallel model is given by

$$P^* = P_1\phi_1 + P_2\phi_2 \tag{9.4}$$

where P is the property of the IPN, P_1 and P_2 are the corresponding properties of components 1 and 2 respectively, and ϕ_1 and ϕ_2 are the volume fractions of components 1 and 2. In a parallel model the components are arranged parallel to one another so that an applied stress elongates each component by the same extent. In case of lower-bound series model, the components are arranged in series with applied stress, and the equation is given by

$$\frac{1}{P} = \frac{\phi_2}{P_2} + \frac{\phi_2}{P_2} \tag{9.5}$$

Another important model, given by Davies *et al.* which [45, 46] accounts for the phase interaction, is

$$P^{1/5} = \phi_1 P_1^{1/5} + \phi_2 P_2^{1/5} \tag{9.6}$$

Finally, the Coran–Patel model [47, 48], which represents a phenomenological adjustment between the parallel and series models and characterized as a one parameter model, is

$$P = f\left(P_u - P_l\right) + P_l \tag{9.7}$$

where $0<f<1$ and its value is given by

$$f = \phi_H^n\left(n\phi_s + 1\right) \tag{9.8}$$

Here n contains the aspects of phase morphology, and Φ_H and φ_s are the volume fractions of the hard and soft phases, respectively. Using $n=1, 2, 3, 4$, etc. arbitrarily for different blend compositions (in terms of % PMMA in the present case), values of P can be determined from Equation 9.7. The storage modulus of PVC/PMMA IPNs at 35°C measured at frequency 1 Hz is plotted against the blend composition in Figure 9.12. Due to very less difference in storage modulus of the pure components (PVC = 14472 and PMMA = 13000 kg/cm^2), almost similar type of prediction about the experimental values was exhibited by different models. Experimental storage modulus values are laid below the predicted values obtained using different models. This may be due to the fact that these equations do not consider the

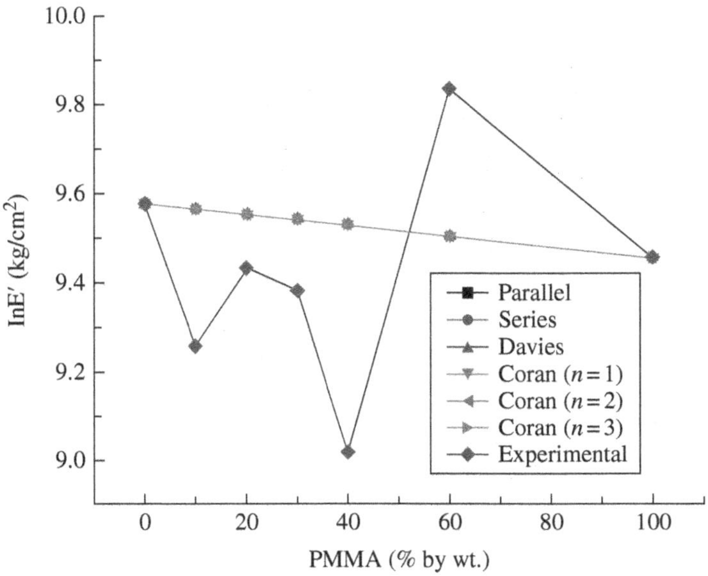

FIGURE 9.12 Plot of theoretical and experimental values of storage modulus against PMMA content in the IPNs measured at 35°C and 1 Hz frequency

specific interaction between the components involved during the IPN formation in contrast to physical blends. In general low value of storage modulus is caused by greater molecular mobility in the low-temperature region (especially below glass transition). Also for the IPNs in this study, glassy to rubbery transition has occurred at a much lower temperature than that predicted from theoretical equations. This shows that plasticization effect is more pronounced due to incorporation of PMMA in the PVC matrix during *in situ* sequential IPN synthesis. Again for MMA46 IPN higher storage modulus is exhibited than theoretical prediction. This may be explained on the basis of higher chain rigidity of crosslinked PMMA and formation of microheterogeneity in the medium.

From the time–temperature superposition (TTS) master curves (Figure 9.13), it was clear that at very long time interval (low frequencies), the IPNs act like elastic solid, whereas at short times they behave like low-modulus rubbers. Microheterogeneous morphology of the IPNs was visible in the scanning electron micrographs.

Physical blends showed bigger size irregular-shaped domains dispersed in the matrix compared to that in the case of IPNs. Also significant variation in the size of the dispersed domains was observed in IPNs with increase in % PMMA, compared to that in the case of blends (Figure 9.14).

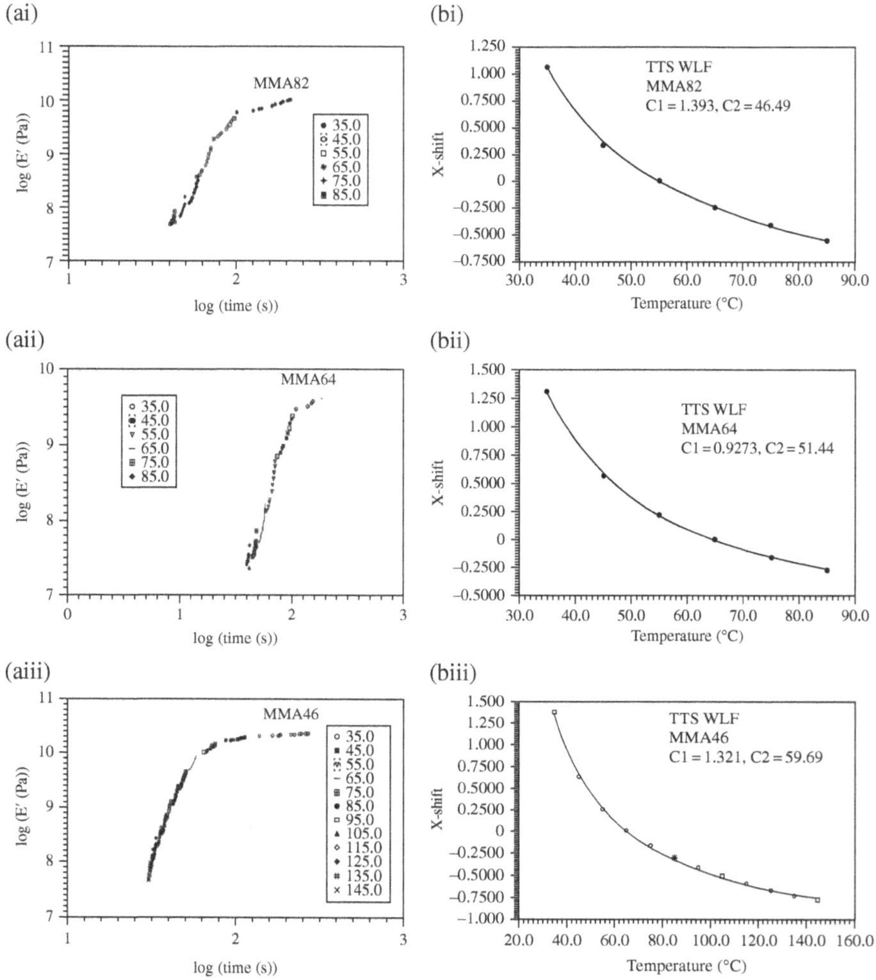

FIGURE 9.13 (a) Master curve of storage modulus for IPNs: (i) MMA82, (ii) MMA64, and (iii) MMA46. (b) Time–temperature shift factor obtained from master curves of storage modulus for IPNs: (i) MMA82, (ii) MMA64, and (iii) MMA46

9.4 OVERALL SUMMARY AND FUTURE SCOPE

DMA of various IPNs involving both thermoplastic and thermosetting polymers has been discussed here considering various aspects. Primarily, two-phase morphology of simultaneous and sequential IPNs can be detected from the two relaxation peaks in tanδ curves corresponding to the relaxations for the individual components or from the two-step decrease in storage modulus related to the extent of flexibility of the different chemical units of the components. Formation

(a)

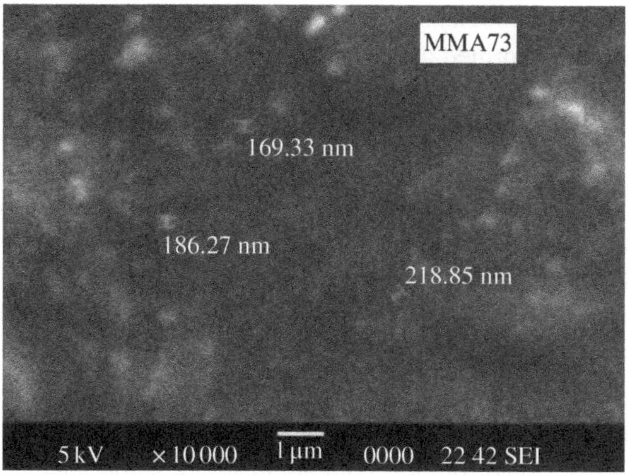

(b)

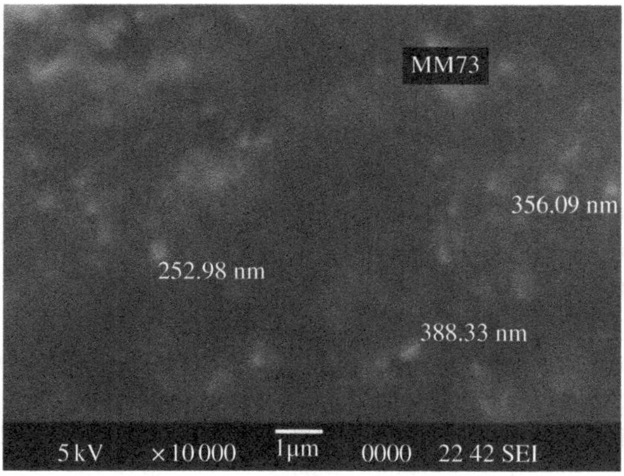

FIGURE 9.14 Scanning electron micrograph of: (a) PVC/PMMA IPNs (MMA73) and (b) blends (MM73) of similar composition

of multiphase structure is closely related to the reaction kinetics, the type of the network formed first, the intranetwork space available for the second monomer polymerization, and subsequent formation of the topological entangled network. Variation in the degree of crosslinking in the components may result in the difference in the peak width of tanδ curves (FWHM) as well as in the storage modulus (E′) values. The observed fact is that the higher the extent of crosslinking (low M_c), the more the broadening in $\tan\delta_{max}$ peak. Quantitative evaluation of overall crosslink density of IPNs is possible from storage modulus data which can be a valuable tool for the prediction of phase morphology.

Therefore variation of the blend ratio, choice of crosslinker system or dosage of polymerization initiator/catalyst, and structural modification of component polymers can be performed accordingly to have the optimum properties even when the morphology determination is not possible due to the unavailability of suitable infrastructure. DMA of IPNs showed an increase in glass transition temperature and storage modulus in the rubbery state with decreasing M_c. The extent of peak height in tanδ curve gives information about the heterogeneity and plasticization effect imposed by the components in IPNs. In the case of sequential full-IPN made of PEO and HTPB, existence and variation of two transition peaks subjected to the composition change give an interesting information about the overlapping tendency of α-relaxation temperature of the two networks which has outweighed the possibility of intermediate transition corresponding to the interpenetrating phase or interface. Generally IFR formation by inclusion of suitable compatibilizer into physical blends is seemed to be beneficial as it may result in gradual transfer of stress from one phase to another, thereby reducing the chance of sudden break in the case of two incompatible polymers. The extent of IFR was varied in the case of PU/PBMA and PU/PS sequential IPNs by simply varying the blend ratio and crosslink density in the PU phase. At a low crosslink density of the PU matrix, its intranetwork volume was sufficient for the phase separation to occur during monomer polymerization at a certain degree of segregation of the components. The higher the crosslink density of the first network, the more is the phase segregation. In some cases the confined network space availability for the growth of the second component may alter the reaction rate and cause very small size of the domains, leading to microphase separation which is not detectable through DMA. In the case of simultaneous IPNs generally phase separation is kinetically controlled, and bigger domains of phase-separated polymers may be formed due to lack of topological restriction. The extent of phase separation in IPNs is further correlated to the extent of segregation (α_{segr}) of the components and the IFR on the basis of viscoelastic properties. The former is applied to both simultaneous and sequential IPNs, while the latter is applied to simultaneous type more often. DMA data can be used to estimate the relative contents of the two phases and fraction of IFR as areas under the corresponding Gaussian maxima in tanδ curves. Use of compatibilizer is effective in reducing the value of degree of segregation as is found with PU/PS simultaneous semi-IPNs. Area under the tanδ curve corresponding to the IFR decreases due to more intimate mixing of the two components through the barrier. This is possible in the case of simultaneous IPNs when the rate of formation of the IPNs (kinetic factor) is outweighed by the *in situ* compatibilization, for example, through complexation in the presence of metal chelates in the case of PU/PMMA IPNs and through copolymerization by the use of comonomer crosslinker for any of the components. The damping properties of the IPNs may be altered by further use of certain solvents in which these materials swell but do not dissolve. Relaxation of swollen samples occurs at a lower temperature than that of the unswollen one due to plasticization effect. Moreover, use of third component (PEDOT) during the IPN formation with PEO/PB is cited here, giving a tailor-made conducting polymer which is sufficiently flexible and can be used as low-temperature actuator. Hence IPNs, formed from various polymers having different types of structural alternatives depending on the blend ratio, crosslinker/catalyst concentration, rate/order of reaction, and compatibilizer nature, can be studied by DMA to predict their properties such as phase morphology, glass transition behavior, damping property, stability, etc. Extensive interpenetration of the components was confirmed from broadening of tanδ curves for the IPNs based on PU/allyl novolac resin in contrast to that of the pure PU.

However, very less attention is paid upon the study of behavior of these materials during long-term use which can also be determined through viscoelastic property analysis from TTS theory as is reported for numerous physical blends of polymers. Similarly the correlation of experimental theoretical data from DMA with various theoretical models known for multiphase polymers has not been attempted. The homogeneity of IPNs can be further checked through Cole–Cole analysis which is usually done for polymer blends. Utilization of viscoelastic properties of sequential and simultaneous IPNs based on thermoplastic polymer components in terms of TTS can be further extended with different other systems apart from PVC/PMMA. Systems composed of components differing widely in their mechanical properties but prescribed for long-term use like actuators, tissue engineering, etc. when combined together may be tried for IPN preparation in the future. Full-IPNs and semi-IPNs can be compared in a separate study in respect to their viscoelastic behavior.

9.5 CONCLUSION

DMA of polymer blends with various thermoplastic and thermoset combinations is reported by several groups of researchers which have given us a broad concept of utilization of viscoelastic properties of IPNs in practical use. The main aspects of IPNs' viscoelastic properties and outcome of the study are listed below:

- DMA gives information about storage modulus (E′), damping properties, and loss modulus of the polymer system.

- Multiphase morphology of the IPNs both for sequential and simultaneous type could be predicted from the DMA data, for example, two tanδ peaks corresponding to the relaxation of the molecules of components present in binary IPNs or two-step decrease in storage modulus with variation in temperature corresponding to the component polymers.
- Width of the $\tan\delta_{max}$ curve varies with the extent of mixing or phase segregation. The broader the peak, the more heterogeneous is the polymer network with a wide distribution of relaxation times. Also homogeneity of the system could be assessed from the shape of the Cole–Cole plot drawn by using the data from DMA. As a rule of thumb, the appearance of semicircular curve when storage modulus is plotted against loss modulus for the system gives indication of homogeneity of the IPNs. Therefore, homogeneity of the PVC/PMMA IPNs may be considered to be confirmed from the simultaneous appearance of single peak on tanδ curve and semicircular nature of Cole–Cole plot in the present work.
- DMA also helps to find the quantitative values of degree of segregation (α_{segr}); therefore optimization of the composition of IPNs can be performed from the storage modulus data. A polymer system is considered to be perfectly homogeneous when $\alpha_{segr}=0$.
- In the present study comparison between simultaneous and sequential IPNs has been shown in terms of the area under IFR (S_m). The greater the IFR, the more is the multiphase separation.
- With increase in crosslinking of the different components in IPNs, T_g, as determined from the temperature corresponding to $\tan\delta_{max}$ for them, increases, but the height of tanδ curves decreases, that is, the material becomes less damping.
- Extensive interpenetration of the components in simultaneous IPNs may be confirmed from the tanδ peak broadening for the IPNs in contrast to that of the pure component.

REFERENCES

[1] Babkina, N.V., Lipatov, Y.S., Alekseeva, T.T. *et al.* (2008) *Polymer Science, Series A*, **50** (7), 798–807.

[2] Lipatov, Y.S. and Alekseeva, T.T. (2007) Chapter 2, Thermodynamics and Phase Separation in IPNs, in Phase Separated *Interpenetrating Polymer Networks*, Springer-Veralg, Heidelberg/New York, p. 15.

[3] Sperling, L.H. (ed) (1981) Interpenetrating polymer networks and related materials, in *Phase Separation and Mechanical Behavior of Multicomponent Polymer Systems*, Plenum, New York, p. 20.

[4] Olabisi, O., Robeson, L.M. and Shaw, M.T. (1979) Properties of miscible polymer systems, in *Polymer-Polymer Miscibility*, Academic Press, New York, p. 295.

[5] Lipatov, Y.S. and Nesterov, A.E. (1997) Thermodynamics of polymer mixing, in *Thermodynamics of Polymer Blends,* vol. 1, Technomic Publishing, Lancaster, PA, p. 81.

[6] Lipatov, Y.S. and Alekseeva, T.T. (eds) (2007) Chapter 3, Heterogeneous Structure and Morphology of IPNs *in Phase Separated Interpenetrating Polymer Networks*, Springer-Veralg, Heidelberg/New York, p. 55.

[7] Gomza, Y.S., Klepko, V.V., Lipatov, Y.S. *et al.* (2008) *Polymer Science, Series A*, **50** (9), 956–964.

[8] Han, X. (1997) Interpenetrating polymer networks based on polyurethane, in *IPNs Around the World Science and Engineering* (eds S.C. Kim and L.H. Sperling), John Wiley, Chichester/New York, p. 75.

[9] Lipatov, Y.S., Alekseeva, T.T., Rosovitsky, V.F. and Babkina, N.V. (1992) *Polymer*, **33**, 610.

[10] Alekseeva, T.T., Grishchuk, S.I., Lipatov, Y.S. *et al.* (2003) *Polymer Science, Series A*, **45**, 721.

[11] Lipatov, Y.S. (2007) *Polymer Bulletin (Berlin)*, **58**, 105.

[12] Sperling, L.H. (1994) PNs around the World. II: Recent advances, in *Advances in Interpenetrating Polymer Networks,* vol. 4 (eds D. Klempner and K. Frisch), Technomic Publishing, Lancaster, PA, p. 1.

[13] Lipatov, Y.S. (1989) Microphase separation in interpenetrating polymer networks and their viscoelasticity, in *Advances in Interpenetrating Polymer Networks,* vol. 1 (eds D. Klempner and K. Frisch), Technomic Publishing, Lancaster, PA, p. 261.

[14] Lipatov, Y.S. and Shilov, V.V. (1984) *Uspekhi Khimii*, **53**, 1197.

[15] Lipatov, Y.S., Alekseeva, T.T., Kosyanchuk, L.F. *et al.* (1999) *Polymer*, **40**, 7083.

[16] Babkina, N.V., Lipatov, Y.S. and Alekseeva, T.T. (2005) *Polymer Science, Series A*, **47**, 1271.

[17] Donatelli, A.A., Sperling, L.H. and Thomas, D.A. (1977) *Journal of Applied Polymer Science*, **21**, 1189.

[18] Jimenez, M.A.G., Armenta, J.L.R., Martinez, A.M.M. *et al.* (2009) *Latin American Applied Research*, **39**, 131.

[19] Tirrell, M. (1996) *Linear Viscoelasticity at the 21st Short Course on Rheometry*, University of Minnesota, Minneapolis.

[20] McCrum, N.G., Read, B.E. and Williams, G. (1967) Dielectric relaxation of polyvinyl alcohol, in *Anelastic and Dielectric Effects in Polymeric Solids*, Dover, New York, p. 350.

[21] Dealey, J.M. and Wissbrum, K.F. (1990) Rheology of multiphase systems, in *Melt Rheology and its Role in Plastics Processing*, Van Nostrand, New York, p. 390.

[22] Menard, K.P. (2000) *Dynamic Mechanical Analysis*, CRC Press, Boca Raton, FL, p. 19.

[23] Worzakowska, M. (2009) *Journal of Materials Science*, **44**, 4059.

[24] Babkina, N.V., Lipatov, Y.S., Alekseeva, T.T. and Yarovaya, N.V. (2009) *Polymer Science, Series A*, **51** (8), 904.

[25] Song, M., Hourston, D.J., Pollock, H.M. *et al.* (1997) *Thermochimica Acta*, **304/305**, 335.

[26] Kosyanchuk, L.F., Babkina, N.V., Yarovaya, N.V. *et al.* (2008) *Polymer Science, Series A*, **50** (4), 434.

[27] Alekseeva, T.T., Lipatov, Y.S., Babkina, N.V. *et al.* (2005) *Polymer*, **46**, 419.

[28] Lipatov, Y.S., Rosovitsky, V.F., Datsko, P.V. and Maslak, Y.V. (1998) *Journal of Applied Polymer Science*, **36**, 1143.

[29] Chaing, W.-Y. and Chang, D.-M. (1995) *European Polymer Journal*, **31** (8), 709.

[30] Qin, C.L., Cai, W., Cai, J. *et al.* (2004) *Materials Chemistry and Physics*, **85**, 402–409.

[31] Tsai, M., Huang, S.L., Liu, S.J. *et al.* (2008) *Desalination*, **233**, 191.

[32] Alekseeva, T.T., Lipatov, Y.S., Babkina, N.V. *et al.* (2008) *Molecular Crystals and Liquid Crystals*, **483**, 191.

[33] Lipatov, Y.S., Alekseeva, T.T., Sorochinskaya, L.A. and Dudarenko, G.V. (2008) *Polymer Bulletin*, **59**, 739.

[34] Brovko, O.O., Sergeeva, L.M., Kuznetsova, V.P. and Lemeshko, V.N. (1998) *European Polymer Journal*, **35**, 2045.

[35] Plesse, C., Vidal, F., Randriamahazaka, H. *et al.* (2005) *Polymer*, **46**, 7771.

[36] Plesse, C., Vidal, F., Gauthier, C. *et al.* (2007) *Polymer*, **48**, 696.

[37] Alekseeva, T.T., Sorochinskaya, L.A., Dudarenko, G.V. and Lipatov, Y.S. (2009) *Polymer Science, Series B*, **51**, 239.

[38] Lipatov, Y.S. (1985) *Pure and Applied Chemistry*, **57** (11), 1691.

[39] Abbasi, F., Mirzadeh, H. and Katbab, A.A. (2002) *Journal of Applied Polymer Science*, **86**, 3480.

[40] Peterson, A.M., Kotthapalli, H., Rahmathullah, M.A.M. and Palmese, G.R. (2012) *Composites Science and Technology*, **72**, 330.

[41] Hourston, D.J., Song, M., Schafer, F.U. *et al.* (1998) *Thermochimica Acta*, **324**, 109–121.

[42] Cole, K.S. and Cole, R.H. (1941) *The Journal of Chemical Physics*, **341**, 9.

[43] Ashaletha, R., Kumaran, M.G. and Thomas, S. (1998) *Polymer Degradation and Stability*, **61**, 431.

[44] Komalam, C., George, K.E., Kumar, P.A.S. *et al.* (2007) *eXPRESS Polymer Letters*, **1** (10), 641.

[45] Ahmad, Z., Al-Awadi, N.A. and Al-Sagheer, F. (2007) *Polymer Degradation and Stability*, **92**, 1025.

[46] George, S., Neelakanthan, N.R., Varughese, K.T. and Thomas, S. (1997) *Journal of Polymer Science Part B: Polymer Physics*, **35**, 2309.

[47] Coran, A.Y. and Patel, R.P. (1976) *Journal of Applied Polymer Science*, **20**, 3005.

[48] Coran, A.Y. and Patel, R.P. (1983) *Rubber Chemistry and Technology*, **56**, 210.

10

INTERPENETRATING AND SEMI-INTERPENETRATING POLYMER NETWORKS OF POLYURETHANE

CHEPURI R.K. RAO, RAMANUJ NARAYAN AND K.V.S.N. RAJU
Polymers and Functional Materials Division, CSIR-Indian Institute of Chemical Technology (IICT), Hyderabad, Telangana, India

10.1 INTRODUCTION

The chemical and physical combination of two or more structurally dissimilar polymers provides a convenient route for the modification of properties to meet specific needs. It facilitates easy processing and, in many cases, may impart flexibility, tensile and impact strength, chemical resistance, and weatherability to the new material [1]. In addition to mechanical blending and copolymerization, IPNs represent a third mechanism by which different polymers can be physically combined.

Normal blending or mixing of two polymers results in a multiphase morphology due to thermodynamic incompatibility which arises from the relatively small gain in entropy of mixing due to contiguity restrictions imposed by their large chain length [2]. On the other hand, if mixing takes place on a lower molecular weight level and polymerization is carried out simultaneously with crosslinking, phase separation may be kinetically

Micro- and Nano-structured Interpenetrating Polymer Networks: From Design to Applications, First Edition. Edited by Sabu Thomas, Daniel Grande, Uroš Cvelbar, K.V.S.N. Raju, Ramanuj Narayan, Selvin P. Thomas and Akhina H.

controlled as the entanglements are made permanent by crosslinking. Thus interpenetrating polymerization is a mode of blending two or more polymers to produce a mixture in which phase separation is very low which is otherwise predominant.

Primarily, phase separation depends on the compatibility of the two polymers, one into another. With highly incompatible polymers, the thermodynamics of phase separation is so powerful that it occurs before crosslinking. With compatible polymers, phase separation can be completely avoided.

The advantage cropped up in strength of IPNs compared with blends is understandable by looking at the mechanical strengths of pure polyurethane (PU), pure polyacrylate (PA), and their blends [3] as shown in Chart 10.1. It clearly shows that IPNs exhibit excellent improved mechanical properties compared with blends.

John Millar was the first to coin the term "interpenetrating polymer network" (IPN) as early as in 1960 [4]. According to Sperling and Mishra [5], interpenetrating polymer networks are a combination of two or more polymers in network form, with at least one polymer polymerized and/or crosslinked in the immediate presence of others. In other words [6], these are unique "alloys" of crosslinked polymers. IPNs are special example of topological isomerism in macromolecules [1]. When broadly classified, these polymer blends diverge into six polymer combinations, namely, polymer blend, grafted copolymer,

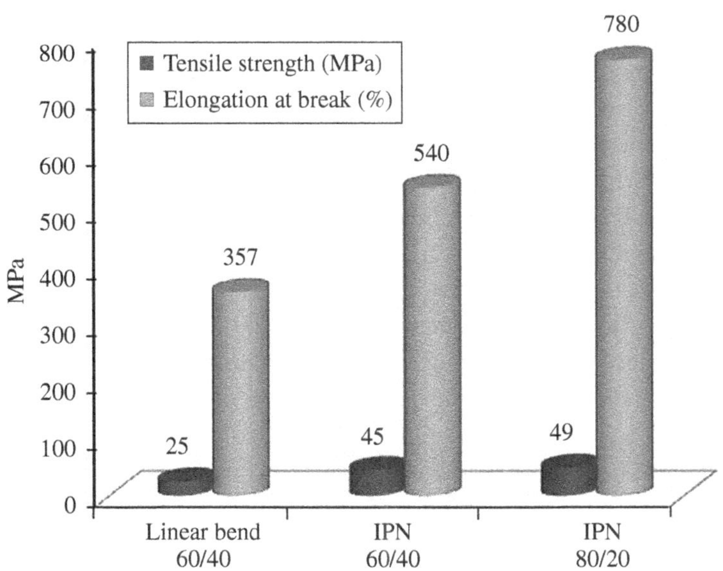

CHART 10.1 Comparison of improved properties for polyurethane-polyacrylate blend and IPNs [3]

block copolymer, AB graft copolymer, semi-IPN, and IPN. In a polymer blend (Scheme 10.1a), the components show no interaction among each other. In graft copolymers (Scheme 10.1b), the end of the side chain is grafted to some part of the backbone chain or at the end. Scheme 10.1b is a single network composed of two kinds of chains known as AB crosslinked copolymer. Here, one of the kinds of chains is bonded to the other at both ends. Scheme 10.1c and d are semi-IPN and IPN, respectively. Systems (a) and (b) of Scheme 10.1 show thermoplastic nature while (c) and (d) are thermosetting in nature. Thermoplastic IPN systems utilize physical crosslinks rather than chemical crosslinks. They may flow at elevated temperatures. In contrast, thermoset IPNs cannot flow without breaking the primary chemical bond.

Several factors such as chemical components of the network, degree of interpenetration of components, and interaction of various functional groups influence the morphology, shape, size, and phase separation behavior of the formed IPNs.

Further, the IPNs are classified as (i) simultaneous IPN, (ii) sequential IPN, (iii) latex IPN, (iv) gradient IPN, (v) thermoplastic IPN, and (vi) semi-IPN.

Simultaneous IPN: In this type of network formation, the monomers I and II along with crosslinkers and activators are mixed and polymerized together. Generally, the two polymerization reactions are noninterfering.

Sequential IPN: In this type of network, the monomer II along with a crosslinker is swollen in already synthesized polymer network I and polymerized.

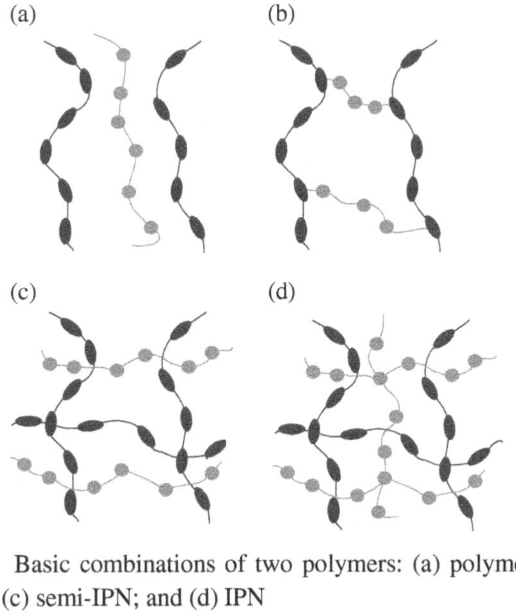

SCHEME 10.1 Basic combinations of two polymers: (a) polymer blend; (b) AB graft copolymer; (c) semi-IPN; and (d) IPN

Latex IPN: These IPNs are in the form of latex, with each particle behaving as micro-IPN. In this type of system, a core–shell structure is often developed and further crosslinking between the particles takes place when film is formed. Thus both networks appear on a single latex which are synthesized by sequential polymerization.

Gradient IPN: In this type of network, crosslinking density is not uniform but varies from location to location.

Thermoplastic IPN: These are considered as materials between a true polymeric blend and an IPN. Hence these materials flow at higher temperatures similar to thermoplastics and remain crosslinked at lower temperature like pure IPN.

Semi-IPN: A network in which one or more polymers are crosslinked and one or more polymers are linear or branched.

Interpenetrating polymer networks (IPNs) provide the possibility of effectively producing advanced multicomponent polymeric systems with new property profiles [7–9]. An IPN has different properties compared with either a copolymer or a polymer blend. It may swell but may not dissolve in solvents. IPNs resist creep or flow to a greater extent than copolymers or blends. IPNs exhibit better physical properties such as tensile strength/elongation compared with polymer blends [1]. The major reason for the differences in properties between polymer blends and IPNs is that the latter have greater adhesion and better mixing.

10.1.1 Polyurethane–Acrylic, Epoxy, and Polyester IPN Systems

Interpenetrating polymer networks (IPNs) of glycerol-modified castor oil polyurethane (GC-PU) and poly(2-hydroxyethylmethacrylate) (PHEMA) were synthesized using *N,N*-methylene-bis(acrylamide) as crosslinker [10]. GC-PU/PHEMA interpenetrating polymer networks were obtained by transfer molding. These were characterized with respect to their resistance to chemical reagents and mechanical properties such as tensile strength, percent elongation, and shore A hardness. These IPNs showed excellent chemical resistance, hardness, elongation, tensile strength properties, and higher compatibility.

Latex IPNs between waterborne PU and poly(acrylic rubber) (PAR) were prepared by sequential polymerization with methyl acrylate, ethyl acrylate, or butyl acrylate [11]. A high degree of interpenetrations was established. Incorporation with acrylate polymers having low glass transition temperatures effectively augmented the hydrophobicity without sacrificing the elasticity of the original PUs. In this investigation, the acrylate monomers were polymerized by UV irradiation to reduce the reaction time.

New kinds of novel PU–PA latex interpenetrating polymer networks (LIPNs) were synthesized [12]. For this purpose, PU dispersions were synthesized first. Using these as seed, PU–PA LIPNs were prepared by the soap-free emulsion polymerization method. It was found that these LIPNs could markedly improve the water resistance and the mechanical properties of PU latex much more than those of PU–PA physical blends due to interpenetrating and entangling between PU and PA latices. The higher the NCO/OH molar ratio in PU dispersions, the larger is the particle size of PU–PA LIPNs, and the average particle size of PU–PA LIPNs becomes larger with an increase in PA content. In another study on PU and PA LIPNs [13], it was observed that the glass transition temperature (T_g) responded to the compatibility between two polymers. When PA content is less than 50%, the LIPN showed a single T_g and this shifted to higher temperature, which indicates higher degree of compatibility between hybrid emulsions. This is caused by high interpenetrating and entangling between the two polymer networks. As a result, the mechanical properties of LIPNs are increased significantly compared to those of physical blends. It was observed that the LIPNs improved water resistance of PU dispersion emulsion.

A study [14] was undertaken on the morphology and mechanical properties of semi-IPNs formed from PU(ester)/methacrylates in different mass ratios. The ester-urethane (PU) prepolymers contained carboxylic groups and methacrylic (PM) prepolymers contained tertiary amine functional groups. The scanning electron micrographs, shifts in glass transition temperature, and mechanical properties showed that their miscibility is enhanced by the interaction of functional groups. The SIPN with functional groups having a PU/PM mass ratio of 0.5/0.5 exhibited the most intense functional group interaction which reflected in enhanced miscibility. Owing to the same reason, Young's modulus and tensile strength of SIPN films were significantly influenced.

Sequential interpenetrating polymer networks (IPNs) of PU from a canola oil-based polyol and poly(methyl methacrylate) (PMMA) were studied [15, 16]. It was observed that the polymerization process of canola oil-based IPNs was accelerated because of the utilization of polyol with primary functional groups, which efficiently lessened the effect of dangling chains and yielded a higher degree of phase mixing. The mechanical properties of canola oil-based IPNs with greater than 75 wt% PMMA were comparable to the corresponding castor oil-based IPNs and were superior to those of the constituent polymers. This is attributed to the finely divided rubber and plastic combination structures of IPNs. At PMMA concentration of less than 65 wt%, canola oil-based IPNs exhibited a mechanical behavior typical of a rigid plastic, whereas castor oil-based IPNs showed a typical mechanical behavior of soft rubber. It is proposed that these new IPN materials with high performance prepared from alternative renewable resources can prove to be valuable substitutes for existing materials in various applications. In another investigation by the same authors [16], a five-phase morphology, that is, sol phase, PUR-rich phase, PUR-rich interphase, PMMA-rich interphase, and PMMA-rich phase, was observed for all the IPNs. Higher degrees of mixing occurred for the IPN with higher NCO/OH molar ratio (2.0/1.0) at PUR concentration of 25 wt%, whereas for the IPN with lower NCO/OH molar ratio (1.6/1.0), higher degrees of mixing occurred at PUR concentration of 35 wt%.

Incorporation of PU into polyacrylamide network in the form of interpenetrating polymer networks may influence to enhance the mechanical and thermal properties of the semi-IPNs due to higher crosslink density imparted by the hard segment content. In a study [17] interpenetrating polymer networks prepared from acrylamide and different NCO-terminated polyurethane prepolymer (PTMEG-PP) showed better thermal stability, compatibility, tensile, and elongation properties compared to the individual polyacrylamide network. This semi-IPN showed low T_g values demonstrating their flexibility and soft characteristics compared to pure polyacrylamide network. Morphological observations through SEM revealed the compatibility of the interpenetrating polymer networks. The semi-IPNs are potential materials for biomedical applications as these were hydrolytically stable in phosphate buffer solution.

Interpenetrated networked adhesives based on PU(urea)/PA were synthesized and evaluated for low energy surfaces [18]. The adhesive morphology took on the IPNs that manifested as a finely dispersed PU(urea) phase in the acrylate phase. Excellent adhesion to low surface energy materials was achieved within a wide range of PU(urea) contents. The IPN adhesives also displayed better flexibility than PA adhesives with HEA-PU as a crosslinking agent.

Two kinds of interpenetrating polymer networks (IPNs) (consisting of two-component PU and vinyl(methacrylic) polymer (PV)), (polyether–castor oil) PU/PV IPN(I) and (polybutadiene–castor oil)PU/PV IPN(II), were synthesized [19]. The former IPN was prepared by polymerization of castor oil, NCO-terminated polyether, and vinyl or methacrylic monomer together, and the latter IPN was obtained by polymerization of castor oil, NCO-terminated polybutadiene, NCO-terminated castor oil, and vinyl or methacrylic monomer together. It is understood that acrylonitrile (AN) is a good monomer for synthesizing IPN(I) but is a poor monomer for preparing IPN(II). At optimum conditions for the synthesis, both the (polyether–castor oil)PU/polyacrylonitrile (PAN) IPNs and the (polybutadiene–castor oil)

PU/polystyrene (PSt) IPNs behave as elastomers with ultimate elongation over 240 and 270%, respectively. TEM micrograph of a (polybutadiene–castor oil)PU/PSt IPN showed a microphase separation in the IPN.

Bismaleimide-modified PU–epoxy IPNs have been studied [20]. The study showed that incorporation of PU into epoxy skeleton led to an increase in mechanical strength and a decrease in T_g. Incorporation of bismaleimide having ester linkages into PU-modified epoxy system increased the thermal stability, tensile, and flexural properties. However, it lead to decrease in the impact strength and glass transition temperature.

Thermoset-type IPNs based on PU and polyester (of phthalic anhydride, maleic anhydride, and propylene glycol) were synthesized and studied [21]. The IPNs are completely miscible before reacting. Once formed, they possess a high degree of conversion which means high chemical stability and overall thermoset behavior. It has been proved by FTIR that the synthesized IPNs are simultaneous and the curing kinetics of PU and polyester are modified during the formation of IPNs.

The kinetics of swelling and the sorption (of chlorobenzene) performance were studied for interpenetrating polymer networks made up of PU and unsaturated polyester [22]. It was found that increasing the amount of PU in the composition reduced crosslinking density in the polyester network that resulted in faster diffusion of the solvent and higher sorption capacity for the solvent. It was also found that the higher the styrene content in the composition, the diffusion of solvent and its sorption inside the polymer network is lowered. Electron microscope analysis on swelled samples showed that no leaching was observed for any phase present in the system, despite phase separation for both the components.

Interpenetrating polymer networks of urethane-modified bismaleimide (UBMI) and the diglycidyl ether of bisphenol A (Ep) (UBMI/Ep graft IPNs) were prepared and studied [23]. The UBMI is introduced and partially grafted to the epoxy by PU graft agents, and then simultaneous bulk polymerization technique is used to prepare the graft IPNs. The morphology of this IPN with PU(PBA) graft agent exhibited a homogeneous one-phase system, while the UBMI/Ep graft IPN with PU(PPG) graft agent showed a two-phase morphology with the UBMI particles dispersed in the epoxy matrix.

10.1.2 PU/Other Polymers

A rigid network such as PSt is useful for chemically linking soft PU networks to get dimensionally stable IPNs [24]. Macromolecular monoliths were synthesized from concentrated emulsions. Matrices with only a PU network were too soft, and it was necessary to add to the formulation a rigid network such as PSt to obtain a material with a good dimensional stability. Either unconnected or interconnected IPNs were prepared. The later IPNs were obtained by using hydroxybutyl methacrylate as a comonomer that chemically links both networks.

The effects of compatibilizing additive (oligo-urethane dimethacrylate (OUDM)) on the kinetics of IPN formation between crosslinked PU and linear PSt and its influence on the microphase separation, viscoelastic, and thermophysical properties have been investigated [25]. It was established that 10, 20 mass% of OUDM introduced into the initial reaction system prevent microphase separation of the system and lead to the formation of compatible IPNs.

Interpenetrating polymer systems based on crosslinked PU and PSt were prepared [26] by a one-shot (*in situ*) method, starting from an initial homogeneous mixture of reagents via noninterfering mechanisms. Despite the difference in refractive index of the components, hazy or optically clear films were obtained, thus indicating various levels of phase separation, also confirmed by glass transition temperature (T_g) measurements. The results suggest that controlling the chemistry and process (crosslink density, composition, and time sequence of events) of *in situ* IPN formation will give various morphologies, and hence properties, ranging from microphase separated materials to larger macrophase separated materials.

In a novel study, a fullerene-based PU star-shaped polymer containing six PU polymer arms was prepared [27]. The polymer is more flexible than that of linear PU. When novolac-type phenolic resin is blended with C60-PU, the intramolecular hydrogen bonding of phenolic resin was destroyed and converted to intermolecular hydrogen bonding between phenolic resin and C60-PU. The crystallinity of C60-PU was destroyed and phenolic resin/C60-PU was amorphous. The phenolic resin blended with different PUs possesses the structure of an IPN. The flexible chain of PU in phenolic resin may improve the toughness of phenolic resin because the flexible chain of PU is able to absorb the energy applied on the phenolic resin instantly.

The simultaneous interpenetrating polymer networks (SINs) of graft vinyl ester resin (VER) and PU were prepared [28]. The grafted VER contained butanol or poly(oxypropylene) side chain. When AIBN was used as initiator for PU/graft VER networks, the morphology and structure of these SINs are mainly controlled by the PU network formed first. The structure of graft VER has almost no influence on the long-range order structure of hard segments. On the other hand, when BPO/DMA was used as an initiator, the side-chain structure of graft VER greatly affected the morphologies of the SINs, although the PU network and graft VER network formed almost simultaneously.

Semi-interpenetrating polymer networks (semi-IPNs) from castor oil-based PU and benzyl starch (BS) were prepared [29]. The studies revealed that the semi-IPN films had good miscibility with BS concentrations of 5–70 wt% because of the strong intermolecular interactions between PU and BS. The ESR data confirmed that the segment volume of PU in the semi-IPNs increased with the addition of BS; that is, the chain stiffness increased as a result of strong interactions between PU and BS macromolecules. It was concluded that starch derivatives containing benzyl groups in the side chains more easily penetrated the PU networks to form semi-IPNs than those containing aliphatic groups, and this led to improved properties.

A cellulose/PU semi-IPN-based electroactive paper (EAPap) actuator has been constructed and studied [30]. For the fabrication of cellulose/PU semi-IPN EAPap actuator, cotton cellulose was dissolved into *N,N*-dimethylacetamide (DMAc) and lithium chloride (LiCl) solvent system. PU prepolymer prepared by poly(di(ethylene glycol) adipate) and hexamethylene diisocyanate (HDI) was mixed with DMAc cellulose solution by stirring. The mixed solution was spin coated on a wafer and cured to form cellulose/PU semi-IPN films using 1,1,1-tris(hydroxymethyl)propane as the crosslinker. The bending actuation performance of the actuator was evaluated in terms of free bending displacement with respect to the actuation frequencies, voltages, and humidity levels. It shows a good bending actuation at room humidity condition.

The effect of NCO/OH ratio on morphology and properties of interpenetrated networks formed between PU and nitrolignin has been investigated [31]. An increase in NCO/OH resulted in an increase in T_g, and solid state ^{13}C NMR study showed that introduction of nitrolignin into PU restricted the motion of PU molecules.

A series of semi-interpenetrating polymer networks (semi-IPN) of PU and poly(vinyl chloride) (PVC) have been obtained by the prepolymer method [32]. It has been found that PVC spherical aggregates are dispersed in the PU matrix, but Cl atoms location indicates partial miscibility of both polymers at the interphase which is probably due to hydrogen bonding and/or dipole–dipole interactions. The PVC component influences the phase behavior of PU's hard segments, as evidenced by DSC results.

Ternary nanocomposites with varying concentrations of nanosized silicon dioxide particles were prepared by adding nanosilica to interpenetrating polymer networks (IPNs) of polyurethane and epoxy resin (PU/EP) [33]. The result from the study showed that adding nanosized silicon dioxide can improve the properties of compatibility, damping, and phase structure of IPN matrices.

Novel nanocomposites with varying contents of organophilic montmorillonite (oMMT) were prepared by intercalating oMMT to interpenetrating polymer networks (IPNs) of PU/EP [34–36]. The results show that oMMT and the IPNs of PU/EP exhibit synergistic effect on the phase structure and morphology of the IPN nanocomposites. The addition of oMMT to the PU/EP IPNs matrix provided twofold benefits to the properties of the IPN nanocomposites. oMMT did not show distinct effect on chemical structure of PU/EP IPNs but promoted the compatibility and phase structure of the IPNs. The "forced compatibility" of PU and EP in interpenetrating process improved the dispersion degree of oMMT. Both the mechanical properties and water resistance of the PU/EP IPN nanocomposites are superior to those of the pure PU/EP IPNs.

A series of graft IPNs were prepared from epoxy (EP) and castor oil-based PU [37]. Results revealed that PU/EP IPNs showed improved damping properties compared with that of EP matrix. In addition, the PU content and testing frequencies had much influence on the damping properties of the IPNs. The IPN composites showed better damping properties with higher PU content or under higher testing frequencies. Mechanical analysis showed that the tensile strength of IPN composites was higher than that of pure EP. It is expected that PU/EP IPNs may be potentially used as structural damping materials.

Poly(urethane-imide) (PUI) membranes, useful for gas separation, were prepared by blending PU prepolymer with different ratio of poly(amide acid) (PAA) [38]. IPN structures are formed through reaction between PU prepolymer and PAA and unsaturated C=C double bond between different hydroxyl-terminated polybutadiene (HTPB) segment of PU. Compositions lower than 50 wt% of PU content (0.5-PUI) possess better compatibility between PU and polyimide(PI) and the glass transition temperature of PU hard segment (T_{gh}) of PUI membranes disappeared. The T_{gh} values, maximum degradation temperature of PUI membranes increased with the increase of PI content. The gas permselectivities of 0.5-PUI are 12 times than that of pure PI.

Tricomponent interpenetrating polymer network (IPN) systems involving castor oil, toluene diisocyanate (TDI), AN, ethylene glycol diacrylate (EGDA), and general-purpose unsaturated polyester resin (GPR) were prepared with various compositions [39]. TGA study indicated that the PU/PAN/GPR IPN underwent single-stage decomposition, showing perfect compatibility at the IPN composition of 10:90 (PU/PAN:GPR). It was found that the tensile strength of the GPR matrix was decreased and flexural and impact strengths were increased after incorporation of PU/PAN networks. The swelling properties in water and toluene were also studied.

Novel semi-interpenetrating polymer network (semi-IPN) membranes were synthesized from 6-*O*-carboxymethylchitosan (6-OCC) and waterborne polyurethanes (WPU) [40]. The growth of a model bacterial—*Escherichia coli*—and the adhesion of platelets on the surfaces of the 6-OCC/WPU semi-IPN membranes indicated that the antibacterial ability and blood compatibility of WPU membranes were improved by the addition of 6-OCC and strongly related to their crosslinking/grafting network structures. It was concluded that the 6-OCC/WPU semi-IPN membranes are good candidates as biomaterial for blood-contacting devices.

10.1.3 PU/Conducting Polymers

10.1.3.1 Polyurethane–Polyaniline IPNs Polyaniline doped by simple mineral acids is not processable owing to its metallike conductivity. In contrast, PU is a processable polymer and a twofold benefit is expected when both are blended. That is, processability will be improved for PAni and conductivity is attained for insulating PU. PAni exists in four oxidation states, as shown in Scheme 10.2.

Transition from one form to another is possible in PAni by simple redox methods (chemically or electrochemically). Emeraldine salt oxidation state (Scheme 10.2b) is the only conducting form of PAni, existing in a lower pH ($pH<2.5$) medium. Because of this reason, for the composite obtained to be conductive, PU should be stable and tolerate lower pH conditions. The best way of blending PAni is to synthesize this polymer in an aqueous dispersion and later blending it into PU water dispersion. Both PU and PAni phases intimately mix into a homogeneous dispersion (Scheme 10.3a). This procedure is widely adopted and there are many reports of such PU–PAni composites.

Apart from the aforementioned methodology, the alternate synthesis of PU–PAni composite seems to be functionalizing the PU backbone with conducting polymer chains. This is equivalent to grafting (Scheme 10.3b). This method is relatively recent and attracting many researchers. In this method NCO-terminated prepolymer is reacted with amine-capped PAni-EB (chain extension) so that the latter is attached to the PU backbone. However, some crosslinking is possible with amine/imine nitrogens in PAni backbone with PU forming networks, as shown in Scheme 10.3b.

In a study [41], oligomeric polyaniline (o-PANI) in emeraldine base (EB) form was blended with epoxy resin to form IPN. The formed film was doped by dipping in an acetic acid solution of *p*-toluene sulfonic acid. The PANI/epoxy

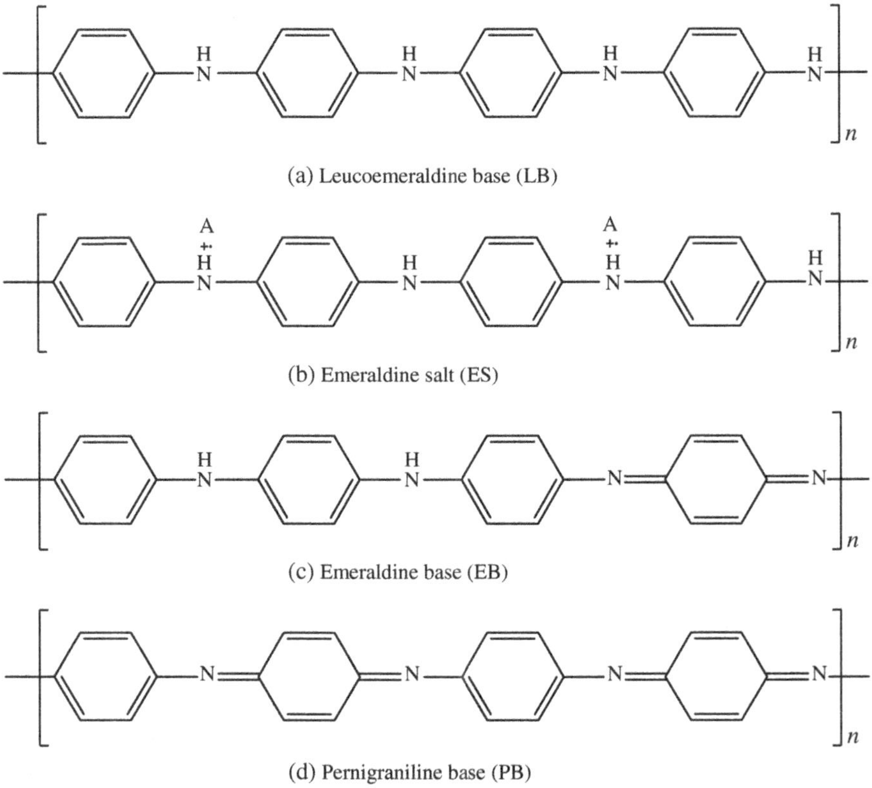

SCHEME 10.2 Various oxidation states of polyaniline: (a) leucoemeraldine base (LB), (b) emeraldine salt (ES), (c) emeraldine base (EB), and (d) pernigraniline base (PB)

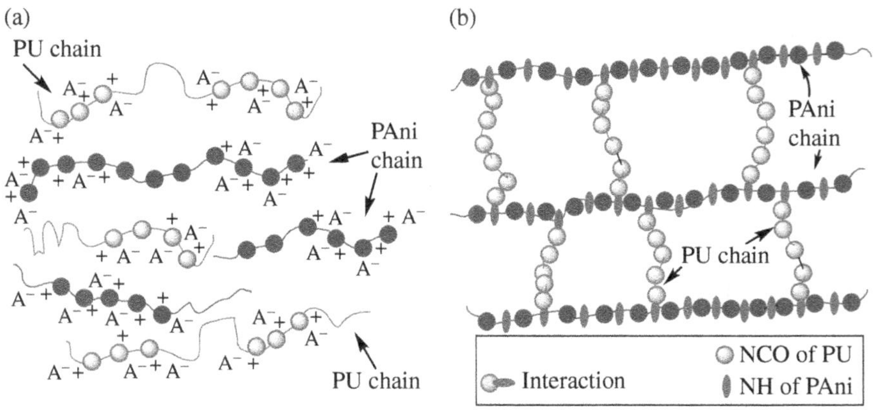

SCHEME 10.3 (a) Polyurethane–polyanline homogeneous blend and (b) polyurethane–polyaniline grafted network

resin composites showed a very low threshold concentration of o-PANI (lower than 1 part o-PANI per 100 parts resin) for electrical conduction (range is 5×10^{-5} to 1×10^{-2} S/cm). The o-PANI/epoxy resin composite contains very fine microscopic phase structure. TEM micrograph for the o-PANI/epoxy resin composite containing 1 part o-PANI per 100 parts resin (1 phr) showed that the o-PANI phase formed a network in the epoxy matrix.

A series of conducting interpenetrating polymer networks (IPNs) are prepared by sequential polymerization of castor oil-based PU with PMMA and polyaniline doped with camphor sulfonic acid ((PAni)CSA) [42]. The effect of different amount of PAni (from 2.5 to 12.5%) on the properties of PU/PMMA (50/50) IPNs such as conductivity, dielectric constant, and dissipation factor has been discussed. (PAni)CSA-filled IPNs showed improved tensile strength than the unfilled IPN system. TGA thermograms of (PAni)CSA-filled PU/PMMA (50/50) IPNs showed a three-step thermal degradation process. SEM micrograms of filled PU/PMMA IPN system showed spherulitic structure at higher concentration of (PAni)CSA.

In order to improve the mechanical performance and water resistance of waterborne conducting polyaniline film, conducting polyaniline/PU–silica hybrid film [43] was prepared in aqueous solution employing silanol-terminated PU and methyltriethoxysilane as sol–gel precursors. The authors believe that a practical route to waterborne processing of conducting polyaniline is achieved. The hybrid film showed surface resistivity of $10^8\,\Omega$ even though the conducting polyaniline loading was only 10 wt% (or 1.5 wt% of polyaniline). The mechanical performance as well as water resistance was significantly improved, making it suitable for antistatic application.

Interconnected networks of PAni chains with PU chains have been established in a study [44a]. Networks were prepared by reaction of NCO-terminated prepolymer with PAni-EB in NMP and blends by simple mixing of two components for 24 h. Shifts of the carbonyl and —NH— absorption coupled with presence of a new relaxation in the DMA spectra suggested the existence of an interphase. The authors forecasted antistatic coatings and corrosion inhibitor applications for this material. Their further study [44b] on UV–Vis results showed that PAni chains were more extended in network than in blends and give shift in visible absorption bands.

The crosslinking effects in composites of carboxylated PU grafted with conducting polyaniline on mechanical and electrical properties were investigated [45]. The crosslinking was effected by reacting ethylenediamine with residual carboxylic groups of PU. The composite showed a marked elastomeric behavior and the conductivity variation during deformation cycles is reversible as the chain relaxation and viscous flow phenomena are drastically suppressed by the crosslinks.

Wang *et al.* studied [46] on blends of PU and aniline-containing poly(urethane-urea) (PUUA) copolymer doped with different protonic acids, namely, camphorsulfonic acid (CSA) and dodecylbenzenesulfonic acid (DBSA). The blends showed a partial compatibility. The blends using CSA as a dopant resulted in better heat-resistant and tensile strength but poorer electronic conductivity than that of DBSA-doped one.

A new strategy for the modification of PU latex with polyaniline was proposed by the Stejskal group [47]. It is based on the introduction of a polymeric steric stabilizer into the latex. The stabilizer prevents the macroscopic precipitation of a conducting polymer during the polymerization. PAni has been prepared for this purpose in the presence of PU latex with an average particle size of 36 nm and poly(*N*-vinylpyrrolidone) (PNVP). The modified latex produced an aqueous colloidal dispersion having the particle size 100–200 nm. The presence of PNVP was essential to obtain a colloid, whose absence led to composite precipitation. The composite coatings obtained after drying contained 5–18 wt% PAni in a protonated emeraldine form with a conductivity up to 10^{-2} S/cm and mechanical properties typical of elastomers.

There has been an attempt to synthesize conducting PAni–PU membrane useful for fuel cells and electrodialysis [48]. The novelty of the preparation is the use of castor oil as renewable resource for PU preparation. Solution membranes were prepared by mixing sulfonated PAni (SPAni) and PAni doped with *p*-toluene sulfonic acid (PAni-PTSA) with castor oil and further HMDI. TGA curve showed a maximum decomposition rate at 350°C, quite different from the polyaniline. Nevertheless, the SPAni/PU membrane showed very good stability until 250°C. The loss modulus curves showed that there was an increase in the PU glass transition temperature with the addition of PAni and SPAni due to their interaction with the PU chains.

Conducting free-standing films of blended PAni.DBSA and thermoplastic polyurethane (TPU) were obtained [49] and were studied for their conductivity versus percolation thresholds. In the casted films, TPU was immiscible with PAni.DBSA, resulting in a gross-phase separation with a low electrical conductivity. The different methods used to prepare the PAni.DBSA/TPU blends significantly affected the conductivity values and percolation threshold. The PAni.DBSA/TPU blends prepared using the *in situ* process showed different morphological characteristics as those obtained by casting. For this preparation process, the electrical conductivities were 6.9×10^{-5} and 1.2×10^{-1} S/cm for 15/85 and 30/70, respectively, approximately 10 times higher than the values observed for the cast blends. The authors suggest potential applications of these materials in electrical processes.

Conducting PU–PAni containing silica composite hybrids were prepared by the chain extension method [50]. Amine-endcapped polyaniline (PAni) was used together with nanosilica bearing silanol groups as chain extenders to prepare the conducting PU hybrids. X-ray diffraction patterns indicated that these samples are typical of semicrystalline/amorphous materials. SEM image of PU-0.5PANI-0.5SiO_2 showed that PAni was dispersed homogeneously and interconnected continuously in the insulating PU–silica matrix. The surface resistivity values of all hybrids were about 10^8–$10^{10}\,\Omega$ and suggested for antielectrostatic application.

A nanotechnological approach was adopted [51] to design nanostructured PAni-blended castor oil-PU coatings. This chapter reported the synthesis of nanostructured MO–PAni (MO = methyl orange, an organic template for nanostructures) and castor oil polyurethane (COPU)-based composites. The effect of loading of nanostructured MO–PAni in COPU on the spectral, physicochemical, and morphological properties has been examined. The dispersion of nanostructured PANI at lower loadings and its intermolecular hydrogen bonding with COPU remarkably enhances the properties of the nanocomposite. The conductivity of pristine MO–PANI was found to be 2.1×10^{-3} S/cm while that of 0.5-MO–PAni/COPU, 1.0-MO–PAni/COPU, and 2.0-MO–PAni/COPU was found to be 5.7×10^{-4} S/cm, 4.3×10^{-4} S/cm, and 3.5×10^{-4} S/cm, respectively. The authors suggested anticorrosion coating application for these materials.

10.1.3.2 PU/Other Conducting Polymers Other conducting polymers such as polypyrrole, polythiophene, and poly(phenylenevinylene) are seldom used for the synthesis of PU blendings or composites and if used the resulting composites are always for a special purpose.

Investigations [52, 53] on conductivity and thermal stabilities of TPU/dodecylbenzenesulfonic acid-doped polypyrrole (PPy.DBSA) nanoblends prepared by solution intercalation (SB) and *in situ* (IS) methods have been carried out. It is observed that the electrical conductivity (σ_{dc}) of nanoblends is influenced by interaction between PPy.DBSA and TPU as well as homogeneous nanolevel dispersion and content of PPy.DBSA in TPU matrix. The maximum value of σ_{dc} has been found at 30 wt% PPy.DBSA for the SB

(0.26 S/cm) nanoblend due to the presence of the hexagonal network with a percolation threshold of 2.5 wt% of PPy.DBSA for IS and SB nanoblends. The temperature variation of conductivity for different nanoblends follows one-dimensional variable range hopping (VRH) model.

IPNs of maleimide-terminated polyurethane (UBMI) and dodecylbenzenesulfonic acid-doped polyaniline (PANIDB) were prepared and studied [54] for thermal, conductivity, and compatibility behavior. Differential scanning calorimetry (DSC) analysis of the IPNs indicated that the compatibility of the polyether-type UBMI with PANIDB is better than that of the polyester-type UBMI with PANIDB. The IPNs exhibited higher conductivity and better mechanical properties when the polyether-type UBMIs were used in the composition. Scanning tunneling microscopy (STM) images of the compatible polyether-type PANIDB/UBMI IPNs indicated that the conducting region arising from PANIDB is distributed as continuous conducting networks in the matrix. This resulted in higher conductivity and very low threshold concentration of the PANIDB for the formation of conducting IPNs.

Using ionomeric PU and polyvinylchloride (PVC), several semi-IPNs have been prepared by varying NCO/OH ratio [55]. Improvement in tensile strength was observed due to an increase in ionic content, increase in physical entanglement, and adhesion at phase boundaries due to the formation of interpenetrated networks. Mechanical properties of IPU/PVC IPNs show improved miscibility. The morphology of the IPNs showed cellular structure.

IPNs of polybutadiene-based PU and PMMA were synthesized [56], and the effect of the incorporation of 2% glycidyl methacrylate (GMA) and 2-hydroxyethyl methacrylate (2-HEMA) on the thermal, mechanical, and morphological properties was investigated. Both 2-HEMA and GMA led to improvements in these properties. However, 2-HEMA-containing IPNs showed somewhat better tensile strength, elongation, and damping characteristics due to better mixing of the components as revealed by SEM studies. DSC showed the presence of a new and third glass transition peak arising from grafting of methyl methacrylate onto PU.

A series of PU and poly(vinylidene chloride) (PVDC) interpenetrating polymer networks (IPNs) were synthesized [57] using the sequential polymerization technique. The physicomechanical and optical properties of the IPNs were investigated. TGA thermograms showed that the thermal degradation of the IPNs proceeded in three steps. Microcrystalline parameters, such as the crystal size and lattice disorder, of the PU/PVDC IPNs were estimated with wide-angle X-ray scattering.

Gradient and conventional semi-interpenetrating polymer networks (semi-IPNs) of PU and PHEMA [58] have been synthesized with possible biomedical applications. The dynamic mechanical analysis and AFM demonstrated that the systems were two-phase systems with incomplete phase separation. It is believed that the improvements in the mechanical properties of the materials through the use of gradient systems would allow the design of a novel materials for biomedical applications.

In a study [59] ternary semi-IPNs of nitrocellulose, poly(vinyl chloride-*co*-vinyl acetate), and PU were synthesized and characterized by FTIR, SEM, and DSC. Semi-IPNs of nitrocellulose–PU are found to be incompatible, whereas the ternary system consisting of nitrocellulose–poly(vinyl chloride-*co*-vinyl acetate) and PU is partially compatible. For a constant NCO/OH ratio, an increase in proportion of NC and Vc–VAc copolymer in the ternary systems shows an increase in tensile strength.

10.1.4 Applications and Concluding Remarks

IPNs have been the subject of extensive study since their advent in 1960 by Millar. Many investigators looked into the synthesis, morphology, properties, and applications. The work by Frisch [60] and Sperling and Friedman [61] showed that IPNs are characterized by topological interlocking of two or more crosslinked chains by two independent polymerization reactions [62]. Later, IPNs were defined as entanglements of polymer networks that are held together by only permanent topological interactions [6, 60–62]. The internetwork entanglements are chemically crosslinked and cannot be pulled apart without breaking one of the bonds.

Many types of IPNs, SINs, semi-IPNs, and so on are available as commercial products [63] and are summarized in Table 10.1. These are useful automotive applications such as gears and tires to medical use such as artificial teeth. The literature search from 2000 to 2011 and some part of 2012 showed that there is a considerable growth in synthesis of different IPNs with PU as one of the main polymers. Table 10.2 summarizes such reports along with their expected or demonstrated applications. These are useful as water and chemical resistance materials [10, 13], adhesive compounds to low energy materials [18], dimensionally stable materials [24], actuators [30], damping materials [33, 37], membranes for fuel cells, dialysis [38, 48], and antistatic and anticorrosion coatings [43, 44, 47, 51].

Polyurethanes are useful biomaterials [65, 66] with products ranging from nasogastric catheters, peritoneal dialysis, infusion pumps to implanted

TABLE 10.1 Some Commercially Known IPNs and Their Applications [63, 64]

S. No.	Trade Name	Composition	Manufacturer	Application
1	Acpol	Acrylic–urethane–polystyrene	Cook composites	Sheet molding compounds
2	ITP	PU–polyester–styrene	ICI Americas Ink	Sheet molding
3	Kelburon	PP-EP/rubber-PE	DSM N.V.	Automotive parts
4	Kraton IPN	SBES–polyester	Shell Co.	Automotive
5	Rimplast	Silicone rubber–PU	LNP plastics	Gears and medical
6	Santoprene	EPDM–PP	Monsanto	Tires, belts, gaskets
7	Silon	PDMS–PTFE	Bio Med Sciences	Burn dressings, scar abatement
8	Somel	EPDM–PP	Du Pont	Outdoor weathering
9	Telcar	EPDM–PP	BF Goodrich	Tubing liners
10	TPR	EPDM–PP	Reichhold Chemical Co.	Auto bumpers, wires, cables
11	Trubyte	Acrylic based	Dentsply	Artificial teeth
12	Vistalon	EPDM–PP	Exxon	Paintable automotive parts

pacemaker parts. In recent years, the chemical composition of PUs was modified by several investigators with the intention of improving blood compatibility and discovering which factors influence most blood-contacting responses. One of such process is making IPNs of PU by carefully designing the ingredient polymers (dental fillings, blood contact materials) [17, 58, 64]. It is well documented [67] that conducting PU materials with conducting polymers can play a vital role as anticorrosion and antistatic coatings and EMI shielding sensors application. It is hoped that the performance of these materials will be greatly enhanced if proper IPNs are synthesized. Similarly, it has been shown that PU-based IPNs are useful in automotive field. Low surface energy materials, such as polypropylene (PP), polyethylene (PE), and polytetrafluoroethylene, have been used in many fields because of their excellent performances. Because of their low energy surfaces, these materials are bonded with great difficulty. The present review also showed report on IPN [18] which is useful as adhesive agent for such materials.

TABLE 10.2 Ingredients and Applications of IPNs of Polyurethanes

S. No.	Type of Polyurethane	Other Polymers	Prominent Measured Property of IPN	Demonstrated/Suggested Application	References
1	Castor oil base	Poly(2-hydroxy methyl methacrylate)	Chemical resistance	Chemical resistance	10
2	TDI	Polyacrylates	Water resistance	—	13
3	Canola oil base	Poly(methyl methacrylate)	Superior mechanical properties	—	15, 16
4	TDI	Polyacrylamide	Hydrolytic stability	Biomedical	17
5	TDI	Polyacrylate	Adhesion to low energy surface	Adhesion to low energy surface	18
6	TDI	Unsaturated polyester	Sorption property	Separation	22
7	HMDI	Polystyrene	Dimensional stability	—	24
8	HMDI	Cellulose	Bending actuation	Bending actuator	30
9	MDI	Epoxy and nano-SiO_2	Damping property	Damping material	33
10	Castor oil base	Epoxy and montmorillonite	Water resistance	—	34–36
11	Castor oil base	Epoxy	Damping property	Structural damping property	37
12	$H_{12}MDI$	Poly(amidic acid)	Gas separation	Membrane for gas separation	38
13	IPDI	Polyaniline–silica	Conductivity	Antistatic application	43
14	TDI	Polyaniline	Conductivity	Antistatic and corrosion inhibitor	45
15	Commercial PU latex	Polyaniline–poly(vinylpyrrolidone)	Conductivity	Conductive coating	46
16	Castor oil base	Polyaniline	Conductivity	Fuel cells, electrodialysis	47
17	IPDI	Polyaniline, silica	Conductivity	Antistatic	43, 48
18	Castor oil base	Nanopolyaniline	Conductivity	Anticorrosion coating	49
19	TDI	Poly(2-hydroxy methyl methacrylate)	Conductivity	Biomedical	40, 56

REFERENCES

[1] D. Klemper, L. Berkowski (1987) *Interpenetrating Networks, Encyclopaedia of Polymer Science and Engineering*, Vol. **8**, p. 282, John Wiley & Sons, New York.

[2] Flory, P.J. (1953) *Principles of Polymer Chemistry*, Cornell University press, Ithaca, NY.

[3] Nicholson, J.W. (2006) Special topics in polymer chemistry. *Journal of Materials Chemistry*, **16**, 3867–3874.

[4] Millar, J.R. (1960) *Journal of the Chemical Society*, **2**, 1311.

[5] Sperling, L.H. and Mishra, V. (1996) *Polymers for Advanced Technologies*, **7**, 197–208.

[6] Suthar, B., Xiao, H.X., Klemper, D. and Frisch, K.C. (1996) A review on kinetic studies on the formation of interpenetrating polymer network. *Polymers for Advanced Technologies*, **7**, 221–233.

[7] Sperling, L.H. (1981) *Interpenetrating Polymer Networks and Related Materials*, Plenum Press, New York.

[8] Klempner, D. and Sperling, L.H. (1994) Interpenetrating polymer networks—an overview, in *Interpenetrating Polymer Networks*, Advanced Chemistry Series No 239 (ed L.A. Utracki), American Chemical Society, Washington, DC, p. 3.

[9] Klempner, D. (1994) IPNs around the world-recent advances, in *Advances in Interpenetrating Polymer Networks*, vol. **IV** (ed K.C. Frisch), Technomic Publishing Co., Lancaster, PA, p. 1.

[10] Prashantha, K., Vasanth kumar Pai, K., Sherigara, B.S. and Prasannakumar, S. (2001) Interpenetrating polymer networks based on polyol modified castor oil polyurethane and poly(2-hydroxyethylmethacrylate): Synthesis, chemical, mechanical and thermal properties. *Bulletin of Materials Science*, **24**, 535–538.

[11] Kim, B.K. and Shin, J.H. (2002) Modification of waterborne polyurethane by forming latex interpenetrating polymer networks with acrylate rubber. *Colloid & Polymer Science*, **280**, 716–724.

[12] Chen, S. and Chen, L. (2003) Structure and properties of polyurethane/polyacrylate latex, interpenetrating networks hybrid emulsions. *Colloid & Polymer Science*, **282**, 14–20.

[13] Chen, L. and Chen, S. (2004) Latex interpenetrating networks based on polyurethane, polyacrylate and epoxy resin. *Progress in Organic Coatings*, **49**, 252–258.

[14] Anzlovar, A. and Zigon, M. (2005) Semi-interpenetrating polymer networks with varying mass ratios of functional urethane and methacrylate prepolymers. *Acta Chimica Slovenica*, **52**, 230–237.

[15] Kong, X. and Narine, S.S. (2008) Sequential interpenetrating polymer networks produced from vegetable oil based polyurethane and poly(methyl methacrylate). *Biomacromolecules*, **9**, 2221–2229.

[16] Kong, X. and Narine, S.S. (2008) Physical properties of sequential interpenetrating polymer networks produced from canola oil-based polyurethane and poly(methyl methacrylate). *Biomacromolecules*, **9**, 1424–1433.

[17] Lynda Merlin, D. and Sivasankar, B. (2009) Synthesis and characterization of semi-interpenetrating polymer networks using biocompatible polyurethane and acrylamide monomer. *European Polymer Journal*, **45**, 165–170.

[18] Li, M., Zheng, Z., Liu, S. *et al.* (2012) Polyurethane (urea)/polyacrylates interpenetrating polymer network (IPN) adhesives for low surface energy materials. *Polymers for Advanced Technologies*, **23**, 1077–1083.

[19] Xie, H.-Q. and Guo, J.-S. (2002) Room temperature synthesis and mechanical properties of two kinds of elastomeric interpenetrating polymer networks based on castor oil. *European Polymer Journal*, **38**, 2271–2277.

[20] Mahesh, K.P.O., Alagar, M. and Ananda Kumar, S. (2003) Mechanical, thermal and morphological behaviour of bismaleimide modified PU-epoxy IPN matrices. *Polymers for Advanced Technologies*, **14**, 137–146.

[21] Ramis, X., Cadenato, A., Morancho, J.M. and Salla, J.M. (2001) Polyurethane-unsaturated polyester interpenetrating networks: thermal and dynamic mechanical thermal behaviour. *Polymer*, **42**, 9469–9479.

[22] Krol, P. and Wojturska, J. (2006) The swelling behavior of interpenetrating polymer networks composed of polyurethane and unsaturated polyester. *Journal of Applied Polymer Science*, **101**, 3511–3519.

[23] Hsieh, K.H., Han, J.L., Yu, C.T. and Fu, S.C. (2001) Graft interpenetrating polymer networks of urethane-modified bismaleimide and epoxy (I): mechanical behavior and morphology. *Polymer*, **42**, 2491–2500.

[24] Lepine, O., Birot, M. and Deleuze, H. (2005) Preparation of macrocellular PU–PS interpenetrating networks. *Polymer*, **46**, 9653–9663.

[25] Alekseeva, T.T., Lipatov, Y.S., Babkina, N.V. *et al.* (2005) Reaction compatibilization in interpenetrating polymer networks II. Polyurethane–polystyrene–oligourethane dimethacrylate system. *Polymer*, **46**, 419–428.

[26] Chenal, J.-M., Colombini, D. and Widmaier, J.-M. (2009) Influence of the process on phase separation induced by radical copolymerization of styrene and divinylbenzene in the presence of reactive polyurethane network precursors. *Journal of Applied Polymer Science*, **99**, 2989–2996.

[27] Ma, C.-C.M., Sung, S.-C., Wang, F.-Y. *et al.* (2001) Thermal, mechanical, and morphological properties of novolac-type phenolic resin blended with fullerenol polyurethane and linear polyurethane. *Journal of Polymer Science Part B: Polymer Physics*, **39**, 2436–2443.

[28] Wang, G.U. and Hu, C.P. (2002) Simultaneous interpenetrating networks consisting of polyurethane and graft vinyl ester resin: effect of different radical initiators. *Journal of Applied Polymer Science*, **84**, 1629–1636.

[29] Cao, X. and Zhang, L. (2005) Miscibility and properties of polyurethane/benzyl starch semi-interpenetrating polymer networks. *Journal of Polymer Science Part B: Polymer Physics*, **43**, 603–615.

[30] Cai, Z. and Kim, J. (2008) Characteristics and performance of electroactive paper actuator made with cellulose/polyurethane semi-interpenetrating polymer networks. *Journal of Applied Polymer Science*, **109**, 3689–3695.

[31] Huang, J. and Zhang, L. (2002) Effect of NCO/OH molar ratio on structure and properties of graft-interpenetrating polymer networks from polyurethane and nitrolignin. *Polymer*, **43**, 2287–2294.

[32] Pielichowski, K. and Janowski, B. (2005) Semi-interpenetrating polymer networks of polyurethane and poly(vinyl chloride): thermal stability assessment. *Journal of Thermal Analysis and Calorimetry*, **80**, 147–151.

[33] Zhang, H., Wang, B., Li, H. *et al.* (2003) Synthesis and characterization of nanocomposites of silicon dioxide and polyurethane and epoxy resin interpenetrating network. *Polymer International*, **52**, 1493–1497.

[34] Jia, Q.M., Zheng, M., Chen, H.X. and Shen, R.J. (2005) Synthesis and characterization of polyurethane/epoxy interpenetrating network nanocomposites with organoclays. *Polymer Bulletin*, **54**, 65–73.

[35] Jia, Q.-M., Zheng, M.-S., Chen, H.-X. and Shen, R.-J. (2006) Morphologies and properties of polyurethane/epoxy resin interpenetrating network nanocomposites modified with organoclay. *Materials Letters*, **60**, 1306–1309.

[36] Jia, Q., Zheng, M., Shen, R. and Chen, H. (2006) Synthesis, characterization and properties of organoclay-modified polyurethane/epoxy interpenetrating polymer network nanocomposites. *Polymer International*, **55**, 257–264.

[37] Chen, S., Wang, Q., Pei, X. and Wang, T. (2010) Dynamic mechanical properties of castor oil-based polyurethane/epoxy graft interpenetrating polymer network composites. *Journal of Applied Polymer Science*, **118**, 1144–1151.

[38] Tsai, M.-H., Huang, S.-L., Liu, S.-J. *et al.* (2008) Synthesis and properties of poly(urethane-imide) interpenetrating network membranes. *Desalination*, **233**, 191–200.

[39] Guhanadan, S., Hariharan, R. and Saroja Devi, M. (2004) Studies on castor-oil based polyurethane/polyacrylonitrile interpenetrating network for toughening of unsaturated polyester resin. *Journal of Applied Polymer Science*, **92**, 817–829.

[40] Yu, S.-H., Mi, F.-L., Shyu, S.-S. *et al.* (2006) Miscibility, mechanical characteristic and platelet adhesion of 6-*O*-carboxymethylchitosan/polyurethane semi-IPN membranes. *Journal of Membrane Science*, **276**, 68–80.

[41] Yang, X., Zhao, T., Yu, Y. and Wei, Y. (2004) Synthesis of conductive polyaniline/epoxy resin composites: doping of the interpenetrating network. *Synthetic Metals*, **142**, 57–61.

[42] Jeevananda, T. and Siddaramaiah (2003) Synthesis and characterization of polyaniline filled PU/PMMA interpenetrating polymer networks. *European Polymer Journal*, **39**, 569–578.

[43] Luo, J., Wang, X., Li, J. *et al.* (2007) Conductive hybrid film from polyaniline and polyurethane silica. *Polymer*, **48**, 4368–4374.

[44] (a) Rodrigues, P.C. and Akcelrud, L. (2003) Networks and blends of polyaniline and polyurethane: correlations between composition and thermal, dynamic mechanical and electrical properties. *Polymer*, **44**, 6891–6899; (b) Rodrigues, P.C., Lisboa-Filho, P.N., Mangrich, A.S. and Akcelrud, L. (2005) Polyaniline/polyurethane networks. II. A spectroscopic study. *Polymer*, **46**, 2285–2296.

[45] Abbati, G., Carone, E., D'Ilario, L. and Martinelli, A. (2003) Polyurethane–polyaniline conducting graft copolymer with improved mechanical properties. *Journal of Applied Polymer Science*, **89**, 2516–2521.

[46] Wang, Y.Z., Hsu, Y.C., Chou, L.C. and Hsieh, K.H. (2004) Study on blends of polyurethane and aniline-containing poly(urethane-urea) copolymer with different protonic acid dopants. *Journal of Polymer Research*, **11**, 127–132.

[47] Sapurina, I., Stejskal, J., Spirkova, M. *et al.* (2005) Polyurethane latex modified with polyaniline. *Synthetic Metals*, **151**, 93–99.

[48] Amado, F.D.R., Rodrigues, L.F., Jr, Forte, M.M.C. and Ferreira, C.A. (2006) Properties evaluation of the membranes synthesized with castor oil polyurethane and polyaniline. *Polymer Engineering & Science*, **46**, 1485–1489.

[49] Vicentini, D.S., Barra, G.M.O., Bertolino, J.R. and Pires, A.T.N. (2007) Polyaniline/thermoplastic polyurethane blends: preparation and evaluation of electrical conductivity. *European Polymer Journal*, **43**, 4565–4572.

[50] Wang, T.-L., Yang, C.-H., Shieh, Y.-T. and Yeh, A.-C. (2009) Synthesis and properties of conducting organic/inorganic polyurethane hybrids. *European Polymer Journal*, **45**, 387–397.

[51] Ahmad, S., Riaz, U. and Alam, J. (2009) Development of sustainable resource-based nanostructured polyaniline/castor oil polyurethane composites. *Advances in Polymer Technology*, **28**, 26–31.

[52] Kotal, M., Srivastava, S.K. and Paramanik, B. (2011) Enhancements in conductivity and thermal stabilities of polypyrrole/polyurethane nanoblends. *The Journal of Physical Chemistry C*, **115**, 1496–1505.

[53] Shenoy, S.L., Cohend, D., Weiss, R.A. and Erkey, C. (2004) Supercritical carbon dioxide aided preparation of conductive polyurethane–polypyrrole composites. *Journal of Supercritical Fluids*, **28**, 233–239.

[54] Liao, D.C., Hsieh, K.H., Chern, Y.C. and Ho, K.S. (1997) Interpenetrating polymer networks of polyaniline and maleimide-terminated polyurethanes. *Synthetic Metals*, **87**, 61–67.

[55] Jaisankar, S.N., Anandprabu, A., Lakshminarayana, Y. and Radhakrishnan, G. (2000) Preparation and properties of semi-interpenetrating polymer networks based on polyurethane ionomer/polyvinyl chloride. *Journal of Materials Science*, **35**, 1065–1068.

[56] Desai, S., Thakore, I.M., Brennan, A. and Devi, S. (2002) Thermomechanical properties and morphology of interpenetrating polymer networks of polyurethane–poly(methyl methacrylate). *Journal of Applied Polymer Science*, **83**, 1576–1585.

[57] Begum, M., Siddaramaiah, Somashekar, R. and Somashekarappa, H. (2007) Synthesis and characterization of polyurethane/poly(vinylidene chloride) interpenetrating polymer networks. *Journal of Applied Polymer Science*, **103**, 1375–1381.

[58] Karabanova, L.V., Mikhalovsky, S.V. and Lloyd, A.W. (2012) Gradient semi-interpenetrating polymer networks based on polyurethane and poly(2-hydroxyethyl methacrylate) for biomedical applications. *Journal of Materials Chemistry*, **22**, 7919–7928.

[59] Natchimuthu, N., Rajalingam, P. and Radhakrishnan, G. (1992) Cellulose nitrate-poly (vinyl Chloride-co-Vinyl acetate)-polyurethane ternary IPNs: FT-IR and morphological studies. *Journal of Applied Polymer Science*, **44**, 981–986.

[60] Frisch, H.L., Klempner, D. and Frisch, K.C. (1969) A topologically interpenetrating elastomeric network. *Journal of Polymer Science. Part B*, **7**, 775.

[61] Myung, D., Waters, D., Wiseman, M. *et al.* (2008) Progress in the development of interpenetrating polymer network hydrogels. *Polymers for Advanced Technologies*, **19**, 647–657.

[62] Sperling, L.H. (1985) Recent advances in interpenetrating polymer networks. *Polymer Engineering & Science*, **25**, 517–520.

[63] Sperling, L.H. (1986) Multicomponent polymer materials, in *Advances in Chemistry*, vol. **211** (eds D.R. Paul and L.H. Sperlin), American Chemical Society Publications, Washington, DC.

[64] Sperling, L.H. (2004) Interpenetrating polymer networks. *Encyclopedia of Polymer Science and Technology*, **10**, 272–311.

[65] Szycher, M. (1999) *Handbook of Polyurethanes*, CRC Press, Massachusetts, US.

[66] Vermette, P., Griesser, H.J., Laroche, G. and Guidoin, R. (2001) *Biomedical Applications of Polyurethanes*, EUREKAH.COM, Georgetown, TX.

[67] Gurunathan, T., Rao, C.R.K., Narayan, R. and Raja, K.V.S.N. (2013) Polyurethane conductive blends and composites: synthesis and applications perspective. *Journal of Materials Science*, **48**, 67–80.

11

SOLID-STATE NMR AND ESR STUDIES OF IPNs

SREĆKO VALIĆ[1,2], MLADEN ANDREIS[2] AND DAMIR KLEPAC[1]

[1] *Department of Chemistry and Biochemistry, School of Medicine, University of Rijeka, Rijeka, Croatia*

[2] *Physical Chemistry Division, Rudjer Bošković Institute, Zagreb, Croatia*

11.1 INTRODUCTION

Interpenetrating polymer networks (IPNs) are a special type of polymer blends, in which two or more polymers are mixed in their network form. However, the usually observed two glass transition temperatures (T_g) indicate that the mixing does not take place at a molecular level and a two-phase morphology is obtained. Since IPNs consist of crosslinked polymers which may or may not be covalently bonded to each other, their morphology is quite complicated, and no general laws have yet been established that correlate synthesis and morphology with the ultimate properties. The combination of rubbery and glassy polymers in IPNs and semi-interpenetrating polymer networks (SIPNs) can produce materials ranging from reinforced elastomers to damping materials and high-impact plastics.

In order to understand the structure and dynamics of these insoluble crosslinked polymer materials, they are usually studied simultaneously with several techniques such as Fourier transform infrared (FTIR) spectroscopy, wide-angle

Micro- and Nano-structured Interpenetrating Polymer Networks: From Design to Applications, First Edition. Edited by Sabu Thomas, Daniel Grande, Uroš Cvelbar, K.V.S.N. Raju, Ramanuj Narayan, Selvin P. Thomas and Akhina H.

X-ray diffraction (WAXRD), X-ray photoelectron spectroscopy (XPS), differential scanning calorimetry (DSC), thermogravimetric analysis (TGA), dynamic mechanical analysis (DMA), and scanning electron microscopy (SEM).

The development of magnetic resonance techniques, particularly solid-state nuclear magnetic resonance (NMR) and electron spin resonance (ESR), in the past few decades enabled the study of structure and dynamics of IPNs at a molecular level. Among most frequently studied types of IPNs by NMR and ESR are SIPNs, materials in which at least one of linear or branched polymer components does not form a network itself but rather interpenetrates into the network of the other component(s).

11.2 THEORETICAL BACKGROUND

11.2.1 Solid-State Nuclear Magnetic Resonance

The broad-line spectra of nuclei with spin $I = 1/2$ in the solid state are mainly a consequence of the dominant contribution of the dipolar Hamiltonian, H_D, which gives rise to a local field B_{loc}. Its magnitude varies as a function of the angle Θ_{ij} between the internuclear vector r_{ij} and the applied magnetic field B_o. Two general types of interactions can be distinguished:

1. Homonuclear dipolar interactions described by the term H_D are present in the spin system of abundant nuclei (I), such as ^{1}H and ^{19}F.
2. In the "diluted" spin system (S) such as ^{13}C and ^{29}Si, homonuclear dipolar interactions are negligible because of the low natural abundance of nuclei under observation; in this case term H_D describes dipolar interactions between unlike spins, which for ^{13}C and ^{29}Si are most frequently the neighboring protons.

One common technique for reducing resonance line widths (used in both cases) is magic angle spinning (MAS). The dipolar Hamiltonian H_D can be reduced to zero if the contribution of the term $(1-3\cos^2\Theta_{ij})$ is zero. Hence, mechanical spinning of the sample at the angle $\Theta_{ij} = 54°44'$ (the so-called magic angle) causes an analogous effect to the rapid molecular motion in solutions, that is, narrowing of lines and high-resolution isotropic spectrum. Furthermore, rotation of the solid sample around the magic angle averages out the chemical shielding tensor to its isotropic value, thus further narrowing the line. However, eliminating H_D by MAS is successful only when the spinning frequency is greater than the interaction to be averaged. MAS high-resolution spectra are obtained for particular spin systems of abundant nuclei, such as protons in swollen crosslinked polymers, which have a certain degree of molecular motion (τ_c of the order 10^{-4} s or longer). But MAS is generally not applicable for removing much larger heteronuclear dipolar interactions of rigid rare spins. In standard techniques limitations arise in obtaining spinning speeds with frequencies faster than the magnitude of the dipole–dipole couplings (typical spinning frequencies are in the range of 4–5 kHz, and they are not sufficient enough to suppress ^{13}C–^{1}H dipolar couplings which are about 50 kHz). Since the dipolar decoupling (DD) cannot remove the broadening caused by chemical shift anisotropy in order to obtain the maximum resolution and maximum sensitivity in the ^{13}C spectra of solid polymers, high-power DD, MAS, and cross-polarization (CP) are used simultaneously (static dipolar interaction can be removed by a high-power DD, and the remaining chemical shift anisotropy can then be averaged to its isotropic value by using MAS). These techniques, as well as use of various multiple pulse sequences and two-dimensional (2D) techniques, give rise to a large number of possibilities for the investigation of structural and dynamical properties of materials in the solid state, particularly for insoluble crosslinked IPNs and SIPNs.

There are two basic approaches to achieve ^{13}C MAS NMR spectra of polymers:

1. Single-pulse excitation (SPE) involves a 90° carbon pulse and the observation of the free induction decay (FID) under the conditions of high-power decoupling. This pulse sequence, usually called a Bloch decay or direct polarization (DP), is successfully applied to polymer systems when there is a sufficient degree of mobility and the carbon T_1 relaxation times are not longer than a few seconds (in solid and crystalline polymers ^{13}C T_1 is usually longer of 1000 s). Thus, the DP pulse sequence is frequently used to study elastomers, but it is also used for characterization of swollen IPNs.
2. CP is a double resonance experiment in which the transfer of magnetization from the proton spin (I) to the carbon spins (S) occurs if the Hartmann–Hahn condition is fulfilled:

$$\gamma_I B_{1I} = \gamma_S B_{1S}.$$

The CP pulse sequence consists of the following: (i) establishing ^{1}H spin polarization in B_0; (ii) cooling of the proton spin system, which is most frequently done by spin-locking in the rotating frame B_1; (iii) bringing ^{13}C and ^{1}H spins in contact during a certain period of time T_C by the Hartmann–Hahn match; (iv) turning off B_{1C} and recording ^{13}C FID; and (v) repeating the pulse sequence following the delay time t_d (Figure 11.1).

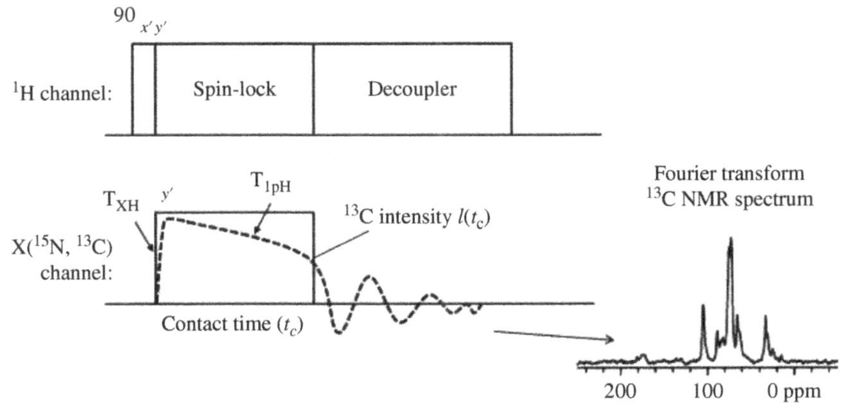

FIGURE 11.1 Cross-polarization pulse sequence. Source: Knicker [1]. Reproduced with permission of Elsevier

Building up the carbon magnetization by the CP process has two major advantages: (i) the relaxation process is much more rapid, because it is determined by the proton T_1, as compared to the carbon T_1 (which allows relatively short recycle delay t_d between the pulse sequences), and (ii) intensity enhancement is four times greater than in the DP experiment ($\gamma_H/\gamma_C = 4$).

However, the major disadvantage of the technique is that the signal intensities observed at certain T_C are not proportional to the concentration of particular groups due to the decrease of signal intensity caused by the proton spin–lattice relaxation time in the rotating frame (^{1}H $T_1\rho$).

This relaxation time is dominated by the spin diffusion process, resulting in a single proton ^{1}H $T_1\rho$ for a whole sample, if it is homogeneous and the protons are strongly coupled, while in the heterogeneous systems such as IPNs it can provide very useful information about molecular dynamics.

Basic principles of NMR and hundreds of pulse sequences with corresponding Hamiltonians regarding both ^{1}H and ^{13}C or ^{29}Si solid-state NMR experiments are presented in numerous textbooks and review articles [2–4]. Therefore, they will not be discussed in detail here. However, it should be mentioned that among most frequently used 2D techniques are correlation spectroscopy (COSY) and nuclear Overhauser effect spectroscopy (NOESY). By COSY experiments it is possible to determine which protons are spin–spin coupled through chemical bonds (2 or 3 bond separation between nuclei), while NOESY experiments detect through-space transfer of nuclear spin polarization. The basic NOESY pulse sequence consists of three 90° pulses. The first pulse creates transverse magnetization which precess during the evolution time t_1 (the evolution time is incremented during the course of the 2D experiment). The second pulse produces longitudinal magnetization equal to the transverse magnetization component orthogonal to the pulse direction, and the third pulse creates the observable transverse magnetization from the remaining longitudinal magnetization.

In the past decades, the development of fast MAS (faster than 20 kHz) [5] and the availability of higher magnetic fields significantly improved the resolution of ^{1}H solid-state MAS NMR spectra. The most important feature of dipole–dipole couplings is that they encode structural information through the $1/r^3$ dependence with respect to the corresponding spin–spin distance. Due to its high sensitivity and high natural abundance, ^{1}H is a favorable nucleus for NMR study of molecular structure; hence, ^{1}H double-quantum (DQ) MAS experiments are commonly employed to determine proton–proton proximities in solid organic compounds. This technique implies the design of homonuclear recoupling pulse sequences capable to efficiently generate DQ correlations between spins of the same nuclear species. Among the large variety of recoupling methods, back-to-back (BABA) recoupling pulse sequence is most frequently used in ^{1}H DQ MAS experiments [6].

11.2.2 Electron Spin Resonance Spectroscopy

Electron spin resonance (ESR) spectroscopy or electron paramagnetic resonance (EPR) spectroscopy is a powerful tool for investigating paramagnetic species like organic or inorganic radicals. The basic principles of this technique are described in detail elsewhere [7–12]. This short introduction to ESR will be focused only on parts which are necessary to understand the application of this technique in the studies of IPNs and SIPNs described in this chapter.

This spectroscopic technique is based on the interaction of an external magnetic field with the unpaired electron in the paramagnetic species. As long as the value of external magnetic field is $B=0$, all unpaired electrons posses the same energy, as shown schematically in Figure 11.2a. When $B>0$, the interaction of electron spin S with the magnetic field B (Zeeman interaction) splits energy of electrons in two levels: electrons with lower energy ($m_s=-1/2$) and higher energy ($m_s=+1/2$). This interaction is described by the Hamiltonian:

$$\mathcal{H} = \mu_B B \mathbf{g} S,$$

where μ_B is the Bohr magneton and $\mathbf{g}$ is the tensor which parameterizes the Zeeman interaction. The energy difference between the two spin states is $\Delta E = E^+ - E^-$, E^+ and E^- being the higher and lower energy values, respectively.

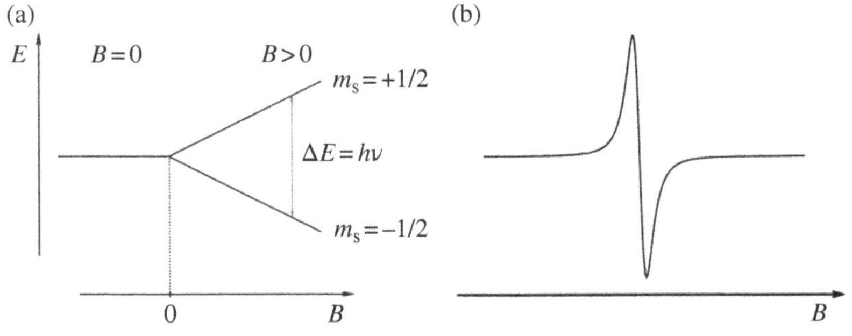

FIGURE 11.2 (a) Energy levels of unpaired electrons placed in an external magnetic field. (b) Schematic presentation of a simple ESR spectrum

An absorption of the photon with the energy $h\nu$ from the microwave field, where h is Planck's constant and ν is the frequency, changes the spin state from $m_s=-1/2$ to $m_s=+1/2$ when $h\nu=\Delta E$. This change results in the absorption line which is usually detected in the form of the first derivative (Figure 11.2b).

IPNs and SIPNs, like many other polymer systems, usually do not possess an intrinsic ESR signal since they have no unpaired electrons. However, an introduction of paramagnetic centers in polymer matrix opens up the possibility for the application of this technique in the study of such systems [7, 10, 11]. Another way is based on the use of high-energy radiation which may cause the formation of radicals in a polymer matrix.

Nitroxyl radicals are known as the most commonly used radicals in studying polymer systems by ESR. Their structures and the way of their application are given in Ref. [13]. ESR spectrum of a nitroxyl radical consists of three lines because of the hyperfine interaction (interaction between the electron spin and the spin of nitrogen nucleus). This interaction additionally splits each Zeeman energy level into three sublevels. Spectral parameters such as the line width, line intensity, or hyperfine splitting value provide useful information about molecular dynamics, matrix structure, free volume, free volume distribution, and dynamic (in)homogeneity [12, 14–21].

Someone who wants to choose a nitroxyl for the study of certain polymer system should take care of its compatibility with the polymer matrix. However, nitroxyl size, shape, polarity, and flexibility are also important parameters [12, 22]. The simplest way of introduction of nitroxyl radicals in a polymer matrix is their mechanical incorporation by diffusion in the presence of a good solvent for both polymer matrix and nitroxyl radical or by a direct diffusion of nitroxyl vapors into polymer at elevated temperature. The latter is possible only when the nitroxyl is able to evaporate at relatively low temperature and can be used for homogeneous systems only. The former leads to more uniform distribution of nitroxyl radicals inside the polymer matrix after solvent removal and is usually applied in the studies of dynamically inhomogeneous systems. Irrespective of the way of mechanical introduction of nitroxyl radicals in polymer matrix, the method is known as spin probe. A second approach implies a covalent binding of a nitroxyl to a polymer chain. This procedure is named spin labeling. The advantage of spin labeling is the precisely known position of the nitroxyl radical in a polymer chain [22]. However, this method demands additional synthetic efforts and is rather limited since it cannot be applied to polymer systems without an adequate functional group. Nitroxyl radicals that are commonly used in polymer investigations are listed in Ref. [13].

Recent investigation indicates that the temperature of probe incorporation process must be strictly controlled, particularly for inhomogeneous systems [11, 23]. In addition, when the spin probe method is used, it is necessary to take care of a possible attachment of the spin probe to polymer chain by hydrogen bond. In such case the effect of hydrogen bond may influence $T_{5\,\mathrm{mT}}$ value (defined in the following text) [24]. An advantage of ESR spectroscopy, when compared to other techniques, is that very low concentrations of nitroxyl radicals (~0.1 wt.% or lower) embedded in a polymer matrix result in a good signal-to-noise ratio. This implies only a slight perturbation of the matrix with no significant influence on its intrinsic properties.

Contrary to the application of nitroxyl radicals, the use of a high-energy radiation, depending on the applied irradiation dose, may provoke important changes in matrix structure and material properties. Besides nitroxyl radicals and radicals generated by high-energy radiation, metallic ions with paramagnetic properties are also used as ESR signal carriers. Some authors apply Cu^{2+} ion for the study of a particular type of IPNs, as described in the following text. The d^9 configuration of Cu^{2+} implies the paramagnetic properties of its compounds, making ESR of Cu^{2+}-containing species a useful tool for structural studies. Cu^{2+} centers typically have tetrahedral or axially elongated octahedral geometry. Their spectra are anisotropic and generally give signals of the axial or orthorhombic type. The ESR spectra of Cu^{2+} can be used for determination of the coordination geometry.

The number of papers reporting ESR results in this field is limited. However, they can be divided into three groups: (i) the studies of molecular dynamics and matrix homogeneity using nitroxyl radicals, (ii) the studies of materials containing high-energy radiation-induced radicals, and (iii) materials with incorporated metallic ions which also give an intrinsic ESR signal.

11.3 NUCLEAR MAGNETIC RESONANCE OF INTERPENETRATING POLYMER NETWORKS AND SEMI-INTERPENETRATING POLYMER NETWORKS

11.3.1 Characterization

Bifunctional IPNs consisting of polystyrene (PS)—PS/poly(4-vinylpyridine), PS/poly(*N*-vinylimidazole), PS/poly(ethyl acrylate), PS/poly(4-vinylpyridine-*co*-ethyl acrylate), and PS/poly(*N*-vinylimidazole-*co*-ethyl acrylate)—have been synthesized and characterized by both ^{13}C CP MAS and ^{13}C DP MAS spectroscopy (Figure 11.3).

DP MAS gives the chemical composition of the lightly crosslinked network under conditions where the more heavily crosslinked network is transparent; the latter is then identified by CP MAS. Combining the spectra with elemental

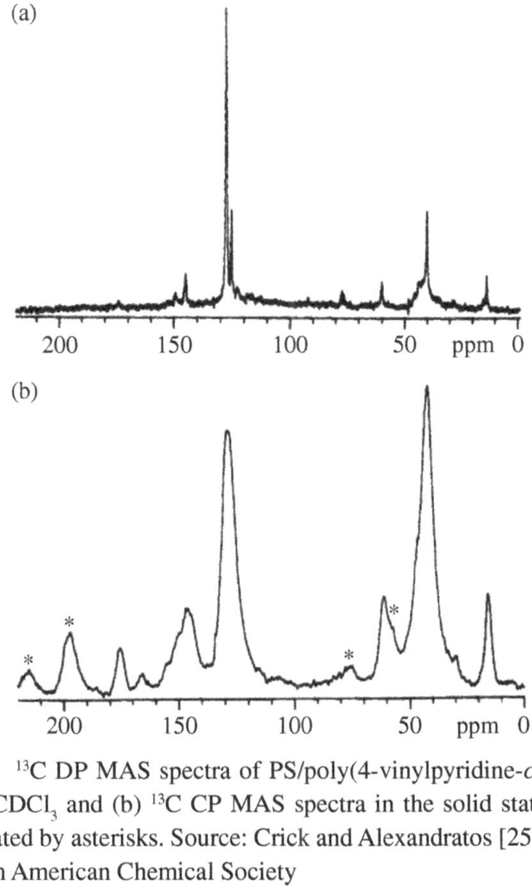

FIGURE 11.3 ^{13}C DP MAS spectra of PS/poly(4-vinylpyridine-*co*-ethyl acrylate) (a) swollen in $CDCl_3$ and (b) ^{13}C CP MAS spectra in the solid state. Spinning side bands are indicated by asterisks. Source: Crick and Alexandratos [25]. Reprinted with permission from American Chemical Society

analyses allows for an unambiguous determination of the ligands present. DP MAS spectra of swollen network show that both monomers are sorbed from solution into the PS beads, while CP MAS spectra confirm that the desired networks have been formed [25].

Due to the excellent thermal stability and unique electrochemical properties, the conducting polymer polyaniline (PANI) has been also studied in various blends, IPNs and SIPNs. Chitosan/polyaniline (CP) SIPNs prepared under the condition of HCl aqueous solution ($pH < 1$) were investigated, among other techniques, also by using solid-state ^{13}C NMR.

Although the crosslinking of chitosan results in less resolved spectra as compared with raw materials and corresponding CP blends, changes of structural properties in SIPNs were detected by the appearance of a new signal at 141 ppm which is attributed to the reaction of amino groups in chitosan with aldehyde groups in glutaraldehyde, that is, the formation of imine bond (—N=CH—) [26].

The spectrum of PANI emeraldine base (PANI-EB) powder has peaks near 124 ppm from the protonated benzenoid unit, peaks near 144 ppm from the nonprotonated benzenoid unit, a peak at 137 ppm from the protonated carbons of the quinoid unit, and a peak at 159 ppm from the imine carbon (unprotonated quinoid unit). The doped PANI emeraldine salt (PANI-ES) powder causes a slight shift of all peaks. The acidic doping of PANI-ES and corresponding CP blend increase the intensity of the peak from the imine carbons with a concomitant increase and shift in the peak from the protonated carbons of the quinoid unit. Thus the NMR spectra show that the chemical nature of the PANI in the IPN is identical to the neat PANI-ES and PANI-EB [27].

Polyaniline/poly(2-acrylamido-2-methylpropane sulfonic acid) (PANI/PAMPS) SIPNs were prepared using the simultaneous method, and the interaction of PAMPS, as a polymeric acid dopant, with PANI was also investigated. As in the aforementioned CP SIPNs, solid-state ^{13}C NMR of these SIPNs exhibited very broad resonance signals, indicating interactions between the components due to the PAMPS sulfonic groups bonding to the benzene rings of PANI, which affected the chemical shift of nearby carbon atoms. Furthermore the SIPN spectrum exhibited a broad asymmetric peak arising from the benzenoid carbons that was shifted downfield at 127.27 ppm. The line broadening is due to the heterogeneity in charge distribution along the polymer chain, resulting from structural disorder of the polymer chains in the solid state [28].

By copolymerization of maleate polysaccharides based on xanthan or gellan with a temperature-sensitive precursor (*N*-isopropylacrylamide), water-swollen

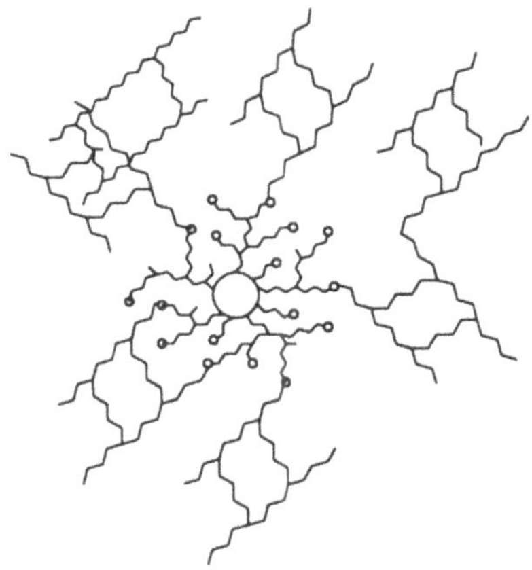

FIGURE 11.4 Schematic structure of graft IPNs based on PU and NL. Source: Huang and Zhang [31]. Reproduced with permission of Elsevier

hydrogels with IPNs or SIPNs were obtained. This new class of unsaturated polysaccharide derivatives was also studied by ^{13}C NMR, and signals are assigned to the components of two comonomers [29, 30].

A series of graft IPNs with different NCO/OH molar ratio was synthesized from castor oil-based polyurethane (PU) and nitrolignin (NL) by changing the content of 1,4-butanediol as chain extender. PU networks are previously formed, and then PU networks were grafted on the NL. In this case, the NL molecule can react with multi-PU network and even with PU entangled network. Interpenetration and entanglement of such networks like a star not only simultaneously enhanced the strength and stiffness of materials but also kept the retractability (Figure 11.4).

Structures of films with and without NL were compared by using solid-state CP MAS ^{13}C NMR (the contribution of NL on the spectra was neglected because of the low content of NL). Chemical shifts of carbons adjacent to urethane in PU–NL films are shifted to higher values, as compared to PU films, indicating that the motion of PU chain was restricted due to the relatively high crosslink network structure [31].

11.3.2 Structure and Dynamics

11.3.2.1 Low-Resolution ^1H Nuclear Magnetic Resonance Low-resolution ^1H NMR operating at 60 MHz was used to monitor spin–spin relaxation time (T_2), and the line shape analysis of FID has been applied to investigate the degree of phase dispersion of IPNs based on elastomeric PU (25 wt.%) and a crosslinked poly(methyl methacrylate) (PMMA). In the so-called in situ sequential IPNs, the two networks are formed after each other, and the network formed first is thought to impede gross phase separation in the final material. This is contrary to the other type of IPNs (in situ simultaneous), in which the formation of both networks is initiated at once and proceeds to completion more or less simultaneously. The results confirm that such IPNs, when prepared sequentially, have a higher degree of phase dispersion than those obtained by the simultaneous synthesis method. Furthermore, in the corresponding neat PU networks, built up from aromatic pluriisocyanate and poly(oxypropylene glycol) (POPG), the rigid crosslink points are not composed of isocyanate only, but include some oxypropylene mers; it appears that the amount of the rigidified part is the same, whatever the molecular weight of POPG [32].

The low-resolution ^{1}H NMR measurements of spin–lattice relaxation times T_1 and $T_1\rho$ (relaxation in the rotating frame) may also give some insight into the domain size because of the presence of spin diffusion in solids, that is, the transport of nuclear spin magnetization. It acts as an averaging process, causing closely adjacent phases with different intrinsic relaxation times to give a common averaged value, and is thus another possible approach to estimate the degree of phase interpenetration in IPNs. These experiments and the Goldman–Shen pulse sequence (also based on spin diffusion) are used to give an estimation of the degree of mixing of two networks in the matrix and of the size of the inclusions of linear PS in the simultaneous IPN. Different SIPNs of PU/PS are examined in order to relate the amount of PS added to the reaction medium with the degree of advancement of the phase separation. Nodules of pure PS are seen for some samples, but they do not correspond to those observed by SEM [33].

11.3.2.2 High-Resolution ^{1}H nuclear Magnetic Resonance From the analytical point of view the limitation of various high-resolution ^{1}H NMR techniques in solids arises from the relatively small chemical shift range (about 10 ppm) as compared with some other frequently studied nuclei. However, solid-state ^{1}H MAS NMR is able to probe local chemical environments and provide useful information concerning proton mobility of conducting polymers on the molecular level, such as in a novel SIPN membranes composed of Nafion117 and crosslinked poly(vinyl pyrrolidone) (PVP) which has been prepared to improve the proton selectivity of perfluorosulfonic acid (PFSA) membrane. The only observed proton peak at 9–10 ppm in the completely dry condition represents the proton exchange between the sulfonic acid group and surrounding water molecules. With the introduction of PVP, the line width becomes broader, indicating that the proton exchange between the sulfonic

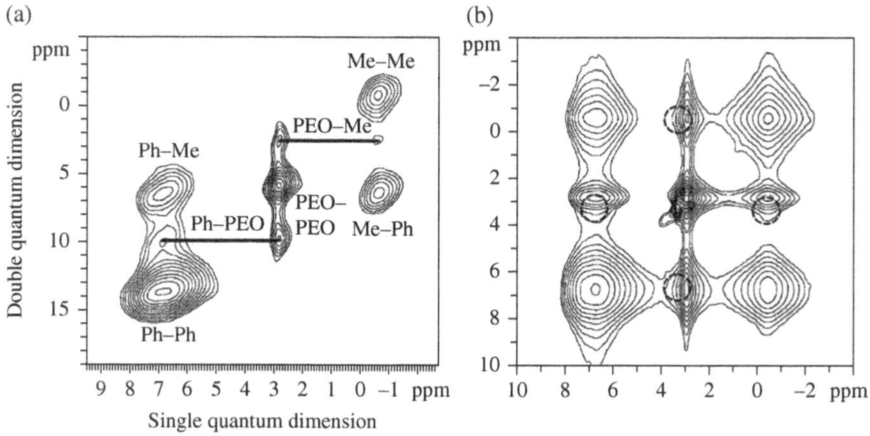

FIGURE 11.5 ^{1}H MAS NMR spectra of PEO/crosslinked silicone 10:90 recorded under MAS at 25 kHz: (a) ^{1}H–^{1}H DQ spectrum (the experiment was performed using the BABA sequence for DQ excitation and reconversion). (b) ^{1}H 2D NOESY spectrum with 50 ms mixing time. The circles in dashed lines indicate that there are no cross-peaks observed between peak II and other protons in the SIPN present. Source: Hou *et al.* [35]. Reproduced with permission of John Wiley & Sons

acid group and the residual water slows down and falls into the intermedium exchange regime. With the increase of humidity, the difference in line width between the Nafion and the Nafion/PVP SIPN membranes rapidly diminishes, indicating that the proton mobility is faster in SIPNs than in Nafion117 [34].

Microstructure and phase behavior of an SIPN consisting of 10% poly(ethylene oxide) (PEO) and 90% crosslinked silicone have been studied using various ^{1}H solid-state NMR methods under fast MAS in combination with well-known polymer characterization techniques. Both ^{1}H–^{1}H DQ MAS NMR spectroscopy and NOESY MAS measurements indicate a mixing of the two components on a molecular level. Thus, ^{1}H solid-state NMR spectra under fast MAS and ^{1}H–^{1}H DQ NMR spectroscopy provide a convenient means to obtain structural and dynamic information. These techniques as well as ^{1}H 2D NOESY MAS NMR are used to investigate the miscibility and the local structure of PEO and the silicone network in a SIPN in which PEO and the crosslinked silicone network are the guest and host polymers, respectively (Figure 11.5).

11.3.2.3 High-Resolution ^{13}C and ^{29}Si CP Magic Angle Spinning Nuclear Magnetic Resonance The chemical shifts and ^{1}H $T_{1}\rho$ from ^{13}C CP MAS experiments are used to investigate molecular interactions between unsaturated polyester (UP) and epoxy (EP). The peaks of IPNs are found to shift downfield, compared with pristine UP. This implies that there is molecular interaction (such as hydrogen bonding) between EP and UP. Moreover, with increasing EP content, the resonance shifted continuously downfield, revealing the gradual increase in hydrogen bonding. Since ^{1}H $T_1\rho$ is sensitive to the short-range spatial proximity of interacting dipole moments of protons, it can be used to determine the homogeneity of the sample. In the EP-rich phase all the values of ^{1}H $T_1\rho$ of the IPNs are lower than that of the pristine EP polymer. This indicates that IPNs possess more rapid molecular motions and faster relaxation than the pristine polymer. This also implies an increase in free volume, which makes the molecular chain easy to move and relax, and consequently the mechanical energy is dissipated effectively. Therefore, these IPN samples are suitable to be used as damping materials. However it should be mentioned that the phase separation was observed for IPNs with UP content higher than 30% [36].

In situ sequential semi-interconnected IPNs based on well-defined unsaturated poly(ester-urethane) (PEU) and PS were synthesized with five PEU/PS ratios including 100/0 (PEU-2), 90/10, 80/20, 70/30 (IS1C), and 50/50 (IS1D) in order to study the effect of length of PS intranetwork bridges on the structure. The ^{13}C CP MAS spectra of three samples are shown in Figure 11.6.

The spectra are recorded with the total sideband suppression (TOSS) technique. Although the intensities of all signals in CP MAS spectra recorded at certain contact time are not a quantitative measure for the concentration of particular carbon atoms, the spectrum of ISlD is similar to that of ISlC. The intensities of the peaks observed for PU were much lower due to the lower concentration of the PU in this IPN. As a result, the signals of quaternary carbons forming carbonyl groups were lost.

Furthermore, the molecular motions of various carbons were studied using the dynamics of the CP via ^{1}H $T_1\rho$ relaxation time, and the quantitative intensities of all the signals were calculated. The ^{1}H $T_1\rho$ values obtained for the PU signals in all the samples were in the range of 1–2 ms, whereas those for the PS signals were in the range of 5–12 ms. Thus, the proton spin diffusion does not pass across PEU and PS, indicating that the IPNs contain heterogeneous cages formed by the chain segments of PEU and PS. This also suggests that microdomains of PS may be present. The ^{1}H $T_1\rho$ values of the PS signals observed in both ISlC and ISlD were longer than those observed for the PS homopolymer, indicating restricted motion of the PS chain segments in the IPNs formed. This shows that the PS intranetwork bridges in the IPNs are constrained. Since the relaxation of PS bridges present in the ISlD is faster than those present in the IPN ISlC, the former is less constrained as

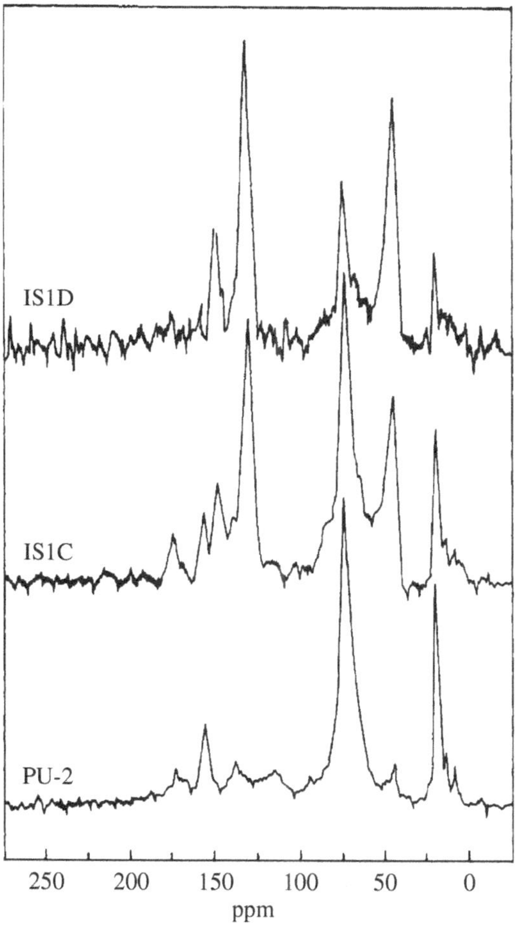

FIGURE 11.6 ^{13}C CP MAS spectra of IPNs PU-2, IS1C, and IS1D. Source: Pandit and Nadkarni [37]. Reproduced with permission of American Chemical Society

compared to the latter, indicating that the PS bridges in the ISlC are in the stretched configuration, while those in the case of ISlD are in the coiled configuration [37].

Novel SIPNs were synthesized on the basis of unplasticized poly(vinyl chloride) (PVC) in three different ways: one route with initiator ("i") and two routes without initiators ("ni"). PVC films containing approximately 10% of a rubbery acrylonitrile–butadiene–styrene (ABS)-derived impact modifier were modified using a vinyl silane from γ-methacryloxypropyltrimethoxysilane (γ-MPTS). A free radical initiator was used to promote reaction of the vinyl groups, and an aqueous acetic acid solution was used to promote the methoxysilane hydrolysis and condensation (HC) reactions for siloxane crosslink formation.

A gel consisting of silane alone was formed prior to the HC process, indicating the formation of an SIPN. The gel content following the HC process far exceeded the silane content, indicating that a significant amount of PVC was entrapped by the silane network. This conclusion is supported by the homogeneous molecular structure and morphology of the films. Several NMR techniques are applied to follow the course of synthesis. Classical liquid-state ^{13}C NMR was used to monitor MPTS after the addition of initiator, while solid-state ^{13}C CP MAS and ^{29}Si CP MAS techniques were used to characterize SIPNs.

It was shown that PVC modification via "i" route yields complete vinyl reaction and complete hydrolysis, yet moderate silane crosslinking (condensation). On the contrary, PVC modified via "ni" routes, in which no vinyl initiator was employed, indicates an incomplete vinyl reaction, complete HC, and a high degree of silane crosslinking. Without an initiator, the vinyl reaction is limited and phase separation results. Highly mobile side chains containing vinyl groups and attached to a siloxy network are found in phase-separated polysilane domains. Furthermore ^{1}H $T_{1}\rho$ relaxation measurements refine these observations by identifying highly mobile and moderately mobile components. A highly mobile component occurs when only HC reactions took place, while a less mobile component was also detected when vinyl reactions also took place [38].

Submicron hybrid particles, where the two components are crosslinked polyethylene (PE) and silica, were prepared by the simultaneous crosslinking of PE and polycondensation of tetraethoxysilane in hot, surfactant-stabilized oil-in-water (O/W) micelles. The previously unknown IPN hybrid particles are made on the basis of simplest organic and inorganic polymers: (i) linear noncrosslinked PE and silica (LPE/silica) and (ii) crosslinked PE and silica (XPE/silica).

The analysis of IPNs was made by solid-state ^{13}C CP MAS, ^{29}Si MAS, and ^{1}H MAS NMR techniques. The deconvolution of ^{29}Si NMR spectra reveals three signals arising from two, three, and four siloxane bonding, respectively, while the ^{13}C spectra of LPE/silica and XPE/silica differ from one other. The three signals in LPE/silica are attributed to amorphous and crystalline PE phase, while the fourth signal observed only in XPE/silica is attributed to the amorphous phase due to the crosslinking. The similar effect is observed in ^{1}H spectra. Furthermore, ^{1}H–^{13}C and ^{1}H–^{29}Si 2D heteronuclear correlation (HETCOR) NMR techniques have been applied to detect connections between ^{13}C or ^{29}Si atoms and the protons directly bonded to them. For example, ^{1}H–^{13}C HETCOR provides information about carbon types. Quaternary carbons have no attached protons and show no correlations, methine carbons show only one peak, methylene carbons can show one or two peaks depending on the chemical shifts of the two protons, while methyl groups generally show a single intense peak or a single multiplet. However, the spectra usually have quite a low resolution, so the fine structure due to multiplets is minimal.

According to ^{1}H–^{13}C HETCOR spectrum (Figure 11.7a), the ^{13}C signal at 32.9 ppm correlates to a very broad ^{1}H signal (300 ppm width at 10 kHz), assigned to the crystalline phase of PE. A second correlation is between ^{13}C

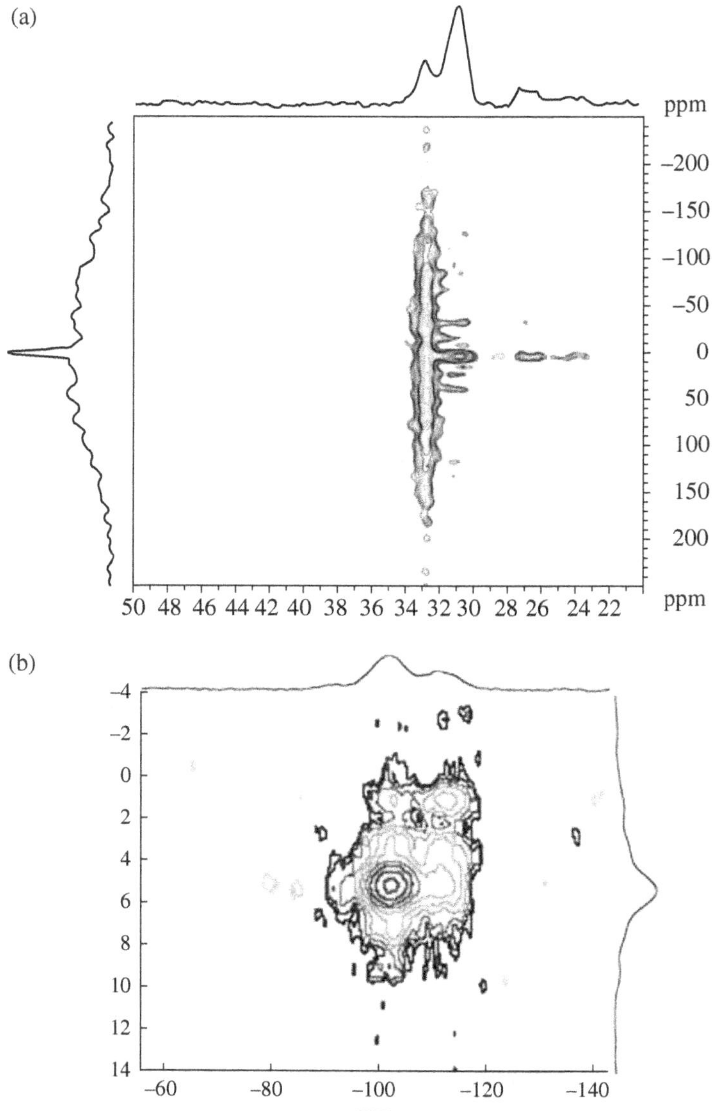

FIGURE 11.7 HETCOR 2D spectrum of XPE/silica: (a) ^{1}H–^{13}C. Vertical axis, ^{1}H; horizontal axis, ^{13}C. (b) ^{1}H–^{29}Si. Vertical axis, ^{1}H; horizontal axis, ^{29}Si. Source: Elimelech *et al.* [39]. Reproduced with permission of Royal Society of Chemistry

signal at 30.1 ppm (amorphous PE) and ^1H at 3.7 ppm (PEG segment of the surfactant), indicating that the surfactant and the amorphous PE are in close proximity to one another. From ^1H–^{29}Si HETCOR spectrum (Figure 11.7b), it is possible to study the interaction between the two components of the composite, the inorganic silica network and the organic crosslinked polymer and the surfactant, and determine the effect of PE crosslinking upon the interaction between these components. It was concluded that inner matrix silica is located relatively far from the amorphous phase of PE and the surfactant (the interaction is similar or lower compared to the linear polymer). However, the surface silica and not fully condensed SiO moieties are in the interaction with the amorphous part of polymer [39].

The modification of porous PVC particles by an in situ stabilizer-free polymerization/crosslinking of a monomer/crosslinker/peroxide solution absorbed within the PVC particles is also performed. PVC particles were modified by XPS (PS crosslinked with divinylbenzene (DVB) solutions), XPMMA (polymethyl methacrylate crosslinked with ethylene glycol dimethacrylate (EGDMA)), and XP(St-*co*-MMA) (styrene/methyl methacrylate 50/50 copolymer crosslinked with 5% DVB solution) using benzoyl peroxide as an initiator. The PVC/modifying monomer mass ratio was 75/25. The modified PVC particles are characterized with various techniques including the solid-state ^{13}C CP MAS technique (Figure 11.8).

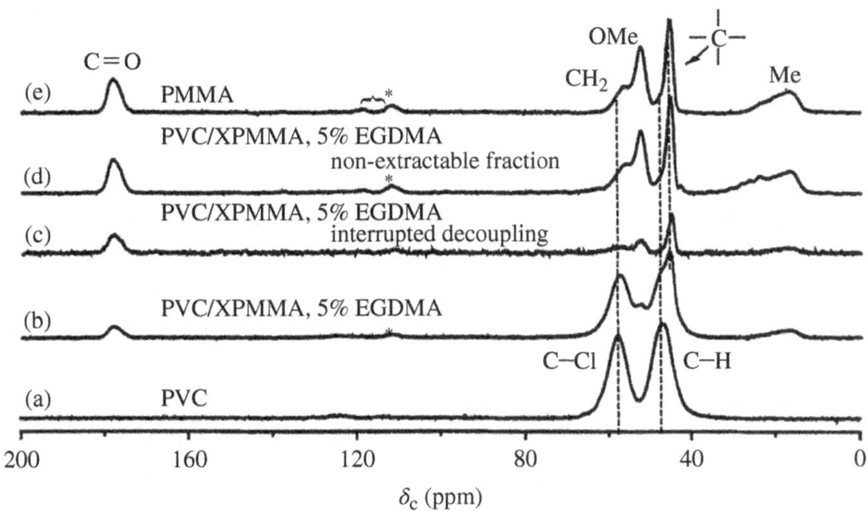

FIGURE 11.8 ^{13}C CP MAS spectra of PVC, PMMA, and PVC/XPMMA = 75/25, 5% EGDMA before and after extraction. Spinning sidebands are indicated by asterisks. Source: Shach-Caplan *et al.* [40]. Reproduced with permission of John Wiley & Sons

The spectra of the nonextractable fraction (spectrum d) and the neat PMMA (spectrum e) are almost identical. Thus, within the accuracy of the NMR measurements, one can conclude that there are no PVC in the nonextractable fraction and thus no indication of grafting of PMMA to PVC. Additionally, interrupted decoupling (ID) experiments were performed. The ID technique results in a substantial intensity attenuation of peaks belonging to nuclei with strong dipolar coupling to protons, leaving intact peaks from quaternary nuclei and from nuclei with motionally averaged dipolar interactions (i.e., methyl groups with fast CH_3 rotations). By employing this technique the ^{13}C CP MAS spectra are further simplified by sorting out individual contributions (PMMA vs. PVC) within the overlapping peaks. The ID emphasizes the presence of the XPMMA in PVC/XPMMA/5% EGDMA (spectrum c). In particular the quaternary carbon peak of PMMA (45.6 ppm), which overlaps the $—CH_2—$ peak of PVC (46 ppm), stands out using the ID pulse sequence. NMR results and the levels of nonextractable fractions found are indicative of low chemical interaction in the SIPN PVC particles [40].

IPNs composed of poly(vinyl alcohol) (PVA) and poly(acrylic acid) (PAAc) are formed by a sequential method, and interpolymer interactions were examined using ^{13}C CP MAS NMR. Evidence on the formation of a PVA–PAAc complex through hydrogen bonds between the hydroxyl groups of the PVA chains and the carbonyl group of the PAAc chains was obtained. The existence of polymer interactions between PVA and PAAc and its effect on the molecular motion of polymer chains was investigated with ^{13}C $T_1\rho$ relaxation times, while the scale of the mixing in the PVA/PAAc IPNs was studied from ^1H $T_1\rho$ relaxation times.

For all the IPNs prepared, the peaks in the ^{13}C CP MAS NMR spectra do not exhibit a single exponential ^{13}C $T_1\rho$ decay behavior due to the presence of a multiplicity of relaxation times (which is in agreement with DSC measurements). Samples with PVA compositions up to 25–30% show a single ^1H $T_1\rho$ value for all the carbons due to an efficient spin diffusion process. Above this concentration there exist different ^1H $T_1\rho$ values for carbons corresponding to PVA and PAAc. This can be attributed to the segregation of PVA from the PAAc network due to phase separation. These data reveal the compatibility between the PVA and the PAAc polymer networks at low PVA concentrations and the occurrence of phase separation at relatively high PVA concentrations [41].

Chitin/poly(2-hydroxyethyl methacrylate) (PHEMA) composites were synthesized via photopolymerization in a gel state of chitin swollen with HEMA monomer as reactive impregnant. The transition behavior and phase structure of the chitin/PHEMA composites were characterized by various techniques including solid-state ^{13}C CP MAS spectroscopy. For compositions containing ≤20 wt.% of chitin, it was assumed that an original network structure of chitin gels was sufficiently perpetuated into the crosslinked PHEMA matrix, resulting in the formation of an IPN-type organization. Generally, a strong interaction on a molecular scale (<1 nm) between two components in a binary blend exists, the electron density around the carbons bearing the interacting groups is perturbed, and some significant changes in chemical shift and/or line shape should be observed in an NMR spectrum compared with those of their respective homopolymers. However in this study these changes are not detected, that is, the spectra of PHEMA-rich composites are a simple superposition of the spectra of both homopolymers, with the apportioned peak intensities. From this observation, it is concluded that no specific intermolecular interaction such as hydrogen bonding takes place in these samples. From $^1H\ T_1\rho$ measurements it was estimated that these IPN-like composites are homogeneous on a scale of less than approximately 2.4 nm [42].

PHEMA and PHEMA interpenetrated with 5% SiO_2 (PHEMA-IPN) were studied by solid-state ^{13}C CP MAS spectroscopy. The similar spectra of two samples consist of six signals assigned to 1-methylene, 2-methylene, β-methylene, 3-methyl, α-quaternary, and carbonyl carbons, respectively. All chemical shifts are nearly independent of temperature. In this study ^{13}C $T_1\rho$ relaxation times are measured as a function of temperature, and the corresponding correlation times and activation energies for each carbon atom have been calculated. The ^{13}C $T_1\rho$ of the α-quaternary and carbonyl in the PHEMA and PHEMA-IPN undergo slow motions, while those of the 1-methylene, 2-methylene, β-methylene, and 3-methyl undergo fast motions. It was found that molecular motions in PHEMA interpenetrated with 5% SiO_2 have higher activation energies than in PHEMA alone. Higher activation energies indicate a higher rigidity for the polymer chains in the PHEMA-IPN sample. Furthermore, the activation energy of the 2-methylene carbon in PHEMA-IPN is distinctly greater than those in PHEMA. This higher activation energy associated with the hydroxy 2-methylene carbon is ascribed to strong interfacial interactions between this group and the SiO_2 phase [43].

The effect of selective crosslinking of poly(styrene-*co*-maleic anhydride) (SMA) in an initially fully miscible blend with poly(styrene-*co*-acrylonitrile) (SAN) is studied as a function of crosslink density. The crosslinking of SMA results in the formation of an SIPN and induces phase separation. For the purpose of ^{13}C CP MAS NMR analysis, SAN was ^{13}C-enriched to 98% at the CN position, and SMA was enriched to 98% at both the CO and the CH_2 positions. The ^{13}C chemical shift of the CN label in SAN is 114 ppm, and those of the CO and CH_2 labels in SMA are 169 and 20 ppm, respectively. The clearly separated responses in the spectra facilitate the discrimination between these polymers.

The CP^3 ^1H spin diffusion pulse sequence [44] was used for the determination of the domain sizes by 2D high-resolution CP MAS NMR COSY or by 1D version using a rotor-synchronized Delay Alternating with Nutation for Tailored Excitation (DANTE) [45] pulse sequence to reverse the ^{13}C magnetization on the CN label of SAN. From these experiments the characteristic scale of the blend morphology was determined on a microscopic level. An approximately linear dependence of the induced domain sizes on the amount of crosslinker was found. An increasing fraction of the blend becomes demixed upon increasing the amount of crosslinker [46].

11.4 ELECTRON SPIN RESONANCE STUDIES OF INTERPENETRATING POLYMER NETWORKS AND SEMI-INTERPENETRATING POLYMER NETWORKS

11.4.1 Nitroxyl Radicals in Studying Interpenetrating Polymer Networks and Semi-Interpenetrating Polymer Networks

The motional transition and heterogeneity of SIPNs based on segmented polyester PU with carboxylic groups and methacrylic copolymer (PM) with tertiary amine groups as a function of concentration of functional groups in both prepolymers were studied using the spin probe method [47]. Nitroxyl radical 2,2,6,6-tetramethyl-1-piperidinyloxy (TEMPO) was incorporated in the polymer networks in toluene solution. Molecular motion of SIPNs with different functional group concentrations (0–0.45 mmol/g) was examined, and the results of motional transition and heterogeneity on a molecular scale were compared with the T_g results measured by DSC. It was shown that the shape of the ESR spectra of spin probe depends on functional group concentration. Moreover, it was found that the temperature interval between the appearance of the fast component and the disappearance of the slow component and the ratio of these two components are influenced by the functional group content. Temperature dependence of the outer extrema separation in the ESR spectra ($2A_{zz}$) is shown in Figure 11.9.

The ESR spectra measured as the function of temperature are sensitive to polymer interactions imposed by functional groups which determine the free volume distribution in the matrix and temperature at which motional transition takes place. The fraction of free volume increases with functional group concentration and reaches its maximum at 0.25 mmol/g, while further increase in the functional group concentration reduces the free volume. These results are discussed in terms of the interference of plasticizing effect of the PU

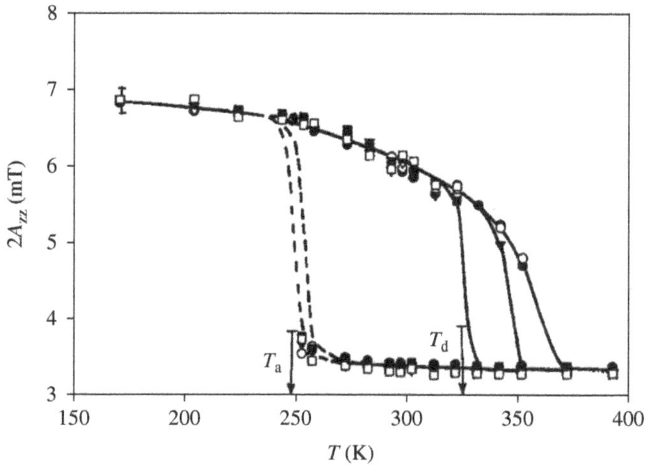

FIGURE 11.9 Temperature dependence of $2A_{zz}$ parameter for the spin-probed SIPN with different concentrations (10^{-2} mmol/g) of functional groups: 0 (●), 5 (○), 15 (▼), 25 (∇), 35 (■), and 45 (□). T_a and T_d are the temperatures of appearance of fast component and disappearance of slow component, respectively. Source: Čulin *et al.* [47]. Reproduced with permission of John Wiley & Sons

component and the formation of possible cluster crosslinks, which restricts segmental motions [47].

The same SIPNs with various functional group concentrations (0–0.45 mmol/g) and corresponding polymer mixtures (1 : 1 mass ratio) were studied by the spin label method [48]. Nitroxyl radical 4-amino-2,2,6,6-tetramethylpiperidine-1-oxyl (TAMINE) was covalently bonded to the PM component [48]. Heterogeneity of segmental motion and transitions due to the additional polymer interactions imposed by complementary functional groups were studied. Results were deduced by analyzing motional behavior of PM-labeled component in a wide temperature range. They show that crosslinking of the PU component and specific interactions between the carboxylic groups bound to the PU hard segments and tertiary amine groups on the PM chains restrict both PU and PM segmental motion. With an increased functional group concentration, more PM segments interact with the PU hard segments, thus increasing mass fraction of PU hard segments in the PM-rich phase. As a result a portion of PM chains interacting with the soft phase is lowered. Spectra shown in Figure 11.10 demonstrate the same trend in motional behavior of both PU/PM mixtures and SIPNs which seems to be dominated by functional group interactions. Considerably large fraction of the slow motion at temperatures between 343 and 373 K in the SIPNs confirm enhanced interpenetration between the PU hard segments and PM chains generated by both

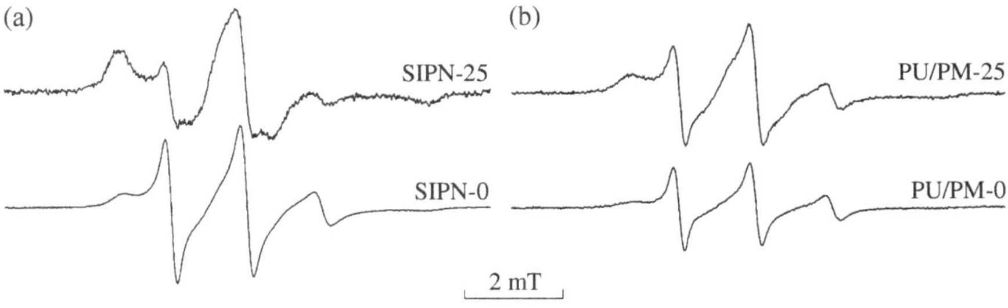

FIGURE 11.10 The ESR spectra of labeled PM in SIPNs (a) and PU/PM mixtures (b) without functional groups (0) and containing 0.25 mmol/g functional groups (25). Spectra were observed at 363 K. Source: Čulin *et al.* [48]. Reproduced with permission of Elsevier

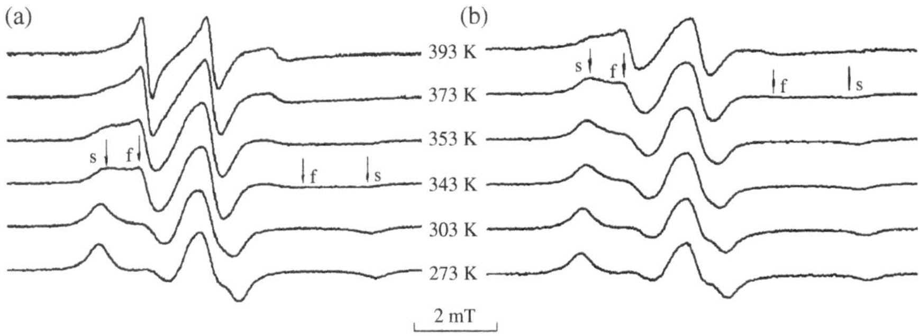

FIGURE 11.11 The ESR spectra of TAMINE covalently attached to the PU chains incorporated in SIPNs (a) and PU/PM mixtures (b) with maximal concentration of functional groups (0.45 mmol/g) observed at different temperatures. Source: Čulin *et al.* [20]. Reproduced with permission of Elsevier

chemical and physical crosslinks. The SIPNs appear to have much better chain interpenetration in comparison with polymer mixtures in which motional contribution of PM segments changed very little from 373 to 393 K. The motional behavior of SIPNs and PU/PM mixtures offers an insight into the generation of local segmental motion correlated to the polymer chain interpenetration and interaction in a wide temperature range.

Additional study was made with the same SIPNs and polymer mixtures (1 : 1 mass ratio) using PU component labeled by TAMINE [20]. It was observed that restriction of segmental motion of PU segments in the PU/PM mixtures increases with the augmentation of functional group content due to the additional noncovalent interactions. The critical concentration of functional groups, above which the motional restriction decreases, was found to be 0.35 mmol/g. Figure 11.11 presents the ESR spectra of labeled PU chains in SIPNs and PU/PM mixtures without functional groups and with their maximal concentration. The effect of functional groups is discussed in terms of the change of local packing density. Consequently, the ESR spectra of samples containing functional groups show higher amount of slow component. This effect is more pronounced in SIPNs than in mixtures. The temperatures of the appearance of the fast component and the disappearance of the slow component depend on the functional group concentration and are shifted to lower values for about 10–15 K, compared to the corresponding temperatures of the pure PU components. Contrarily, when these temperatures are compared to those of SIPNs with the same composition, shifts to higher values are observed. In fact, covalent bonding between the polymer components in SIPNs restricts segmental motions of PU chains. In addition, carboxylic groups incorporated into the PU hard segments and tertiary amine groups into the PM chains increase the possibility of complex interactions in the network. Those complex interactions influence the change of local free volume and segmental motion, which is reflected in the line shapes of temperature-dependent ESR spectra [20].

The morphology and miscibility of SIPNs based on poly(styrene-*co*-methacrylic acid) [P(S-*co*-MAA)] of different carboxylic acid contents and poly(ε-caprolactone) (PCL) have been studied using nitroxyl radical TEMPOL as a spin label. While the ESR spectra of spin-labeled PCL exhibit one motional component at any specific temperature, two spectral components are observed for all SIPNs over a temperature range ΔT ($\Delta T = T_d - T_a$) arising from two microphases: a PCL-rich microdomain and a P(S-*co*-MAA)-rich microdomain. The miscibility can be improved by increasing the carboxylic acid content, which could enhance the hydrogen bonding interactions between the ester groups of PCL and carboxylic acid groups in P(S-*co*-MAA). It was also found that the microphase separation became more remarkable with increasing intracomponent crosslinking of the SIPNs [49].

SIPNs with controlled hydrogen bonding interactions based on [P(S-*co*-MAA)] and poly(butyl methacrylate) (PBMA) were studied by the spin probe method [50]. The carboxylic acid groups in [P(S-*co*-MAA)] act not only as proton donor groups but also as crosslinking points. The study was focused on the effect of the hydrogen bonding between constitutive polymers and the crosslinking of [P(S-*co*-MAA)] on the motional dynamics and local heterogeneities. The coexistence of the two spectral components was ascribed to the probe molecules in a rigid [P(S-*co*-MAA)]-rich microdomain and in a flexible PBMA-rich microdomain, due to local composition fluctuations. The T_a values of SIPN were stabilized at 260 K, irrespective of the carboxylic acid content. A complex correlation between T_d and the carboxylic acid content was found, while the same T_a values for all the SIPN samples indicated that the appearance of the fast component was not sensitive to the polymer segmental motion. It was assumed that at temperatures below 260 K the spin probe molecules rotated in the "holes" of dimensions comparable to that of a probe in the polymer matrix, which resulted from the integrative effect on the microphase structure of the intracomponent crosslinking and the counterpart intercomponent hydrogen bonding [50]. When the SIPN sample contained 29 mol% of carboxylic acids, the plasticizing effect made the motion of [P(S-*co*-MAA)] segments less restricting. With further increase in the content of carboxylic acids, the extensive hydrogen bonding interactions acted as a crosslinking agent in the formation of point-like and possible physical crosslinks, which hindered the segmental motion of [P(S-*co*-MAA)]. It was also found that an increase in the carboxylic acid content to 38 mol% increases the $T_{5\,mT}$, indicating the interpenetration of the PBMA chains in the [P(S-*co*-MAA)] networks. The enhanced hydrogen bonding interactions made the interpenetration between the soft PBMA chains and the hard [P(S-*co*-MAA)] networks more and more pronounced, and consequently, the polymer matrix rigidity decreased. The improved matrix rigidity in the SIPN containing 38 mol% carboxylic acids was ascribed to the intensive interactions acting as physical crosslinking [50]. On the base of described results, the authors assumed that some type of competition exists in the microphase structure between the intercomponent hydrogen bonding interactions, which improved the miscibility of samples, and the intracomponent crosslinking, which might lead to phase separation.

A series of SIPNs from castor-oil-based PU and benzyl starch (BS) was investigated with nitroxyl radical TEMPOL. Two methods were applied: nitroxyl radicals were dispersed in SIPN matrices (spin probe) and covalently bonded to PU chains (spin label). Information about the segment size and interaction between the PU and BS components were obtained on the base of the ESR spectra. The spectra of both spin-probed and spin-labeled SIPNs, consisting of a single motional component, confirmed that the segment volume of PU in the SIPNs increased with the addition of BS, that is, the chain stiffness increased as a result of strong interactions between PU and BS macromolecules [51].

TEMPOL was applied as a spin probe in the study of molecular motions and compatibility of a thermodynamically incompatible polypropylene/polyester (PP/PES) SIPN system. The effects of PP network modification on its compatibility with PES were investigated. PP as a nonpolar polymer has no strong intermolecular interactions with other polymers. Therefore, by modification with isocyanuric acid triallylester (IATE), polar functional groups were introduced in PP chains. Their active sites (–NCO) can react with adequate functional groups in the polymer chains. The ESR spectra show isotropic slow motions with rising temperature and indicate the presence of two motional phases above 270 K. At temperatures above 330 K, spectra consist of three narrow lines indicating fast probe motions. The best interpenetrating and entangling of two phases are found for 87/13 PP/PS composition [52].

Main chain and segmental dynamics of polyisoprene (PI) and PMMA chains in SIPNs were systematically studied over a wide range of temperatures as a function of composition, crosslink density, and molecular weight [53]. The ESR spectra of TEMPOL in SIPNs with different compositions (35 and 50 wt.% of PMMA), crosslink density (2 and 4%), and molecular weight of PI in the low-temperature region are similar to the spectra of pure spin-probed PI. At higher temperatures, spectra show a central line of higher amplitude when compared to pure PI spectra, suggesting a dynamic anisotropic state of rotation. In the range from 293 to 390 K, the spectra are composed of two components differing in their outer peak separation and line shapes. These two spectral components correspond to spin probes embedded in two different motional environments, an indication of microphase separation. A shift of T_{5mT} values to higher temperatures with increasing PMMA content is observed due to the decreased mobility. The slow motion component is predominantly connected with spin probes embedded in phase resembling restricted molecular motion of the PMMA and PI chains close to rigid PMMA chains. The narrow component is due to the spin probes located in the mobile PI phase and flexible PMMA chains. At higher temperatures, the mobility of probe molecules in each component becomes comparable, and only fast motional spectra are obtained. It is assumed that the system has a range of chain dynamics centered on the two types of sites detected in ESR experiments. The significance of ESR spectroscopy as a potential tool that allows a close look into structural and dynamic heterogeneities in multicomponent polymer systems on a length scale too small (<5 nm) to be observed by DSC, DMA, or electron microscopy was clearly demonstrated. The shifting of the appearance of narrow component to higher temperature is an indication of the degree of interpenetration of rigid PMMA chains into highly mobile PI chains. The presence of rigid PMMA chains stiffens nearby highly mobile PI chains and imparts restrictions on its motional behavior. This gives rise to dynamic heterogeneities within the PI matrix. The PI chains located far from PMMA chains are undergoing molecular motions above its T_g, and those close to or intimately mixed with rigid PMMA chains are subjected to restricted molecular motion which resulted in the observed shift.

The characteristic molecular architecture developed during IPN synthesis, such as permanent physical entanglement of chains of both components and penetration of PMMA chains into entangled PI network, leads to the interaction and cooperative motional behavior of both components at higher temperature. Hence, the narrow component is at higher temperatures. Additionally, it is found that PI chains in close proximity to PMMA chains are stiffened to some degree, which in turn results in dynamic heterogeneity within PI and PMMA phases. The ESR observation confirms the degree of maximum miscibility at the concentration of PMMA less than 35 wt.%, sample M in Figure 11.12. Spectra of highly crosslinked SIPN with 50/50 composition (sample O) show a considerable decrease in contribution from the rigid phase with an increase in narrow component. The increased crosslink density was found to enhance mixing between the two phases and consequently induce some flexibility in the highly rigid PMMA chains. SIPN with 50/50 composition (sample H) show complex spectra with still a considerable proportion of spin probe motion arising from PMMA-rich domains located in highly mobile environment

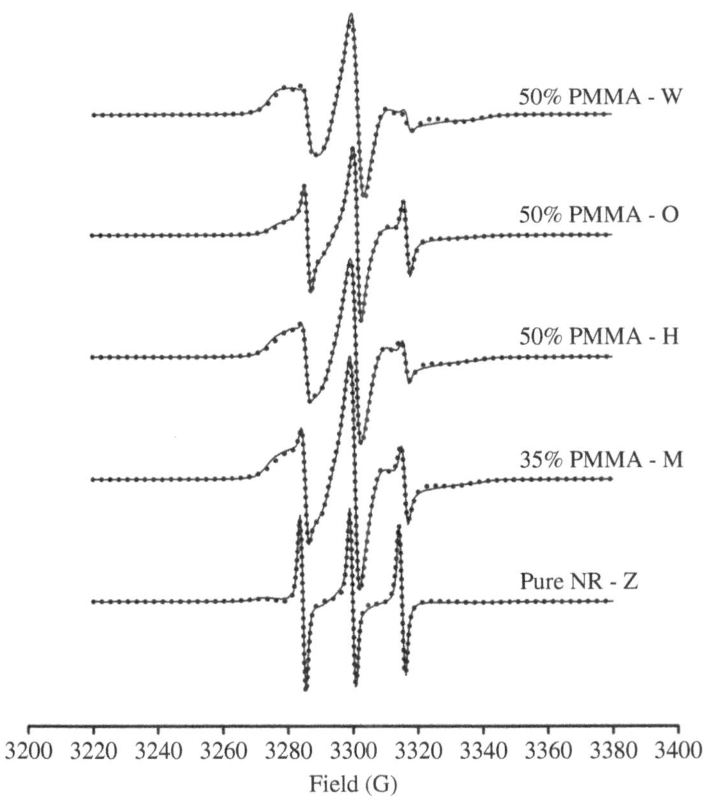

FIGURE 11.12 Experimental (full lines) and simulated (dotted lines) ESR spectra of SIPNs observed at 363 K. Source: John *et al.* [53]. Reproduced with permission of Elsevier

(slow component 87.4% and fast component 12.6%). For this to happen, the PMMA domains have to be considerably large and the probe must go deep into the domains. Lower crosslinking and higher PMMA content facilitated the development of larger PMMA domains in this SIPN. The low molecular weight SIPN with 50/50 composition (sample W) shows major contribution from rigid component (92.9%) even at higher temperature unlike other SIPNs. Here the ESR spectral line shape and simulation data reflects the continuity of two phases.

11.4.2 Radicals Induced by High-Energy Radiation

IPN hydrogel was made by the grafting of acrylamide (AAm) and acrylic acid (AAc) onto the preirradiated PP film. AAm and AAc were grafted onto the preirradiated PP film by two times grafting reaction. The ESR spectra show that the radiation-induced free radicals are still active after the first grafting. Trapped radicals on samples with different storage conditions and reaction conditions were determined by ESR [54].

The same types of hydrogels were prepared by grafting of AAm, *N*-methylolacrylamide (NMA), and AAc onto the preirradiated PP membrane. In the different stages of grafting under different reaction conditions, trapped radicals in the membrane samples were probed by ESR. The ESR spectra (Figure 11.13) show that the radiation-induced free radicals remained active after the first and second grafting reaction. Highest grafting degree of the IPN was obtained for AAc in the first grafting, second grafting of AAm was high,

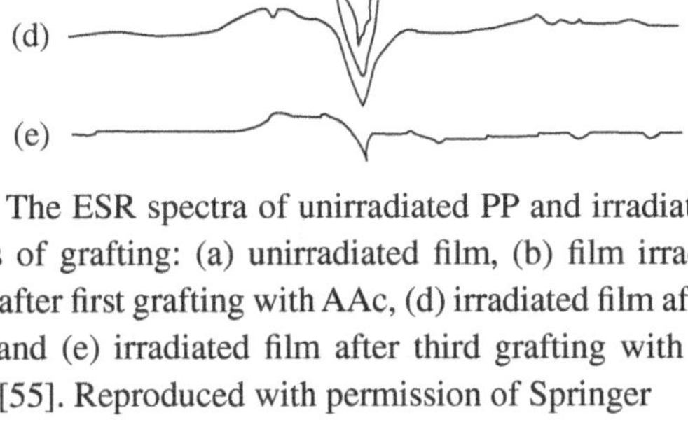

FIGURE 11.13 The ESR spectra of unirradiated PP and irradiated PP membranes at different stages of grafting: (a) unirradiated film, (b) film irradiated at 150 kGy, (c) irradiated film after first grafting with AAc, (d) irradiated film after second grafting with AAc/AAm, and (e) irradiated film after third grafting with AAc/AAm/NMA. Source: Wu *et al.* [55]. Reproduced with permission of Springer

and third grafting of NMA indicated that the remaining radicals after two grafting steps were still active. The grafting degree of PP-g-AAc hydrogel from the first grafting carried out at 70 °C for 2 h was 117.8%. The grafting degree of AAc/AAm onto PP-g-AAc was 266.8%. The degree of grafting of NMA in the third step at the temperature of 80 °C onto PP-g-(AAc-AAm) hydrogel varied with the time. With the increasing of reaction time from 1 to 5 h, the grafting degree of NMA onto PP-g-(AAc-AAm) was 31.2, 57.8, 88.3, 103.5, and 122.7%, respectively. It was concluded that using this method, the NMA components of the grafted copolymer can be easily adjusted by controlling the reaction time [55].

Chen *et al.* studied the grafting reactivity of the electron beam preirradiated PP film with AAc and AAm repeatedly or intermittently. The preirradiated PP film could be grafted several times intermittently, and the free radicals or peroxides on the samples could be determined after several times grafting reaction. The effects of storage time, reaction time, and repeated reaction times on the degree of grafting were investigated. The trapped radicals, peroxy radicals, and peroxides on the preirradiated and reacted PP films were determined by using ESR and 1,1-diphenyl-2-picrylhydrazyl (DPPH), respectively. When the samples were stored at −20 °C, the decrease of free radicals was slow, and a clear free radical signal could be still detected by ESR even after 20 days. At room temperature, however, the decrease was very fast, and almost no free radical signal could be found by ESR with 20-day storage after irradiation (Figure 11.14). The trapped radicals may react with O_2 during the storage and

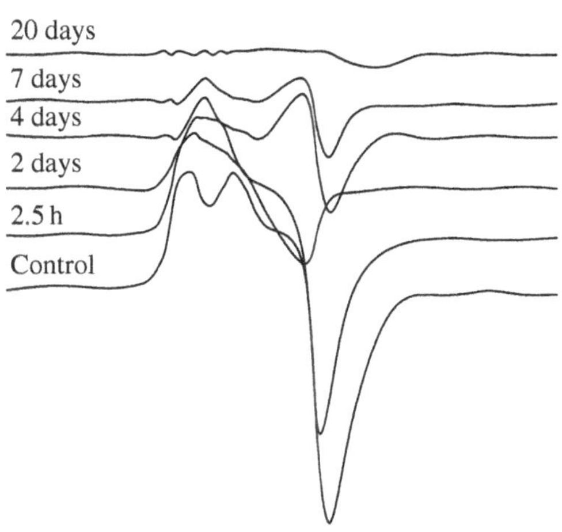

FIGURE 11.14 The ESR spectra of the preirradiated PP film at different storage times. Sample was stored at room temperature. Source: Chen *et al.* [56]. Reproduced with permission of Elsevier

form peroxy radicals quickly, and then they may be transformed into relatively stable peroxides at last. At room temperature, the transformation of peroxy radicals into peroxides was faster. Therefore, the amount of peroxide on the preirradiated PP films increased with the storage time, and the increase is faster when the samples were stored at room temperature [56].

11.4.3 Copper(II) Ions

A new mixed-valence copper(I,II) coordination complex with $[Cu_2O_8]$ paddle-wheel dicopper units has been hydrothermally synthesized and characterized by single-crystal X-ray diffraction, IR, XPS, ESR, and TGA. It exhibits a 2D interpenetrating grid framework, in which five-coordinated Cu(II) and three-coordinated Cu(I) environments are established. The oxidation state of Cu ions has been confirmed by XPS and ESR. The powder ESR spectra observed at room temperature supported the existence of Cu(II). Owing to the Cu(II)–Cu(II) interaction bridged by carboxylate groups, the ESR spectrum demonstrates two sets of signals for effective S states $S=1$ and $S=1/2$, respectively. The experimental spectrum was well reproduced using the parameters in an axial symmetry ($S=1$, $g_{\parallel}=2.4126$, $g_{\perp}=2.1180$, $D=0.376\,cm^{-1}$; $S=1/2$, $g_{\parallel}=2.2866$, $g_{\perp}=2.0903$; $E=0\,cm^{-1}$). These results illustrate well that reaction conditions, especially metal-to-ligand ratio and temperature, play very important roles in the formation of different coordination frameworks, and this also offers the possibility to control the formation of such network structures by varying these factors [57].

The synthesis and structural characterization of the 2D Cu(II) coordination polymer, namely, $\{[Cu_2(BPP)_4(NCS)_4]\}_n$, where BPP is the nitrogen ligand 1,3-bis(4-pyridyl)propane, are described in Ref. [58]. Single-crystal diffraction analysis shows that the asymmetric unit consists of two crystallographically independent Cu(II) ions that adopt a distorted octahedral geometry (Figure 11.15). Each Cu(II) center is coordinated by four nitrogen atoms from different BPP ligands and by other two nitrogen atoms from isothiocyanate groups. The BPP ligands link the metal centers generating an undulated 2D net of (4,4) topology. Two sets of 2D sheets interlock each other in the same plane, giving rise to a twofold parallel interpenetrating network.

Figure 11.16a shows ESR spectrum of a polycrystalline sample of $\{[Cu_2(BPP)_4(NCS)_4]\}_n$ measured at room temperature. The spectrum exhibits broad lines, which are explained by two different isolated Cu^{2+} ions consistent with the structure determined from the X-ray diffraction data. The ESR spectrum is due to electron Zeeman and hyperfine interaction of two different $S=1/2$ and $I=3/2$ spin systems. Hyperfine structure is not well resolved. However,

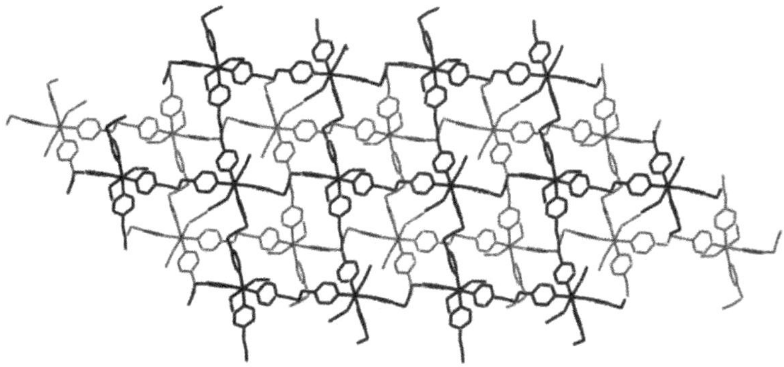

FIGURE 11.15 Two-dimensional interpenetrated network in $\{[Cu_2(BPP)_4(NCS)_4]\}_n$. Source: Farani *et al.* [58]. Reproduced with permission of Elsevier

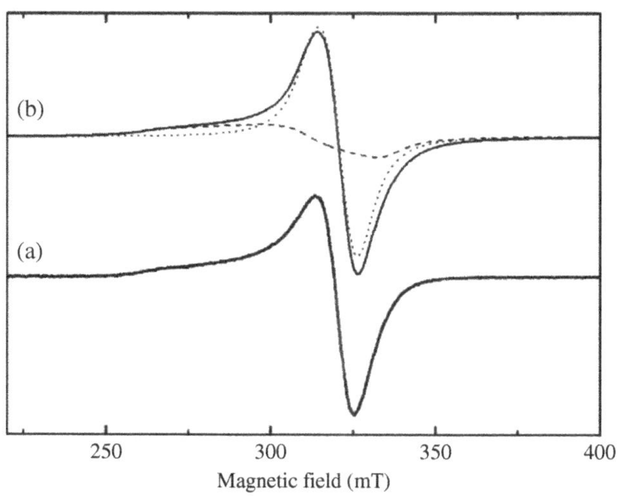

FIGURE 11.16 The EPR spectra of $\{[Cu_2(BPP)_4(NCS)_4]\}_n$. (a) Experimental and (b) calculated: two isolated Cu(II) centers (traced line) and the sum of both (solid line). Source: Farani *et al.* [58]. Reproduced with permission of Elsevier

it contributes to the line width and asymmetric line shape. Figure 11.16b shows two calculated Cu^{2+} centers (traced lines) together with the sum of both showing good agreement with the experimental spectrum (Figure 11.16a). The calculation reveals that both centers have nearly axial **g** tensors. ESR measurements down to low temperatures (10 K) do not modify significantly the line shape and spin Hamiltonian parameters. No magnetic coupling of the two individual Cu^{2+} centers is observed. From the previous results authors conclude that Cu^{2+} ions are bound in two nearly axial symmetry sites, one of them more distorted than the other. These observations are consistent with the crystal structure for $\{[Cu_2(BPP)_4(NCS)_4]\}_n$, in which the two individual Cu^{2+} sites have a distance of about 1.3 nm. The ESR spectra do not reveal the superhyperfine structure because the nitrogen interactions contribute to line broadening.

11.5 CONCLUSION

This chapter presents an overview of application of NMR and ESR spectroscopies in studies of IPNs and SIPNs. Reviewed articles clearly demonstrate the high potential of magnetic resonance techniques for investigation of structure and motional heterogeneity at the segmental level, usually not detected by other techniques. The applied solid-state NMR methods include relaxation measurements and low- and high-resolution ^1H, ^{13}C, and ^{29}Si CP MAS, while ESR experiments cover nitroxyl radicals used as spin probe or spin label, radicals induced by high-energy radiation, and copper(II) ions. These magnetic resonance techniques together with FTIR spectroscopy, WAXRD, XPS, DSC, TGA, DMA, and SEM offer the possibility to obtain a more complete picture of these complex polymer systems and details that are important for their application.

REFERENCES

[1] Knicker, H. (2011) Solid state CPMAS C-13 and N-15 NMR spectroscopy in organic geochemistry and how spin dynamics can either aggravate or improve spectra interpretation. *Organic Geochemistry*, **42**, 867–890.

[2] Dybowski, C. and Lichter, R.L. (1987) *NMR Spectroscopy Techniques*, Marcel Dekker, New York.

[3] Andreis, M. and Koenig, J.L. (1989) Application of NMR to crosslinked polymer systems. *Advances in Polymer Science*, **89**, 69–160.

[4] Schmidt-Rohr, K. and Spiess, H.W. (1994) *Multidimensional Solid-State NMR and Polymers*, Academic Press, San Diego, CA.

[5] Jakobson, H.J. (2002) A Retrospect on the Development of High-Speed Sample Spinners for Solids Since 1980, in *Encyclopedia of Nuclear Magnetic Resonance* (eds D.M. Grant and R.K. Harris), Wiley, Chichester, p. 5704.

[6] Feike, M., Demco, D.E., Graf, R. *et al.* (1996) Broadband multiple-quantum NMR spectroscopy. *Journal of Magnetic Resonance, Series A*, **122**, 214–221.

[7] Berliner, L.J. (ed) (1976) *Spin Labeling II: Theory and Application*, Academic Press, New York.

[8] Marsh, D. (1981) Electron Spin Resonance: Spin Labels, in *Membrane Spectroscopy* (ed E. Grell), Springer-Verlag, Berlin, p. 51.

[9] Weil, J.A. and Bolton, J.R. (2007) *Electron Paramagnetic Resonance: Elementary Theory and Practical Applications*, Wiley-Interscience, Hoboken, NJ.

[10] Schlick, S. (ed) (2006) *Advanced ESR Methods in Polymer Research*, Wiley-Interscience, Hoboken, NJ.

[11] Valić, S. (2010) Electron Spin Resonance in Studying Nanocomposite Rubber Materials, in *Rubber Nanocomposites: Preparation, Properties and Applications* (eds S. Thomas and S. Ranimol), John Wiley & Sons, Singapore, p. 391.

[12] Veksli, Z., Andreis, M. and Rakvin, B. (2000) ESR spectroscopy for the study of polymer heterogeneity. *Progress in Polymer Science*, **25**, 949–986.

[13] Valić, S., Andreis, M. and Klepac, D. (2011) ESR Spectroscopy of Multiphase Polymer Systems, in *Handbook of Multiphase Polymer Systems* (eds A. Boudenne, L. Ibos, Y. Candau and S. Thomas), John Wiley & Sons Ltd., Chichester, p. 551.

[14] Berliner, L.J. (1998) *Spin Labelling; The Next Millennium*, Plenum Press, New York.

[15] Bordignon, E. and Steinhoff, H.J. (2007) Membrane Protein Structure and Dynamics Studied by Site-Directed Spin Labeling ESR, in *ESR Spectroscopy in Membrane Biophysics* (eds M.A. Hemminga and L.J. Berliner), Springer Science+Business Media, New York, p. 129.

[16] Marinović, T., Valić, S., Andreis, M. and Veksli, Z. (1991) Study of matrix inhomogeneity of natural-rubber and synthetic polyisoprenes by a spin probe method. *Polymer*, **32**, 2519–2522.

[17] Valić, S., Andreis, M., Veksli, Z. and Charlesby, A. (1991) Matrix inhomogeneity in crosslinked rubber and rubber emulsions. *Radiation Physics and Chemistry*, **37**, 257–261.

[18] Kovarskii, A.L., Kulish, E.I., Placek, Y. *et al.* (1993) Application of the spin-probe method for the study of 2-phase polymer systems—poly(vinyl chloride) polyethylene blends [Russian]. *Vysokomolekulyarnye Soedineniya, Seriya A*, **35**, A1669–A1673.

[19] Hloušková, Z., Tiňo, J. and Chodák, I. (1994) Study of PE-based composites by the spin-probe method. *European Polymer Journal*, **30**, 175–178.

[20] Čulin, J., Andreis, M., Veksli, Z. *et al.* (2005) ESR-spin labelling study of semi-interpenetrating networks and polymer mixtures based on functionalized polyurethanes and polymethacrylates. *European Polymer Journal*, **41**, 1874–1882.

[21] Schlick, S., Harvey, R.D., Alonsoamigo, M.G. and Klempner, D. (1989) Study of phase-separation in interpenetrating polymer networks using nitroxide spin labels. *Macromolecules*, **22**, 822–830.

[22] Jeschke, G. (2002) Determination of the nanostructure of polymer materials by electron paramagnetic resonance spectroscopy [Review]. *Macromolecular Rapid Communications*, **23**, 227–246.

[23] Didović, M., Klepac, D. and Valić, S. (2008) ESR-spin probe method in studying natural rubber: an effect of the probe incorporation temperature. *Macromolecular Symposia*, **265**, 144–147.

[24] Pace, M.D. and Snow, A.W. (1995) Nitroxide spin probe/label study of hydrogen bonding and probe size effects in a linear epoxy polymer. *Macromolecules*, **28**, 5300–5305.

[25] Crick, D.W. and Alexandratos, S.D. (1993) Characterization of bifunctional interpenetrating polymer networks via solid-state ^{13}C NMR-spectroscopy. *Macromolecules*, **26**, 3267–3270.

[26] Kim, S.J., Shin, S.R., Spinks, G.M. *et al.* (2005) Synthesis and characteristics of a semi-interpenetrating polymer network based on chitosan/polyaniline under different pH conditions. *Journal of Applied Polymer Science*, **96**, 867–873.

[27] Shin, S.R., Park, S.J., Yoon, S.G. *et al.* (2005) Synthesis of conducting polyaniline in semi-IPN based on chitosan. *Synthetic Metals*, **154**, 213–216.

[28] Kim, S.J., Lee, N.R., Yi, B.J. and Kim, S.I. (2006) Synthesis and characterization of polymeric acid-doped polyaniline interpenetrating polymer networks. *Journal of Macromolecular Science, Part A: Pure and Applied Chemistry*, **43**, 497–505.

[29] Hamcerencu, M., Desbrieres, J., Popa, M. and Riess, G. (2009) Stimuli-sensitive xanthan derivatives/N-isopropylacrylamide hydrogels: influence of cross-linking agent on interpenetrating polymer network properties. *Biomacromolecules*, **10**, 1911–1922.

[30] Hamcerencu, M., Desbrieres, J., Popa, M. and Riess, G. (2012) Original stimuli-sensitive polysaccharide derivatives/N-isopropylacrylamide hydrogels. Role of polysaccharide backbone. *Carbohydrate Polymers*, **89**, 438–447.

[31] Huang, J. and Zhang, L.N. (2002) Effects of NCO/OH molar ratio on structure and properties of graft-interpenetrating polymer networks from polyurethane and nitrolignin. *Polymer*, **43**, 2287–2294.

[32] Parizel, N., Meyer, G. and Weill, G. (1993) Nuclear magnetic resonance lineshape studies of interpenetrating polymer networks. *Polymer*, **34**, 2495–2502.

[33] Parizel, N., Meyer, G. and Weill, G. (1995) Morphologies of semi and full interpenetrating polymer networks by nuclear magnetic resonance relaxation times. *Polymer*, **36**, 2323–2330.

[34] Li, T., Zhong, G.M., Fu, R.Q. and Yang, Y. (2010) Synthesis and characterization of Nafion/cross-linked PVP semi-interpenetrating polymer network membrane for direct methanol fuel cell. *Journal of Membrane Science*, **354**, 189–197.

[35] Hou, S.S., Graf, R., Spiess, H.W. and Kuo, P.L. (2001) An investigation into PEO/crosslinked-silicone semi-interpenetrating polymer network using ^{1}H solid-state NMR spectroscopy under fast MAS. *Macromolecular Rapid Communications*, **22**, 1386–1389.

[36] Shih, Y.F. and Jeng, R.J. (2004) IPNs based on unsaturated polyester/epoxy: IV. Investigation on hydrogen bonding, compatibility and interaction behavior. *Polymer International*, **53**, 1892–1898.

[37] Pandit, S.B. and Nadkarni, V.M. (1994) Sequential interconnected interpenetrating polymer networks of polyurethane and polystyrene. 1. Synthesis and chemical structure elucidation. *Macromolecules*, **27**, 4583–4594.

[38] Sluszny, A., Silverstein, M.S., Kababya, S. *et al.* (2001) Novel semi-IPN through vinyl silane polymerization and crosslinking within PVC films. *Journal of Polymer Science Part A: Polymer Chemistry*, **39**, 8–22.

[39] Elimelech, H., Nedelec, J.M., Hardy-Dessources, A. *et al.* (2010) Cross-linked polyethylene@silica: the first full interpenetrating network hybrid particles. *Journal of Materials Chemistry*, **20**, 9515–9522.

[40] Shach-Caplan, M., Narkis, M. and Silverstein, M.S. (2002) Modification of porous suspension-PVC particles by stabilizer-free aqueous dispersion polymerization of absorbed monomers. *Polymer Engineering & Science*, **42**, 911–924.

[41] Hernandez, R., Perez, E., Mijangos, C. and Lopez, D. (2005) Poly(vinyl alcohol)-poly(acrylic acid) interpenetrating networks. Study on phase separation and molecular motions. *Polymer*, **46**, 7066–7071.

[42] Miyashita, Y., Kobayashi, R., Kimura, N. *et al.* (1997) Transition behavior and phase structure of chitin/poly(2-hydroxyethyl methacrylate) composites synthesized by a solution coagulation bulk polymerization method. *Carbohydrate Polymers*, **34**, 221–228.

[43] Lim, A.R., Schueneman, G.T. and Novak, B.M. (1999) The T-l rho C-13 spin-lattice relaxation time of interpenetrating networks by solid state NMR. *Solid State Communications*, **109**, 465–470.

[44] Mulder, F.M., Heinen, W., van Duin, M. *et al.* (1998) Spin diffusion with C-13 selection and detection for the characterization of morphology in labeled polymer blends with MAS NMR. *Journal of the American Chemical Society*, **120**, 12891–12894.

[45] Bodenhausen, G., Freeman, R. and Morris, G.A. (1976) Simple pulse sequence for selective excitation in fourier-transform NMR. *Journal of Magnetic Resonance*, **23**, 171–175.

[46] Mulder, F.M., Heinen, W., van Duin, M. *et al.* (2000) Cross-linking induced phase separation in SAN/SMA semi-interpenetrating polymer networks observed by solid state NMR and site specific isotope enrichment. *Macromolecules*, **33**, 5544–5548.

[47] Čulin, J., Veksli, Z., Anžlovar, A. and Žigon, M. (2003) Spin probe study of semi-interpenetrating polymer networks based on polyurethane and polymethacrylate functional prepolymers. *Polymer International*, **52**, 1346–1350.

[48] Čulin, J., Šmit, I., Andreis, M. *et al.* (2005) Motional heterogeneity and phase separation of semi-interpenetrating networks and mixtures based on functionalised polyurethane and polymethacrylate prepolymers. *Polymer*, **46**, 89–99.

[49] Qiu, F.R., Chen, S.M., Ping, Z.H. and Yin, G.M. (2005) ESR spin-label study of poly(styrene-co-methacrylic acid)/poly(epsilon-caprolactone) semi-interpenetrating polymer networks with controlled hydrogen-bond interactions. *Magnetic Resonance in Chemistry*, **43**, 918–925.

[50] Qiu, F.R., Chen, S.M. and Ping, Z.H. (2005) Study of the miscibility and segmental motion of STMAA-PBMA polymer blends and semi-interpenetrating polymer networks by an ESR spin probe method. *Magnetic Resonance in Chemistry*, **43**, 411–416.

[51] Cao, X.D. and Zhang, L.N. (2005) Miscibility and properties of polyurethane/benzyl starch semi-interpenetrating polymer networks. *Journal of Polymer Science Part B: Polymer Physics*, **43**, 603–615.

[52] Wang, J.Y., Li, F., Li, Y.X. *et al.* (2000) Study on molecular movement and compatibility of polypropylene/polyester semi-interpenetrating polymer networks. *Journal of Polymer Engineering*, **20**, 201–211.

[53] John, J., Klepac, D., Didović, M. *et al.* (2010) Main chain and segmental dynamics of semi interpenetrating polymer networks based on polyisoprene and poly(methyl methacrylate). *Polymer*, **51**, 2390–2402.

[54] Chen, J., Yang, L.M., Wu, M.H. *et al.* (2000) Preparation of interpenetrating polymer networks by two times grafting of monomers onto preirradiated polypropylene film. *Radiation Physics and Chemistry*, **59**, 313–316.

[55] Wu, M.H., Chen, J., Qian, Q. and Bao, B.R. (2002) Thermo-responsive interpenetrating polymer networks composed of AAc/AAm/NMA by radiation grafting. *Journal of Radioanalytical and Nuclear Chemistry*, **252**, 531–535.

[56] Chen, J., Yang, L.M., Chen, L.Q. *et al.* (2004) An interesting grafting reactivity of EB preirradiated polypropylene film. *Radiation Physics and Chemistry*, **69**, 149–154.

[57] Wen, Y.H., Cheng, H.K., Zhang, H. *et al.* (2004) A new mixed-valence copper(I,II) coordination polymer with 2-D interpenetrating framework based on paddle-wheel dicopper units. *Inorganic Chemistry Communications*, **7**, 1120–1123.

[58] Farani, R.D., Teles, W.M., Pinheiro, C.B. *et al.* (2008) Spectroscopic and structural analyses of the copper(II) 2-D coordination polymer $\{[Cu_2(BPP)_4(NCS)_4]\}_n$ (BPP=1,3-bis(4-pyridyl)propane) comprising interpenetrated layers of (4,4) topology. *Inorganica Chimica Acta*, **361**, 2045–2050.

12

DIFFUSION, TRANSPORT, AND BARRIER PROPERTIES OF IPNs

RUNCY WILSON[1], S. ANIL KUMAR[2], MIRAN MOZETIČ[3], UROŠ CVELBAR[3] AND SABU THOMAS[1,4]

[1] *School of Chemical Sciences, Mahatma Gandhi University, Kottayam, Kerala, India*
[2] *Department of Chemistry, N.S.S. College, Ottapalam, Kerala, India*
[3] *Department of Surface Engineering, Jozef Stefan Institute, Ljubljana, Slovenia*
[4] *International and Interuniversity Center for Nanoscience and Nanotechnology, Mahatma Gandhi University, Kottayam, Kerala, India*

12.1 INTRODUCTION

An interpenetrating polymer network (IPN) is a polymer comprising two or more networks which are at least partially interlaced on a polymer scale but not covalently bonded to each other. The network cannot be separated unless chemical bonds are broken. The two or more networks can be envisioned to be entangled in such a way that they are concatenated and cannot be pulled apart, but not bonded to each other by any chemical bond. Simply mixing two or more polymers does not create an interpenetrating polymer network. The interpenetrating polymer network (IPN) is an assembly of interpenetrating networks of two crosslinked polymers (having or without having mutual chemical interaction), at least one of which is synthesized and crosslinked in

Micro- and Nano-structured Interpenetrating Polymer Networks: From Design to Applications, First Edition. Edited by Sabu Thomas, Daniel Grande, Uroš Cvelbar, K.V.S.N. Raju, Ramanuj Narayan, Selvin P. Thomas and Akhina H.

the presence of the other [1]. If only one component of the assembly is crosslinked leaving the other in linear form, the system is termed a semi-interpenetrating polymer network (SIPN) [2]. Many SIPNs are generally formed from water-soluble polymers by crosslinking them either using radiation or chemicals or by polymerizing hydrophilic monomers in the presence of a crosslinker. SIPNs have been studied with a particular emphasis being put on their reversible volume changes in response to external stimuli, such as temperature, pH, solvent composition, ionic strength, and electric field [3, 4]. Because of their drastic change of properties in response to environmental stimuli, these SIPNs may be used in controlled drug delivery [5], molecular separation [6], tissue culture substrates [7], and several technical applications [8].

The IPN polymeric systems result from interpenetration of a few polymer networks, when one of them is formed by polymerization or crosslinking in the presence of the other. In most cases, the processes that form the polymer systems follow different polymerization mechanisms, for example, free radical chain polymerization, typical for vinyl monomers, and step-growth polymerization polyaddition or polycondensation in the case of polyurethanes (PUR) or epoxy resins. The polymer systems produced in that way form separate but interpenetrating spatial structures, with basically no chemical bonds observed between them [5, 9–12]. Chemical composition of the network, degree of interpenetration of system components, and possible physical interactions between functional groups affect morphology of IPN-forming polymers. These factors influence the physicomechanical properties of IPNs and create new possibilities for employing them as a modern organic material in many applications like membranes, gel permeation chromatography, drug release control, and so on [13–15].

12.2 BACKGROUND OF INTERPENETRATING POLYMER NETWORKS

The literature shows that IPNs were invented in the beginning of 1914. The Aylsworth was first in inventing an IPN in 1914 by combining the then new phenol–formaldehyde compositions with natural rubber and sulfur [16, 17]. In effect, by adding natural rubber and sulfur, Aylsworth made the world's first rubber-toughened plastic, decades before rubber-toughened styrenics became commercial. The Aylsworth's composition was used for toughening phonograph records and other materials from 1914 to 1929. The next known discovery of IPNs was by Staudinger and Hutchinson as revealed in a patent with a 1941 application date [18]. To smooth the surfaces of transparent plastic sheeting for esthetic reasons, Staudinger and Hutchinson took sheets of cross-linked polystyrene (PS) or polymethyl methacrylate (PMMA) and swelled the sheets with the same monomer mix as network I. After polymerization, the swelled sheets were stretched to reduce the waviness or surface imperfections of the original material. The next independent invention of IPNs was by Solt in 1955 [19]. Solt developed cationic–anionic ion exchange resins using suspension-sized particles, and both networks were oppositely charged. The idea of both charges on the same particle (rather than different particles) was that the exchange efficiency would be improved if both charges were in juxtaposition, but still separated in space. Indeed, the charges must be separated, even by a fraction of a nanometer, to prevent coacervation. The term "interpenetrating polymer networks" was coined by Millar [20] in 1960. Several investigating teams found IPNs independently in the 1960s. Early work by Shibayama and Suzuki [21], Lipatov and Sergeeva [22], Frisch *et al.* [23], and Sperling and Friedman [24] set the pace for modern development, founding the field of IPNs.

12.3 TRANSPORT PROPERTIES: THEORETICAL AND PRACTICAL ASPECTS

The transport behavior of various organic solvents and gases through polymer membranes is of great importance, because they are widely used for various barrier applications. These transport studies become of utmost importance while designing barrier material or tubes for transporting liquids and gases. Molecular transport through polymers has become a subject with a variety of challenges and opportunities for practitioners of chemical engineering and applied chemistry in recent years [25–34]. The knowledge of the performance of polymers in the environment of hazardous solvents, vapors, and temperature is essential for their successful applications as structural engineering materials. Researchers are more and more enthusiastically adopting the use of small molecules as microstructural probes for polymers [35].

In biomedical applications also polymeric membranes are extensively used. Polymer films serve, by selective diffusion, to remove certain components from the blood when a damaged kidney can no longer do so. The components to be removed are urea, uric acid, water, creatine, phosphates, and excess chlorides from the blood through the membrane [36]. In recent years, new polymers have entered the area of controlled drug release. Many of these materials are designed to degrade within the body; among them are polylactides, polyglycolides, poly(lactide-*co*-glycolides), polyanhydrides, and so on.

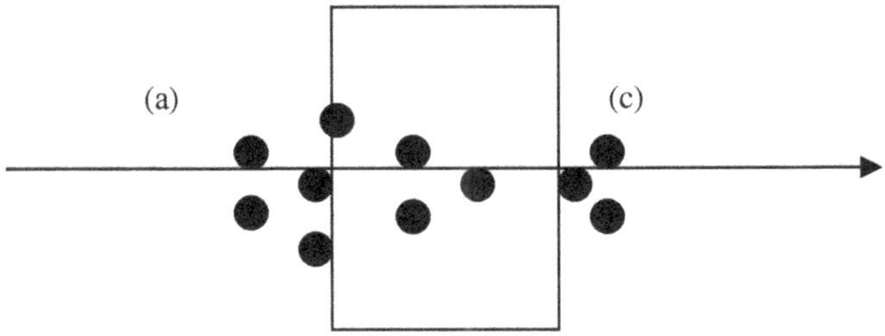

FIGURE 12.1 Schematic representation of transport mechanism

There are three primary mechanisms by which active agents can be released from a delivery system: diffusion, degradation, and swelling, followed by diffusion. Any or all of these mechanisms may occur in a given release system. An ideal drug delivery system should be inert, biocompatible, mechanically strong, comfortable for the patient, capable of achieving high drug loading, safe from accidental release, simple to administer and remove, and easy to fabricate and sterilize. The schematic representation of the processes involved in the transport mechanism is as follows (Figure 12.1):

(a) Adhesion of species on the surface of the matrix
(b) Diffusion of species through the matrix
(c) Desorption of the species from the matrix

The transport process slowly tries to equalize the concentration difference or the chemical potential of the penetrant in the phases separated by the membrane. This process can be described by Fick's first and second laws of diffusion. If a concentration gradient is established across some arbitrary reference section in the polymer, a net transport of penetrant occurs in the direction of decreasing concentration. This phenomenon can be described in terms of Fick's first law of diffusion according to which the diffusive flux J (the amount of penetrant passing through a plane of unit area normal to the direction of flow during unit time) in the x-direction of flow is proportional to the concentration gradient:

$$J = -D\frac{\partial c}{\partial x} \qquad (12.1)$$

where D is the diffusion coefficient and c is the concentration of the diffusing molecule. This equation is applicable when the diffusion is in the steady state.

On the other hand, Fick's second law describes the unsteady-state transport process, which is given by the rate of change of the penetrant concentration at a plane within the membrane as

$$D\frac{\partial^2 c}{\partial x^2} = \frac{\partial c}{\partial t} \tag{12.2}$$

This is an ideal case in which the membrane is isotropic and D is independent of distance, time, and concentration. Depending on the boundary conditions, there are many solutions available for Equation 12.2.

When the diffusion rate J achieves a steady value (no longer varies with time), a material balance requires that it also be independent of x, in which case Equation 12.1 can be integrated to give

$$\int_0^h J dx \int_{c=c_1}^{c=c_2} D(dc) \tag{12.3}$$

$$J = -\frac{1}{h}\int_{c=c_1}^{c=c_2} D(dc) \tag{12.4}$$

Equation 12.4 indicates the steady-state diffusion rate which is inversely proportional to the overall membrane thickness h for a given set of boundary conditions. To complete the integration, the variation of D with c must be known. In some systems, the diffusion coefficient does vary with concentration because the van der Waals forces are high between the diffusing molecules themselves and between the diffusing molecules and the polymer chains.

When there are relatively weak intermolecular forces of attraction between the diffusing molecules and the polymer chains, the solubility is low, and the polymer structure is not significantly altered. Under these circumstances c can vary without affecting D, and Equation 12.4 then becomes

$$J = -\frac{D}{h}(c_2 - c_1) \tag{12.5}$$

where c_2 and c_1 are, respectively, the concentrations of the diffusing molecules dissolved at the downstream and upstream polymer-membrane faces. In many cases, it is the pressures or partial pressures p of a gas or vapor above the faces of the polymer film, rather than the surface concentration, which are known. These quantities are related by Henry's law, which states that

$$C = S \cdot p \tag{12.6}$$

where S is the solubility constant for a given gas–polymer system.

The combination of Equations 12.5 and 12.6 gives the well-known permeation equation

$$J = DS\left[\frac{p_1 - p_2}{h}\right] \tag{12.7}$$

The product *DS* is called the permeability coefficient, *P*, so that

$$P = D \cdot S \tag{12.8}$$

12.4 TRANSPORT MECHANISM

The mol uptake (Q_t) for the solvent by 0.1 kg of the polymer for a specific time, t, under solvent effects, is calculated using the expression

$$Q_t\left(\text{mol\%}\right) = \frac{\text{Mass of solvent sorbed by the polymer}}{\text{Molecular mass of the solvent / initial mass of polymer}} \times 100 \tag{12.9}$$

The transport properties of polymeric membranes can be followed by using the empirical equation

$$\log\frac{Q_t}{Q_\infty} = \log k + n \log t \tag{12.10}$$

The slope of the plot of $\log Q_t / Q_\infty$ versus log t gives the value of n which indicates the mechanism of transport and its y-intercept gives the value of k which depends upon the structural significance of polymer as well as its interaction with the solvent. According to the n values obtained from the above equation, three basic modes of transport are distinguished. If $n = 0.5$ the diffusion mechanism is Fickian; in that case the rate of diffusion of permeant molecules is much less than the polymer segment mobility. If $n = 1$, the mechanism is non-Fickian; this may be considered in systems in which permeant diffusion rates are much faster than polymer relaxation process. If n lies between $0.5 < n < 1$, the diffusion mechanism is non-Fickian and is anomalous, as it occurs when the penetrant mobility and polymer segment relaxation rates are similar. Research in this field showed that the sorption and diffusion of penetrants through polymers are controlled by the polymer structure, its cross-link density, temperature, and so on [37]. The transport of liquids through polymers generally follows Fick's law of diffusion. However, there are also cases where diffusion deviates from Fick's law [38].

The knowledge of the performance of polymers in the environment of hazardous solvents, vapors, and temperature is essential for their successful applications as structural engineering materials. The sorption and diffusion of gases and organic solvents in polymers have been a topic of great relevance, both from the theoretical and technological point of view [39–44]. When a crosslinked polymer is placed in contact with a suitable solvent, imbibition of the solvent takes place with consequent swelling of the specimen. As a result, the dimensional network gets extended until equilibrium swelling is achieved at which the rate of imbibition of solvent equals the rate of expulsion. The sorption behavior gives an idea about the permeability and diffusion coefficient of penetrant through polymers, which are used in various applications such as membranes, ion exchangers, and so on. The transport properties of various engineering polymers like PUR, crosslinked PS, natural rubber [45–48], and various synthetic rubbers have been extensively studied using organic solvents with various polarity and interacting groups and also using aqueous salt solutions of various acids by many researchers [49, 50]. However, the transport properties of interpenetrating polymer networks (IPNs) are rarely found in the literature even though they are widely used in many applications such as membranes, gel permeation chromatography, control release of drugs, and enzyme immobilization [51–53].

With increasing importance of polymer blends in engineering areas [54–58], studies have been directed to develop interpenetrating polymer network (IPN) polymers. In IPNs, the transport behavior depends on the composition of components and their chemical interaction, as well as morphology [59]. Hence, transport study could be a very good technique to follow the morphology and structure–property relationships of IPNs. The swelling behavior of any polymer network depends upon the nature of the polymer, polymer–solvent compatibility, and degree of crosslinking. However, in the case of ionic networks, swelling kinetics depends upon mass transfer limitations, ion exchange, and ionic interaction [60]. Swelling kinetics of hydrogels can be classified as diffusion controlled (Fickian) or relaxation controlled (non-Fickian). When diffusion into the hydrogel occurs much faster than the relaxation of the polymer chains, the swelling kinetics is said to be diffusion controlled.

Krusic *et al.* [61] prepared SIPNs based on polyacrylamide/poly(itaconic acid) and demonstrated that poly(itaconic acid) interacts with the polyacrylamide gel mainly by hydrogen bonding, resulting in the formation of an interpolymer complex. Nghiem *et al.* [62] have studied nanofiltration (NF) with membranes of different permeability. It was demonstrated that NF membranes can be used to remove pesticides and that the removal mechanism is mainly influenced by steric and electrostatic interactions [63, 64]. When the interaction between membrane and adsorbing solute is important for separation, then the nature of solute has something to reveal about membrane. Van de Bruggen *et al.* [65] noticed that the physicochemical properties of solute and membranes influence the separation behavior. It was found that the physical and chemical structure of polymers and their morphology, crosslink density, the shape and size of solvent molecules, and temperature influence the molecular transport of the solvents through polymers.

12.5 SORPTION AND DIFFUSION OF SOLVENTS

Ajith *et al.* [66] reported the equilibrium swelling experiments of interpenetrating polymer networks (IPNs) based on castor oil-based PUR and unsaturated polyester (UPE). The sorption behavior found to vary with the hydroxyl number (molecular weight) of the polyol employed in the preparation of PUR. Studies revealed that the sorption is anomalous in nature in all the cases and the diffusion coefficient is found to increase with both increase in NCO/OH ratio and UPE content, whereas sorption coefficient (S) shows a reverse trend in all the samples. From the chlorobenzene sorption curves of PUR, UPE, and various IPNs (NCO/OH.1.2), it is clear that the homopolymer networks 100% PUR (100 : 00) and 100% UPE (00 : 100) show the maximum and minimum values of Q_t, respectively, at a particular time. The uptake capacity of IPNs was intermediate between that of the constituents. With increase in the UPE content and NCO/OH ratio, sorption decreases and reaches a minimum value irrespective of the type of PUR used in all the samples. All the IPNs studied show high crosslink densities and hence low diffusion and sorption values compared to the common crosslinked networks. For a given crosslink density, the sorption decreases with an increase in the polyester content. It has been shown that the increase in the crosslink density also reduces the sorption for a given composition.

Piotr *et al.* [67] studied the kinetics of swelling and the sorption performance for the polymer compositions with interpenetrating polymer networks made up of polyurethane and UPE with chlorobenzene at 25°C. It was found that the rates for solvent transport and solvent absorption processes were controlled by the chemical composition of the formulation studied. It was also found that the increasing share of polyurethane in the composition reduced crosslinking density in the polyester network that resulted in faster diffusion of the solvent and higher sorption capacity for the solvent. The higher the styrene contents in the composition, the higher the crosslinking density in the system, and hence the diffusion of solvent and its sorption inside the polymer network was much more difficult. From the morphological investigation, it is clear that the chlorobenzene increases the volume of the superficial layer that can result in gradual breaking away of the fragment located at the sample edges. This is a similar behavior that can be observed in the case of "neat" UP resin that has been crosslinked with styrene. No changes have been observed in the structure of sample core. The solvent is swelling the whole sample volume in the case of IPN samples with two-component matrices. After a time needed to reach equilibrium, changes in the appearance of the matrix become apparent, but there are no differences visible in the morphology of condensed phases located deep inside, away from the surface. Although the two-component matrix is not leached out completely, the matrix becomes less compact and numerous voids are present. The structure is formed which is nearly openwork. The more compact the spatial structure of the formed IPNs is, the slower the solvent diffusion if the achieved levels of equilibrium swelling are lower. Increasing amounts of ST in IPN compositions slow down the sample swelling process. The opposite effect is exhibited by the presence of PUR in the system. It should be pointed out that the method of analysis of IPN, which was described earlier, is a universal research procedure and it could be successfully used to understand the structure of other multiphase crosslinked polymer materials. It can be noted that the swelling process became more intense after a considerable growth of Q_t and then quick sample decomposition. That can be explained by differences in chemical affinity of chlorobenzene to structures that are formally crosslinked and only physically tangled up, just as present in IPN-type materials. The destruction of network will also lead to inferior cohesion of the material. Thus, the shear stresses between two layers, swollen gel coating and glassy (rigid) core, result in cracks of the sample surface layer [68]. What type of behavior is specific for the systems is responsible for the so-called anomalous swelling [69]. When cracks are formed, they can speed up the swelling process within the glassy core since the solvent finds it easier to migrate inside the core through the cracks in the outer shell. The equilibrium degree of swelling in those cases was assumed to be equal to the value Q_t just before the scratches and cracks appeared on the sample surface since any further diffusion of solvent took place under conditions which were no longer, since the solvent–sample contact surface was increased considerably.

Muneera *et al.* [70] reported the molecular migration of n-alkane penetrants such as hexane, heptane, octane, nonane, and decane through PUR/PBMA (50/50) membranes at 25, 40, and 60°C by the weight gain method. Sorption results of n-alkanes at 40 and 60°C for the TDI-based PUR/PBMA IPN membranes were studied. At all the temperatures, transport follows a systematic trend with the size of n-alkanes. For instance, sorption values for n-hexane are much smaller than those observed for n-decane. Sorption behavior at 40°C is almost complete, but at 60°C sorption tendencies for nonane and decane are widely different than those observed for lower n-alkanes, namely, hexane, heptane, and octane. It is observed that sorption values of all the alkanes increase systematically with increasing temperature.

An increase in polymer–solvent interactions leads to increased sorption and transport of liquids such that diffusion process often becomes concentration dependent. The Fickian model has been used to estimate the diffusion coefficient. The lower diffusion coefficients are observed for MDI-based PUR/PBMA IPNs compared with TDI. Diffusion coefficients decrease with increasing size of the penetrant. The temperature dependence of transport coefficients has been used to compute the activation parameters from the Arrhenius plots. The ΔH values are positive, whereas ΔS values are negative for both TDI- and MDI-based PUR/PBMA IPNs. It can be seen that the D values decrease with an increase in the molecular weight of the penetrants. The sequence of variation of D with respect to penetrants is hexane > heptane > octane > nonane > decane. The D values also increase with an increase in temperature. Huang *et al.* [71] investigated the sorption and transport properties of hydrocarbon membranes based on poly(vinyl alcohol) (PVA) network and poly(styrenesulfonic acid-*co*-maleic acid) (PSSA-MA). The water/methanol solvent uptake characteristics of Nafion 115 and SIPN membranes were studied. The behaviors of both the membranes were clearly distinguishable. This study systematically investigated the sorption and transport properties of PVA-based SIPN membranes and compared them with Nafion 115. A significant difference in the sorption behavior of Nafion 115 and SIPN-80 membranes was identified by solvent uptake and k values. Nafion 115 showed no preference in sorption of water and methanol. In contrast, SIPN membranes exhibited excellent selectivity of sorption of water over methanol. The solvent uptake of Nafion 115 increased as the methanol concentration increased up to a methanol mole fraction of 0.63 and then decreased. On the other hand, the solvent uptake of the SIPN-80 membrane decreased sluggishly as the methanol concentration increased. Moreover, the ratios of $kCH_3OH{=}kH_2O$ for SIPN membranes were only about one-third that of Nafion 115 (Figure 12.2).

The Arrhenius plot of water self-diffusion coefficients for Nafion 115 and SIPN membranes in a fully hydrated state exhibited a positive temperature dependence of self-diffusion coefficient. All membranes exhibited positive temperature dependence of self-diffusion coefficient. The slopes of the SIPN-80 and Nafion 115 membranes are quite close and are slightly higher than that of the SIPN-60 membrane. This result can be explained by the

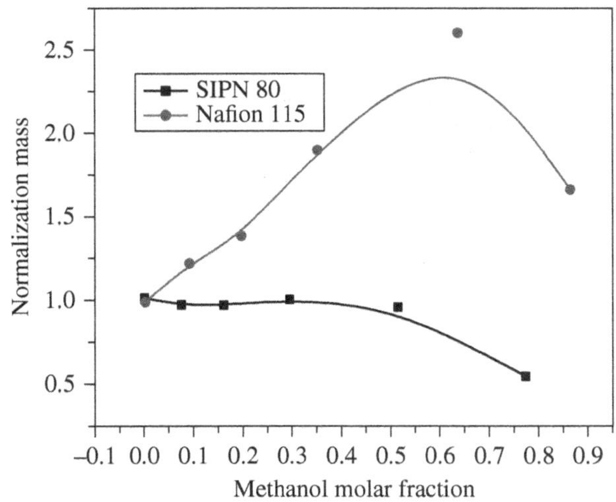

FIGURE 12.2 Solvent uptakes of Nafion VR 115 and SIPN-80 membranes as a function of methanol concentration

distribution of water states in these membranes by the evaporation of free water that occurs faster than bound water in a vaporized measuring condition. This means that membranes with a higher content of free water will exhibit greater degrees of dehydration at higher temperatures. The Arrhenius plot of the methanol self-diffusion coefficients for SIPN-80 and Nafion 115 membranes under a fully methanol uptake state at different temperatures noted that the methanol self-diffusion coefficient in SIPN-80 sluggishly decreased when the temperature is over 50°C. However, it still linearly increased in Nafion 115 during the measured temperature range, even up to 70°C. This result implied that methanol is well retained in Nafion 115 at temperatures up to 70°C.

The transport properties of the SIPN and Nafion 115 membranes were also identified by water and methanol self-diffusion coefficients. Compared with Nafion 115, SIPN membranes showed lower water and methanol self-diffusion coefficients. Moreover, the ratio of $DCH_3OH{=}DH_2O$ for SIPN-80 was approximately only one-seventh that of Nafion 115. Evidently, there is a good correlation between the sorption behavior and transport behavior for Nafion 115 and SIPN-80 membranes. The SIPN-80 membrane exhibited excellent selectivity of water over methanol in both sorption and transport properties. With a methanol permeability of $4.1\times10^{-7}\,cm^2/s$, the SIPN-80 membrane promises to be a good candidate as a proton-conducting membrane in DMFCs. Moslem *et al.* [72] reported on the swelling behavior of the cross-linked PVA/poly(vinylpyrrolidone) (PVP), and they found a relatively fast rate of swelling at the beginning of the process. The amount of absorbed water determined from the immersion water uptake experiment was of greater than 1400%. Because of the neutrality of the hydrogels, the maximum of swelling was achieved at pH = 7. Maximum extent of swelling was achieved at 25°C, and the swelling of hydrogels was found to decrease at higher temperatures. The results show that the water absorption of crosslinked PVP/PVA hydrogels tend to increase with increasing PVP ratio in the copolymer but decrease with increasing potassium peroxodisulfate concentration. In fact, hydrogen bonding interactions of PVA hydroxyl groups tend to crystallize the PVA chains through physical crosslinking. On the other hand, the steric hindrance of the bulky pyrrolidone groups restricts the chains and decreases the crystallinity of PVA segments. Also, PVP is more hydrophilic than PVA, as it was observed when dissolving the polymer. The amorphous nature of PVP and its higher affinity for water make the network (which is rich in PVP) swell to a greater extent. However, in the ratios lower than 28 molar percent of PVP, the water absorption increases smoothly due to local concentration increment of hydroxyl groups. However, the swelling character of this hydrogel did not differ significantly from the PVA-rich hydrogel. It was also noticed that the water absorption of the gel gradually diminished at higher potassium peroxodisulfate concentration. The rate of water absorption was significantly higher at the immersion times less than 5 h, whereas after it, the rate of absorption only slightly increased.

Wen-fu lee *et al.* [73] prepared SIPN polymer gels and IPN polymer gels with thermosensitivity by introducing a biodegradable polymer, chitosan, into the *N*-isopropyacrylamide (PNIPAAm) gel system. The swelling ratios (function of time) of the PNIPAAm, SIPN, and IPN gels at 25°C in deionized water show that the key factor in swelling is the crosslinking of chitosan, and the equilibrium swelling ratios of IPN gels decrease with an increase of glutaraldehyde content, and it was also clear that the PNIPAAm/chitosan SIPN and IPN hydrogels deswell more quickly at their gel transition temperatures and also shows that the gel transition temperature for the present gels is not affected by the addition of the chitosan component in the gels. The scanning electron microscopic images show that the hydrogels are porous and the pore size of SIPN and IPN gels is smaller than that of the PNIPAAm gel. The pore distribution of the SIPN gel also is nondirectional, but the network of IPN gel exhibits a lamellar structure, supporting the fact that the IPN structure provides smaller domains of phase separation. The microphotographs of fractured surfaces show that the pore walls of SIPN and IPN gels become thicker than those of the PNIPAAm gel and make the gel strength increase.

Kumar *et al.* [74] evaluated the transport behavior of polyethylene glycol (PEG)-based polyurethane/PMMA (50/50) SIPN with aromatic probe molecules. The effects of solubility parameters of solvents on the transport phenomenon have been studied. It is found that as the difference in solubility parameter of solvents and polymer membrane is less, there is an increase in sorption and permeation coefficients. This indicates that there are more interactions between solvent and polymer, which leads to salvation, followed by diffusion into membrane. The diffusion coefficient result suggests that when PUR/PMMA membrane used as barrier for aromatic probe molecules, penetration of probe molecules depends on several factors, such as nature of membrane, size, temperature, solubility parameter and its compatibility with probe molecules, and also their chemical characteristics. Desorption behavior of IPN has been measured, and the rate of evaporation of benzene during desorption process is found to be highest compared to other penetrants. The effects of temperature on sorption and desorption cycles have been measured. It is found that as temperature increases the mass uptake values during sorption and mass loss values during desorption increase. The activation energy for diffusion E_D and permeation E_P was calculated for these systems. The ΔS for sorption and resorption lies in the range of 0.70–3.16 and 12.83–23.45 kJ/mol K, respectively. Similarly ΔH for sorption and resorption processes are 0.70–2.37 and 4.80–9.20 kJ/mol, respectively. These values did not show any regular trend. This may be due to the complicated structure and morphology of IPN. For all IPN penetrant systems, positive ΔH values were observed. The positive values of ΔH indicate that the sorption is an endothermic process and is dominated by Henry's mode, that is, the sorption proceeds through creation of new sites or pores in the polymer. The experimental diffusion coefficients were compared with theoretical predictions.

Aji *et al.* [75] reported the transport of aromatic hydrocarbons through SIPN and full-interpenetrating polymer network membranes (IPNs). The interpenetrating polymer network membranes were synthesized from natural rubber and PS by the sequential polymerization technique. The sorption was carried out in a series of aromatic solvents. It was found that in all cases, the uptake value increased by about 50% as the PS content decreased from 70 to 30%. The diffusion, sorption, and permeation coefficients were evaluated. As the crosslink density was increased, the uptake decreased by 40%.

12.6 GAS BARRIER PROPERTIES OF INTERPENETRATING POLYMER NETWORKS

Flexible packaging films for oxygen-sensitive products (such as foods and electronics) are generally made of multilayer structures. Each of these layers brings a specific property to the material such as high oxygen barrier and good adhesion to a substrate. The packaging of fruits and vegetables requires inward permeation of CO_2 and respiration of oxygen without the loss of water. In meat-wrap and cheese-wrap applications, the polymer films chosen are those which will allow at least certain minimum oxygen diffusion from air and minimize flavor loss, respectively. The dimensional stability and integrity of polymeric materials in an aggressive liquid environment are highly essential for their successful technological and engineering applications.

The current trend is to reduce the number of these layers while keeping the same mechanical and technical characteristics. The development of more "ecological" systems (absence of halogenated compounds such as poly(vinylidene chloride) and of solvents) is also a current trend. Generally, organic polymers which exhibit high oxygen barrier properties present crystalline domains and a compact structure. So, IPNs with high crosslink densities of both networks and which have the highest possible degree of interpenetration would be the best candidates.

Simic *et al.* [76] have examined the oxygen barrier and adhesion (to the substrate) properties of IPNs composed of poly(methacrylate) and epoxy–amine networks made in situ between two oriented polypropylene sheets. The oxygen transmission rate (OTR) values for (*N,N′*-bis (2-methacryloyloxyethyl)-1,9-nonylene biscarbamate) diUdiMA/triethylenetetramine-*N,N*-diglycidyl-4-glycidyloxyaniline (diUdiMA/TETA-DGGOA) films irradiated after different waiting times, noticed that high crosslinking density inside the cured layer can be responsible for a better O_2 barrier effect. Based on their investigation it can be concluded that cross-reactions in the IPNs prepared from a free radically photocured bis(methacrylate) compound and epoxy resins (cured with plurifunctional amines) were absent. Thin films containing a diUdiMA/TETA-DGGOA IPN as the central layer present rather good gas barrier properties. In the studied systems, no influence of the curing sequence (modulated by the moment of irradiation) on IPN gas barrier properties has been shown. This result indicated that either the moment of irradiation does not affect IPN structure or that the change of structure has no effect on IPN permeability.

12.7 PERVAPORATION CHARACTERISTICS OF INTERPENETRATING POLYMER NETWORKS

Pervaporation is a membrane-based process for liquid separation. Unlike the conventional distillation where the separation depends on the vapor–liquid equilibrium, the pervaporation separates the liquid mixtures by the difference in the affinity (or solubility) and diffusivity of each liquid component with the membrane. This makes the pervaporation appropriate for breaking up of the azeotropes. In order to achieve a good separation, the membrane must contain the active sites that interact strongly with the separated species.

12.8 PRINCIPLES OF PERVAPORATION

In pervaporation, the liquid mixture is in direct contact with one side of the membrane, and the permeated product is removed as vapor at the downstream side by applying a low pressure. The material transported undergoes a phase change. The minimum energy utilized is the enthalpy of vaporization of the permeate. The permeate composition is mainly determined by the relative affinities of the components in the feed to the membrane and their unequal mobilities through the membrane. Therefore permeate composition will differ from that of feed composition.

There are two approaches to describe the mass transport in pervaporation: (i) the solution–diffusion model and (ii) the pore flow model. The solution–diffusion model is accepted by several researchers [77–79]. According to this mechanism, pervaporation consists of three consecutive steps: (i) sorption of the permeant from the feed liquid to the membrane, (ii) diffusion of the permeant in the membrane, and (iii) desorption of the permeant to the vapor phase on the downstream side of the membrane [80].

Amnuaypanich *et al.* [81] investigated the pervaporation dehydration of water–ethanol mixtures using the mixed matrix (MM) membranes prepared from natural rubber and crosslinked PVA SIPN embedded with the zeolite 4A. The membranes were employed in the pervaporation dehydration of water–ethanol mixtures. Natural rubber limits the swelling of the membranes in water, while PVA, on the other hand, improves the water absorption due to its hydrophilicity. Zeolite 4A with a uniform porous structure provides an alternative route for the transport of water molecules through the membranes and, with the molecular sieve, enhances the water selectivity of the membranes.

Pervaporation dehydration of water–ethanol mixtures was performed using the MM membranes, and the results showed that for low water concentration of the feed, increasing the zeolite loading can improve both the total permeation flux and the water separation factor. However, as further increasing the feedwater concentration, although the total permeation flux increased, this parathion factor was reduced because the membranes were swollen substantially and cause the loose structure of the surrounding polymer networks. Increasing the feed temperature was found to affect the permeation flux and separation factor of the MM membranes in a similar manner as increasing water concentration of the feed mixtures. The enhancement of flux with increasing of the feed temperature is due to the decrease of interactions between permeates and the permeates with a membrane at higher temperature. Following the study of feed temperature effect, the Arrhenius plots were constructed, and the activation energies for water (*EJ*W) and ethanol (*EJ*E) permeation through the MM membranes were estimated. The lower *EJ*W values imply that water molecules are transported through the membranes with less restrictive compared to the transport of ethanol molecules.

Lu *et al.* [82] prepared a novel interpenetrating polymer network membranes from PVA/PVP blends of different compositions. The two polymer components were independently crosslinked chemically with glutaraldehyde and photochemically with 4,4′-diazostilbene-2,2′-disulfonic acid disodium salt. The membrane performances were studied in pervaporation of tetrahydrofuran (THF)/water and THF/methanol mixtures. It was found that the membranes were excellent in THF dehydration, but much less efficient for the separation of THF/methanol mixtures.

12.9 VAPOR SORPTION

Dynamic vapor sorption (DVS) is a gravimetric technique that measures how quickly and how much of a solvent is absorbed by a sample: such as a dry powder absorbing water. It does this by varying the vapor concentration surrounding the sample and measuring the change in mass which this produces. Water vapor is most commonly used, but it is also possible to use a wide range of organic solvents. DVS was originally developed to replace the time- and labor-intensive desiccators and saturated salt solutions to measure water vapor sorption isotherms. Water sorption is important for biocompatibility because of skin contact, body insertion, food packing, and so on. Many researchers have studied the relationship between sorption and polymer films. Sanopoulou *et al.* [83] reported on non-Fickian sorption kinetics in polymer films. Vachoud *et al.* [84] reported on the sorption and desorption of solute in chitin, showing that the sorption of the solute depends on electrostatic interactions. Mackin *et al.* [85] measured the content of amorphous or crystalline regions by moisture sorption. Wadoso *et al.* [86] studied and compared three methods to determine the vapor activity during hydration and showed that all methods are useful for measuring hydration step activities. Shi and James [87] studied moisture transport on copolyester thin films. Kim *et al.* [88] reported the vapor sorption studies on the PVA/chitosan interpenetrating polymer networks (IPN) prepared by UV irradiation. In their studies they noticed that the vapor sorption behavior is more affected by the density of water vapor than by hydrophilic properties with increasing temperature. Equilibrium water uptake increases as humidity increases, and the increase is more noticeable at high humidity. Their vapor sorption behavior was more affected by the density of water vapor than by their hydrophilic properties with increasing temperature.

12.10 CONCLUSION

Interpenetrating polymer networks and related materials represent one of the fastest-growing fields in polymer science. It is widely used in the transportation, consumer, industrial, hygiene, and medical sectors where they provide decorative finish, high resilience, high tear strength, and so forth. Toward the end of the last century, great attention has been focused on the diffusion, sorption, and permeation of solvents into IPNs since these basic phenomena play vital roles in several important areas of engineering and industry. The IPNs show excellent properties toward the gas barrier properties, pervaporation, and vapor sorption properties.

12.11 APPLICATIONS, CHALLENGES, DIFFICULTIES, AND FUTURE DIRECTIONS

In the area of IPNs, the sorption and transport mechanisms have not been well understood. The design of porous networks based on IPNs has received a lot of attention recently. The porous IPN materials may find potential applications in separation techniques as chromatography support as well as membrane catalysis and pervaporation. There exists much interest for the development of biopolymer-based IPNs for various applications. IPN-based drug delivery system is designed to deliver drugs at a predetermined rate for a designed period of time with minimum fluctuation which is not explored in detail. Its physical and biological characteristics such as enhanced solubility of hydrophobic drugs, imparting drug stability in the formulation, good swelling capacity, biodegradability, biocompatibility, and targeting of drug in a specific tissue make it suitable for drug delivery as well as biomedical applications.

REFERENCES

[1] Sperling, L.H. (1981) *Interpenetrating Polymer Networks and Related Materials*, Plenum Press, New York.

[2] Shin, M.S., Kim, S.I., Kim, I.Y. *et al.* (2002) *Journal of Applied Polymer Science*, **84**, 2591.

[3] Kim, S.J., Park, S.J., Shin, M.S. *et al.* (2002) *Journal of Applied Polymer Science*, **85**, 1956.

[4] Lee, S.J., Kim, S.S. and Lee, Y.M. (2000) *Carbohydrate Polymers*, **41**, 197.

[5] Frisch, H.L., Du, Y. and Schultz, M. (1998) *Polymer Networks. Principles of Their Formation, Structure and Properties*, Blackie Academic and Professional, London.

[6] Okano, T., Bae, Y.H., Jacobs, H. and Kim, S.W. (1990) *Journal of Controlled Release*, **11**, 255.

[7] Feil, H., Bae, Y.H., Feijen, J. and Kim, S.W. (1991) *Journal of Membrane Science*, **64**, 283.

[8] Von Recum, H., Okano, T. and Kim, S.W. (1998) *Journal of Controlled Release*, **55**, 121.

[9] Osada, Y. and Kajiwara, K. (2001) *Gel Handbook*, Academic Press, San Diego, CA.

[10] Kro´l, P. and Wojturska, J. (2002) *Polimery (Warsaw)*, **47**, 242.

[11] Kro´l, P. and Wojturska, J. (2003) *Journal of Applied Polymer Science*, **88**, 327.

[12] Kro´l, P. and Wojturska, J. (2004) *Polimery*, **49**, 790.

[13] Kro´l, P. and Wojturska, J. (2005) *Journal of Applied Polymer Science*, **97**, 1855.

[14] Lee, Y.K., Tak, T.M., Lee, D.S. and Kim, S.C. (1990) *Journal of Membrane Science*, **52**, 157.

[15] Lee, Y.K., Sohn, I.S., Jeon, E.J. and Kim, S.C. (1991) *Polym Journal (Tokyo)*, **23**, 427.

[16] Aylsworth, J.W. (1914) U.S. Patent 1,111,284.

[17] Sperling, L.H. (1987) *Polymer News*, **12**, 332.

[18] Staudinger, J.J.P. and Hutchinson, H.M. (1951) U.S. Patent 2,539,377.

[19] Solt, G.S. (1955) British Patent 728,508.

[20] Millar, J.R. (1960) *J. Chem. Soc.*, **263**, 1311.

[21] Shibayama, K. and Suzuki, Y. (1967) *Rubber Chemistry and Technology*, **40**, 476.

[22] Lipatov, Y.S. and Sergeeva, C.M. (1967) *Russian Chemical Reviews*, **45**, 63.

[23] Frisch, H.L., Klempner, D. and Frisch, K.C. (1969) *Journal of Polymer Science Polymer Physics Edition*, **7**, 775.

[24] Sperling, L.H. and Friedman, D.W. (1969) *Journal of Polymer Science*, **7**, 425.

[25] Lee, M.-J., Choi, Y.S., Kang, Y.S. *et al.* (2012) *Journal of Materials Chemistry*, **22** (18522).

[26] Lee, W.-F. and Lu, H.-C. (2013) *Journal of Applied Polymer Science*, **127** (5), 3663.

[27] Stephen, R., Varghese, S., Joseph, K. *et al.* (2006) *Journal of Membrane Science*, **282**, 162.

[28] Kim, Y.K., Park, H.B. and Lee, Y.M. (2005) *Journal of Membrane Science*, **251**, 159.

[29] Khurma, J.R., Rohindra, D.R. and Nand, A.V. (2005) *Polymer Bulletin*, **54**, 1959.

[30] Frank, A., Rath, S.K. and Venkatraman, S.S. (2005) *Journal of Controlled Release*, **102**, 333.

[31] Fress, K., Sipek, M., Hynek, V. *et al.* (2004) *Journal of Membrane Science*, **240**, 179.

[32] Ruthven, D.M. (2004) *Chemical Engineering Science*, **59**, 4531.

[33] Matsui, S. and Paul, D.R. (2004) *Journal of Membrane Science*, **235**, 25.

[34] Tsujita, Y. (2003) *Progress in Polymer Science*, **39**, 201.

[35] Vieth, W.R. (1990) *Diffusion in and through Polymers-Principles and Applications*, Hanser Publishers, New York.

[36] Friedlander, H.Z. and Rickles, R.N. (1966) *Chemical Engineer*, **73**, 111.

[37] Mathai, A.E., Singh, R.P. and Thomas, S. (2003) *Polymer Engineering & Science*, **43**, 704.

[38] Park, G.S. and Crank, J. (1968) *Diffusion in Polymers*, Academic Press, London.

[39] Huelck, V., Thomas, D.A. and Sperling, L.H. (1972) *Macromolecules*, **5**, 340.

[40] Poh, B.T., Adachi, K. and Kotaka, T. (1987) *Macromolecules*, **20**, 2563.

[41] Berens, A.R. and Hopfenbergh, H.B. (1978) *Polymer*, **19**, 489.

[42] Hopfenbergh, H.B. (1974) *Permeability of Plastic Films and Coatings to Gases, Vapours and Liquids*, Plenum Press, New York.

[43] Errede, L.A. (1986) *Macromolecules*, **19**, 654.

[44] Harogoppad, S.B., Aminabhavi, T.M. and Balundgi, R.H. (1992) *Journal of Polymer Science*, **44**, 1687.

[45] Shanthamurthy, U.A. and Aminabhavi, T.M. (1990) *Polymer*, **31**, 1757.

[46] Mathew, A.P., Packirisamy, S., Kumaran, M.G. and Thomas, S. (1995) *Polymer*, **36**, 4935.

[47] Vittoria, V. and Ruvolo Filho, A. (1992) *Journal of Applied Polymer Science*, **44**, 2127.

[48] Harogoppad, S.B. and Aminabhavi, T.M. (1990) *Polymer*, **31**, 2346.

[49] Frisch, H.L. and Frisch, K.C. (1979) *Progress in Organic Coatings*, **7**, 105.

[50] Frisch, H.L., Cifaratti, J., Palma, R. *et al.* (1977) *Polymer Alloys*, Plenum Press, New York.

[51] Marti, G.C., Enssani, E. and Shen, M. (1981) *Journal of Applied Polymer Science*, **26**, 1465.

[52] Kolarz, B.M. (1980) *Angewandte Makromolekulare Chemie*, **90**, 167.

[53] Pozniak, G. and Trochimczuk, W. (1982) *Angewandte Makromolekulare Chemie*, **104**, 1.

[54] Miao, T., Miller, E.J., McKenzie, C. and Oldinski, R.A. (2015) *Journal of Materials Chemistry B*, **48** (9242).

[55] Hopfenberg, H.B. and Paul, D.R. (1976) *Polymer Blends*, Academic Press, New York.

[56] Kolarik, J. and Gueskens, G. (1997) *Polymer Networks and Blends*, **71**, 3.

[57] Aminabhavi, T.M. and Phayde, H.T.S. (1995) *Journal of Applied Polymer Science*, **55**, 1335.

[58] George, S.C., Varguhese, K.T. and Thomas, S. (2000) *Polymer*, **41**, 579.

[59] Frank, S. and Lauterbur, P.C. (1993) *Nature*, **363**, 334.

[60] Peppas, N.A. and Colombo, P. (1997) *Journal of Controlled Release*, **45**, 35.

[61] Krusic, K.M., Dzunuzovic, E., Trifunovic, S. and Filipovic, J. (2003) *Polymer Bulletin*, **51**, 159.

[62] Nghiem, L.D., Schafer, A.I. and Elimelech, M. (2004) *Environmental Science & Technology*, **38**, 1888.

[63] Van der Bruggen, B., Schaep, J., Wilms, D. and Vandecasteele, C. (1999) *Journal of Membrane Science*, **156**, 29.

[64] Berg, P., Hagmeyer, G. and Gimbel, R. (1997) *Desalination*, **113**, 205.

[65] Van der Bruggen, B., Schaep, J., Maes, W. *et al.* (1998) *Desalination*, **117**, 139.

[66] Ajithkumar, S., Patel, N.K. and Kansara, S.S. (2000) *European Polymer Journal*, **36**, 2387.

[67] Krol, P. and Wojturska, J. (2006) *Journal of Applied Polymer Science*, **101**, 3511–3519.

[68] Tanigami, T., Yano, K.K. and Matsuzawa, Y.S. (1995) *Polymer*, **36**, 2941.

[69] Ghi, P.Y., Hill, D.J.T., Maillet, D. and Whittaker, A.K. (1997) *Polymer*, **38**, 3985.

[70] Begum, M., Siddaramaiah, Kumar, H. and Aminabhavi, T.M. (2003) *Journal of Applied Polymer Science*, **90**, 739–746.

[71] Huang, Y.F., Hwang, B.J. and Lin, C.W. (2009) *Journal of Applied Polymer Science*, **113**, 342.

[72] Lakouraj, M.M., Tajbakhsh, M. and Mokhtary, M. (2005) *Iranian Polymer Journal*, **14**, 1022.

[73] Lee, W.-F. and Chen, Y.-J. (2001) *Journal of Applied Polymer Science*, **82**, 2487.

[74] Kumar, H. and Siddaramaiah (2005) *Polymer*, **46**, 7140.

[75] Aji, P.M., Packirisamy, S., Stephen, R. and Thomas, S. (2002) *Journal of Membrane Science*, **201**, 213.

[76] Simic, V., Boileau, S., Bouteiller, L. *et al.* (2002) *European Polymer Journal*, **38**, 2449.

[77] Katoka, T., Tsuru, T., Nakao, S. and Kimura, S. (1991) *Journal of Chemical Engineering of Japan*, **24**, 334.

[78] Katoka, T., Tsuru, T., Nakao, S. and Kimura, S. (1991) *Journal of Chemical Engineering of Japan*, **24**, 326.

[79] Lee, C.H. (1975) *Journal of Applied Polymer Science*, **19**, 83.

[80] Wijmans, J.G. and Baker, R.W. (1995) *Journal of Membrane Science*, **107**, 1.

[81] Amnuaypanich, S., Patthana, J. and Phinyocheep, P. (2009) *Chemical Engineering Science*, **64**, 4908.

[82] Lu, J., Nguyen, Q., Zhou, J. and Ping, Z.-H. (2003) *Journal of Applied Polymer Science*, **89**, 2814.

[83] Sanopoulou, M. and Petropoulos, J.H. (2001) *Macromolecules*, **34**, 1400.

[84] Vachoud, L., Zydowicz, N. and Domard, A. (2001) *International Journal of Biological Macromolecules*, **28**, 93.

[85] Mackin, L., Zanon, R., Pack, J.M. *et al.* (2002) *International Journal of Pharmaceutics*, **231**, 227.

[86] Wadso, L. and Markova, N. (2001) *European Journal of Pharmaceutics and Biopharmaceutics*, **51**, 77.

[87] Shi, F.F. and James, E. (1998) *Journal of Polymer Science. Part B*, **36**, 1025.

[88] Kim, S.J., Lee, K.J., Kim, S.I., Lee, K.B. and Park, Y.D. (2003) *Journal of Applied Polymer Science*, **90**, 86.

13

AGING OF INTERPENETRATING POLYMER NETWORKS

SELVIN P. THOMAS[1] AND MOHAMMED N. ALGHAMDI[2]

[1] *Yanbu Research Center, Royal Commission for Yanbu Colleges and Institutes, Yanbu Al-sinaiah, Saudi Arabia*

[2] *Mechanical Engineering Technology Department, Yanbu Industrial College, Royal Commission for Yanbu Colleges and Institutes, Yanbu Al-sinaiah, Saudi Arabia*

13.1 INTRODUCTION

In the 1970s, interpenetrating polymer networks (IPNs) were developed as a kind of novel polymeric materials, and they were extensively applied in reinforced rubbers, toughened plastics, damping materials, coatings, and functional materials due to their unique forced compatibility [1, 2]. Additionally, IPN technology was proven an excellent polymer modification method for stable integration of two polymer networks with different properties or different functions by physical entanglements [3–5]. IPNs are a combination of two or more polymers in network form, with at least one such polymer polymerized and/or crosslinked in the immediate presence of the other(s) [6, 7]. The idea is to effect a molecular interpenetration of the networks; in practice, most IPNs form immiscible compositions, usually phase separating during some stage of the synthesis. The IPNs are part of the broad class of polymer blends. People define polymer blends in two ways: as any combination of two or more polymers (the broad definition) and as a

Micro- and Nano-structured Interpenetrating Polymer Networks: From Design to Applications, First Edition. Edited by Sabu Thomas, Daniel Grande, Uroš Cvelbar, K.V.S.N. Raju, Ramanuj Narayan, Selvin P. Thomas and Akhina H.

combination of two or more polymers without any substantial quantities of crosslink, graft, or block junctions among the several polymer chains (the narrow definition).

IPNs can be classified into different ones according to the synthesis methods and properties. Broadly they are sequential, simultaneous, latex, gradient, thermoplastic, and semi-IPNs. All these types are important in the case of different applications in drug delivery, membrane separation, and others. The aging behavior of IPNs is very important in designing many applications. A brief account on the aging characteristics of IPNs is described in this chapter.

13.2 AGING OF INTERPENETRATING POLYMER NETWORKS

Aging of IPNs can be done by different methods such as heat, ozone, oxygen, and flame resistance. The various mechanisms governing the aging process will also be discussed in the following sections.

13.2.1 Thermal Aging

Thermal aging is done by the application of heat, and it can be measured by various means. Usually thermogravimetric analysis (TGA) is employed to analyze the aging properties. A few examples are discussed.

13.2.1.1 Poly(Ethylene Oxide) (PEO)/Poly(3,4-Ethylenedioxythiophene) (PEDOT) Semi-Interpenetrating Polymer Networks PEO/PEDOT IPNs belong to the semi-IPN category. Semi-IPNs are a promising solution to associate two or more polymers and lead to new tailor-made materials. They are defined as a combination of at least one crosslinked polymer and one linear polymer synthesized in the presence of each other. Verge *et al.* synthesized the PEO/PEDOT IPNs by preparing PEO IPN first and followed by swelling in PEDOT [8].

The PEO network is obtained by free radical copolymerization of poly(ethylene glycol) dimethacrylate and methoxy poly (ethylene glycol) methacrylate (50 wt%) in the presence of EDOT (2 wt%). Conducting semi-IPNs are prepared by dipping the 2 wt% EDOT-swollen PEO films at room temperature into an iron chloride solution for various times. The experimental conditions lead to a preferential polymerization just under the PEO network surface rather than in the bulk. The color changes from pale blue to dark blue depending on the amount of PEDOT in the network. The self-supported PEO/PEDOT conducting semi-IPN films can be handled easily and remain flexible.

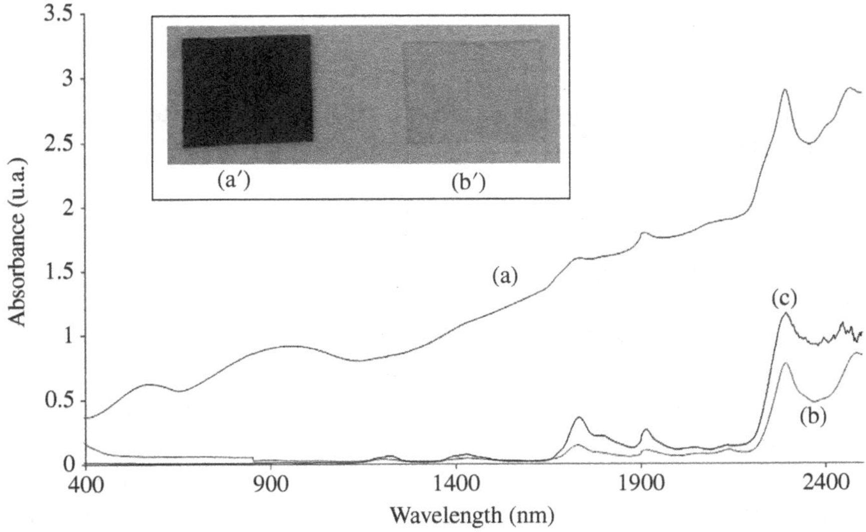

FIGURE 13.1 Visible–NIR spectra of PEO/PEDOT semi-IPN: (a) before aging test; (b) after 58 h of aging test at 80°C in open air; (c) single PEO network. The insert shows the pictures of a PEO/PEDOT semi-IPN: (a′) before aging test; (b′) after 48 h of aging test at 80°C in open air. Source: Verge *et al.* [8], pp. 3864–3870. Reproduced with permission of Elsevier

The PEO/PEDOT semi-IPN films were then submitted to a thermal aging at 80°C in open air for 8 h. The color turns from blue to pale yellow meaning that a degradation of PEDOT occurs. Figure 13.1 shows the absorbances of the film in the 400–2500 nm range before (a) and after (b) thermal aging at 80°C. The insert shows the pictures of a PEO/PEDOT semi-IPN before (a′) and after (b′) the aging test. The absorption spectrum of a native PEO/PEDOT semi-IPN presents three main transition bands (Figure 13.1a). The first one is localized around λ_{max} = 590 nm, corresponding to the PEDOT π–π* transition. This transition indicates that some nondoped PEDOT is present in the conducting semi-IPN. The second one is centered at λ_{max} = 920 nm and the third band covers the full near-infrared (NIR) range and remains difficult to be assigned with a λ_{max}. These two last absorptions correspond to the polaronic bands characteristic of the oxidized PEDOT. Both reduced and oxidized forms of PEDOT simultaneously exist in the film after methanol washings. After the thermal treatment, the absorbance of the PEO/PEDOT semi-IPN strongly collapses in the entire visible–NIR range (Figure 13.1b). The π–π* transition as well as the polaronic bands have totally disappeared. Clearly, the resulting spectrum is very similar to that of a single PEO network (Figure 13.1c); in

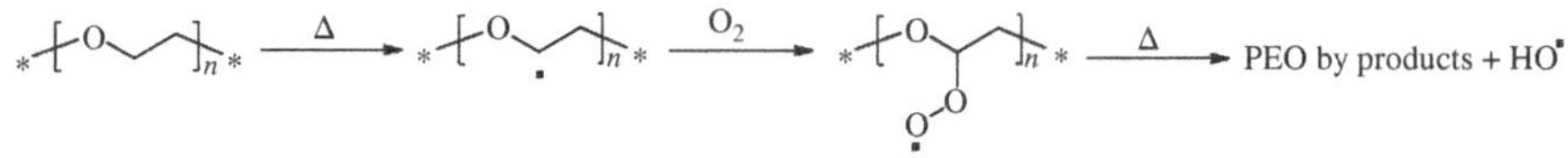

SCHEME 13.1 Degradation mechanism of PEO according to Ref. [9]. Source: Morlat and Gardette [9]. Reproduced with permission of Elsevier

particular, the absorption bands at 1910 and 2285 nm can be assigned to the PEO network rather than the PEDOT. Such results indicate that both oxidized and reduced PEDOT forms are damaged during the degradation process. Although the thermal stability of PEDOT is as high as 150°C, this study shows that PEDOT embedded within the PEO matrix is not stable any more even for temperatures close to 80°C. The PEO network could be responsible of this degradation.

Morlat and Gardette [9] have described the PEO degradation mechanism under thermal aging at 50°C in the presence of O_2. The authors demonstrated that hydroxyl radicals can be generated by the oxidation of PEO chains according to the mechanism described in Scheme 13.1.

In the present investigation, it has been understood that the thermal stability of the IPNs is poor and it is mainly due to degradation of the PEO network in the system. A complete PEDOT degradation after 8 h at 80°C has been observed by UV–visible spectroscopy and by cyclic voltamperometry. As previously shown by Gardette *et al.*, experiments confirm that under thermo-oxidation conditions in open air, the formation of hydroxyl radicals (HO) occurs by oxidation of PEO. It was further demonstrated that the HO react with the PEDOT leading to the rupture of the π-conjugation in the backbone, hence, to the loss of electroactivity. The introduction of a radical scavenger in the semi-IPN confirms the role of HO radicals in the mechanism of the PEDOT degradation. In these conditions and in open air, the lifetime of the film has been significantly improved.

It is also possible to confirm these results by TGA. Thermal experiments were performed at 80°C under air versus heating time. The thermal stability of PEO/PEDOT semi-IPNs has been evaluated on sample containing or no 4TBC. When there is no radical scavenger in the sample, the TGA curve shows a weight loss process corresponding to the all semi-IPN degradation. In these experimental conditions, the weight loss is around 15 wt% after 24 h (Figure 13.2a). A rather different behavior is observed if the TGA is carried out in the presence of 4TBC in the semi-IPN. The weight loss is less than 1 wt% after 24 h (Figure 13.2b) and does not exceed 9 wt% for a period as long as 3 days. This result is indicative of the higher thermal stability of the semi-IPN/4TBC compared with the semi-IPN alone. As shown in Figure 13.2c, the

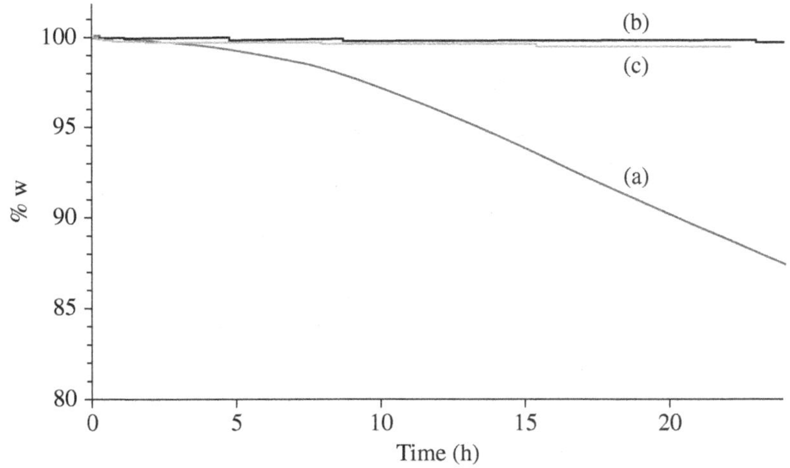

FIGURE 13.2 Thermogravimetric analysis at 80°C as a function of time of a PEO/PEDOT semi-IPN: (a) under synthetic air; (b) in the presence of 4TBC and under synthetic air; (c) under argon. Source: Verge *et al.* [8], pp. 3864–3870. Reproduced with permission of Elsevier

thermal stability of the semi-IPN is significantly improved under argon atmosphere, that is, the weight loss is below 1 wt% after 24 h, nullifying the effect of oxygen in the system.

13.2.1.2 Semi-2-Interpenetrating Polymer Networks from Polyetherimide and Citraconimide A series of semi-2-interpenetrating polymeric networks (semi-2-IPNs) were prepared through blending in solution using two different polyimides: biscitraconamic acid as a precursor of biscitraconimide (MBMI) with various proportions of polyetherimide (PEI) to achieve optimum properties [10]. Biscitraconamic acid was prepared by reacting citraconic anhydride (CA), 3,3′,4,4′-benzophenone tetracarboxylic dianhydride (BTDA) and bis(3-aminopropyl)phenyl phosphine (BAPPP), and it was characterized by differential scanning calorimetry (DSC), Fourier transform infrared spectroscopy (FTIR), and 1H-NMR spectroscopy. Both biscitraconamic acid and PEI were blended in *N,N*-dimethylacetamide (DMAc) solution, casted, and thermally cured up to 3000°C to give semi-2-IPNs. Semi-IPN polyimide samples were placed in an air-circulating oven at 3000°C for 200 h and further extended for 400 h. The samples were weighed at the scheduled aging time intervals. Semi-IPNs lost 9–14% and 17–25% of its original weight after isothermal aging, respectively. Neat MBMI resin showed 25.7% of weight loss for 400 h which reduces to 24.6 and 20.5 with incorporation of

10 and 50 wt% of PEI. This increase in thermo-oxidative stability is attributed to the presence of phenyl rings in the backbone of PEI.

13.2.1.3 Ethylene-co-Vinyl Acetate (EVA)/Polypyrrole (PPy) Interpenetrating Polymer Network Thermally stimulated depolarization current measurements and aging effects on poly(ethylene-*co*-vinyl acetate) and polypyrrole IPN composites were carried out in the temperature range between 283 and 343 K [11]. Thermally stimulated depolarization current appears as a very valuable technique to study slow mobility in the amorphous state. The results of thermal aging of the studied IPN composites were beneficial to explain the mechanism of the polarization in the material. The use of thermal aging enhances the sensitivity of thermally stimulated depolarization current (TSDC) and enables one to resolve the structure of distributed and overlapping relaxations. The increase of the conductivity in IPN composites at temperature below the relaxation temperature is mainly due to positive carriers in the material. The apparent activation energy of the relaxation of these IPN composites obtained experimentally is affected greatly by the aging time and the poling temperature.

Figure 13.3 shows several global TSDC data obtained in the aforementioned IPN composite (31% PPy content) for different times of aging (t_w's; 300–1200 s). From the inspection of Figure 13.3, the TSDC peaks obtained

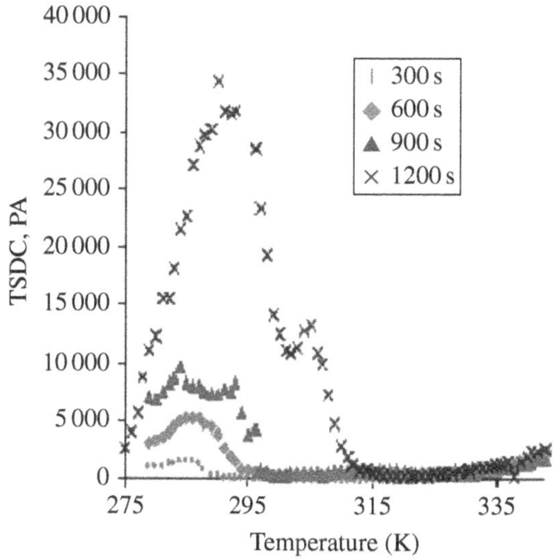

FIGURE 13.3 Effect of aging for different waiting times t_w (s). Source: Nasr *et al.* [11], pp. 301–305. Public Domain

for large waiting times have wider widths and appear at higher temperatures than those obtained for shorter aging times, and a shoulder appears at higher temperature (305 K). From the figure, various parameters such as relaxation time and activation energies were successfully calculated.

13.2.2 UV Radiation Aging

Photochemical stability of IPNs can be included in the aging behavior of the same. It includes the irradiation with sunlight or UV radiation and the effect of it on different properties. These data are important for assessing the lifetime of IPNs exposed to sunlight and to determine the possibility of modifying the surface properties of such polymeric materials during UV irradiation.

13.2.2.1 Polyurethane–Epoxy Resin Interpenetrating Polymer Networks

Rosu *et al.* synthesized polyurethane (PU)–epoxy resin (ER)-based semi-IPNs according to a sequential procedure by mixing the PU with variable amounts of ER in the presence of ethylene diamine [12]. A 40 mm^2 film sample was irradiated with a medium pressure mercury lamp, in the field 240–370 nm, and source power was 100 W. Irradiation of samples was performed in air, using a rotating hexagonal prism-shaped device. Irradiation was done for 100 and 200 h, and various properties such as color, gloss, weight loss, contact angle, and retained water content were measured.

L* decrease and red and yellow chromophore accumulation (a* and b* increase) on the film surfaces determined a change in color from gray to brown for all studied samples. These changes were lesser for semi-interpenetrating polymer networks (S-IPN) structures than for PU films due to high opacity of S-IPN films. Therefore, the UV radiation penetrated deeper in the PU film than in the S-IPN structures. The loss of gloss was highest in the PU sample due to surface roughness increase. Epoxy resin networks (ERN) content increase did not seem to significantly affect the variation of gloss retention. Contact angle variation significantly decreased during the first 100 h of irradiation time and increased later, due to accumulation of photodegradation products on the surfaces. This aspect could also be related to roughness increase during irradiation. No correlation between ERN content and contact angle variation could be found. Photodegradation of both PU and ER structures was investigated by FTIR technique. It was concluded from the FTIR spectra that photodegradation of the PU occurred via Norrish-type reactions and photo-Fries rearrangements, slower in the S-IPN structures. The presence of photo-oxidation processes was also identified. There also exists the possibility of photochemical protection of the PU for the networks containing up to 20%

ERN due to films opacity increase. This aspect is possible due to the ER's light retention capacity. Phase separation occurred for sample S-IPN-5 which contains more than 20% ERN. The mass losses were similar for all samples. The variation of retained water content upon irradiation showed an increase for all samples. The epoxy content had a normal influence on water content.

13.2.2.2 Polyisobutene/Polystyrene Interpenetrating Polymer Networks

In order to improve polyisobutene (PIB) mechanical properties, Vancaeyzeele *et al.* combined a PIB network with a polystyrene (PS) one into an IPN architecture [13]. The PIB network is formed by alcohol addition between the hydroxyl end groups of a telechelic dihydroxypolyisobutene and a pluri-isocyanate. The PS network is synthesized by free radical copolymerization of styrene with divinylbenzene. The material aging study was performed by UV exposure (500 W) at 358°C (Helios Italquartz). The absorbances of single networks and IPNs were measured with a Jasco V570 spectrometer at 350 nm every 24 h and the absorbance variation (ΔA) was calculated as

$$\Delta A = \frac{A_t - A_0}{A_0} \tag{13.1}$$

where A_0 and A_t are the sample absorbance before and after UV exposure, respectively. The IPN architecture should lead to a protection of each individual network from the physical and/or chemical attacks. The single PIB and PS networks and the PIB/PS IPNs are exposed to UV in order to promote their aging. The material absorbance is measured at 350 nm every 24 h in order to estimate the aging behavior.

The single PS network is much more sensitive to UV irradiation than the single PIB one or PIB/PS IPNs. Indeed, the PS single network absorbance is multiplied by 9 after 120 h and the material turns yellow contrary to single PIB network and IPNs whose absorbance hardly doubled. Aromatic compounds are well known to lead to yellowish material following UV irradiation. The difference of absorbance variations between the single networks and IPNs is not due to a dilution effect as IPN absorbance variations are lower than those of the single networks. The PS phase inside the IPN architecture is thus protected from irradiation.

13.2.2.3 Compatible IPNs Based on Epoxy and Methacrylated Epoxy Resins

Lin *et al.* synthesized IPNs from methacrylated diglycidyl ether of bisphenol A (MADGEBA) by blending with DGEBA in different weight ratios [14].

The materials were cured simultaneously with dicumyl peroxide (DICUP) and hexahydrophthalic anhydride (HHPA). The samples were irradiated with ultraviolet in a Q-UVA weatherometer for various periods to study their aging behavior. Experimental results indicated that the methacryloyl moiety thus incorporated confers significant stabilization against photodegradation. The accelerated aging study of DGEBA by FTIR agreed with the degradation mechanisms proposed earlier. Aging experiments revealed that incorporation of the methacryloyl moiety into the DGEBA either at chain ends or by means of IPN resulted in significant protection of the epoxy from UV damage. Mechanical test of specimens confirmed the result of FTIR study.

13.2.3 Water Aging

Measurements of water uptake on IPNs were performed after drying the samples on suitable agents and then immersing them in distilled water. The hydration can be calculated by specific equations after leaving the samples immersed in water for various times. The times can be minutes to days.

13.2.3.1 Polyurethane/Poly(2-Hydroxyethyl Methacrylate) Semi-Interpenetrating Polymer Networks Stamatopoulou *et al.* prepared and investigated the nanocomposites based on sequential semi-IPN of crosslinked PU and linear poly(2-hydroxyethyl methacrylate) with 0.25 and 3 wt% of nanosilica filler [15]. Semi-IPNs, which were used as the matrix to create the nanocomposites, were obtained as follows. First, the PU network was synthesized by a two-step method. The first step was the preparation of adduct of trimethylol propane (TMP) and toluylene diisocyanate (TDI). Adduct of TMP/TDI was synthesized by reaction of 1 g-equiv. of TMP with 1.5 g-equiv. of TDI at 65°C. The reaction was performed to achieve the theoretically calculated isocyanate group content, which was determined by titration method. The second step was the synthesis of three-dimensional PU. The PU network was obtained from a mixture of poly(oxypropylene glycol), POPG, of MM 2000 and adduct TMP/TDI (ratio 1 : 2 g-equiv.) at 80°C in a nitrogen atmosphere. Semi-IPNs were obtained by a sequential method. Films of PU network were subjected to swelling in 2-hydroxyethyl methacrylate (HEMA) monomer containing photoinitiator. Swelling was performed to equilibrium. Then the photopolymerization of monomer HEMA in the matrix of PU was performed.

Aging measurements of water uptake were performed after drying the samples over P_2O_5 and then immersing them in distilled water. The hydration was calculated using Equation 13.2 after leaving the samples immersed in water for 6, 35, and 130 days:

$$h = \frac{m_{\text{water}}}{m_{\text{dry,sample}}} = \frac{m_{\text{hydrated,sample}} - m_{\text{dry,sample}}}{m_{\text{dry,sample}}} \tag{13.2}$$

where $m_{\text{dry,sample}}$ is the mass of the sample equilibrated at 0.02 RH.

In Figure 13.4, the results of immersion experiments are shown: the content of water in the samples of PU, the semi-IPN, and the nanocomposites after 6, 35, and 130 days of immersion in water, to follow effects of aging and hydrolysis. For comparison, results of water content at ambient relative humidity and of sorption from the vapor phase at 0.98 RH are also presented. For neat PU, the apparent water uptake from immersion is significantly smaller than from the vapor phase already after 6 days of immersion and decreases further with increasing time of immersion. The results showed that the hydrophilicity of the semi-IPN is more or less preserved in the various nanocomposites and even enhanced in the case of unmodified silica as filler. Effects of water on polymer dynamics were found to be particularly

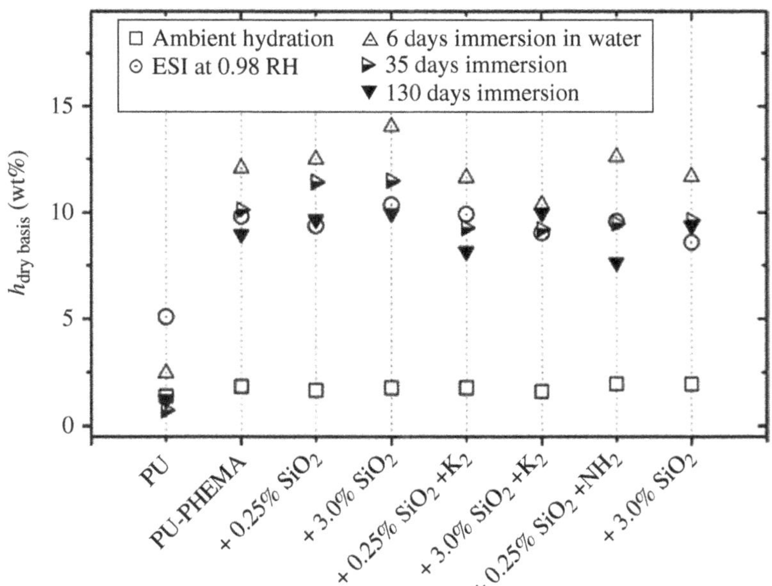

FIGURE 13.4 Comparative hydration results for neat PU, the PU–PHEMA semi-IPN, and its nanocomposites. Results are shown for samples equilibrated at ambient humidity and at 0.98 RH and samples immersed in water for 6, 35, and 130 days. Source: Stamatopoulou *et al.* [15], pp. 397–408. Reproduced with permission of Wiley

pronounced for the secondary $\beta_{sw,PHEMA}$ and the β_{PU} relaxations, in agreement with interpretations in terms of hydrogen bonding interactions with specific groups in the structure of the two polymers.

13.2.3.2 Gelatin–Alginate Interpenetrating Polymer Network Hydrogels

The gelatin–alginate biopolymer hydrogels were synthesized via a sequential interpenetrating network method [16]. Figure 13.5 shows the schematic diagram of the gelatin–alginate hydrogel formation.

The typical procedure is as follows: 15% (w/v) gelatin was dissolved indifferent concentrations of alginate solution (0–3%, w/v) by gentle heating to obtain a polymer solution with various compositions. Microbial transglutaminase (mTG) was then added to the polymer solution to achieve a final concentration of 20 U/ml. Then, the solution was decanted into a homemade mold consisting of a pair of glass plates separated by 2 mm thick glass slides and kept at 40°C for 30 min to form an enzymatic crosslinked, semi-interpenetrated (semi-IPN) hydrogel. Subsequently, the semi-IPN hydrogel was immersed in a 50 mM aqueous solution of $CaCl_2$ for another 1 h at room temperature to allow for

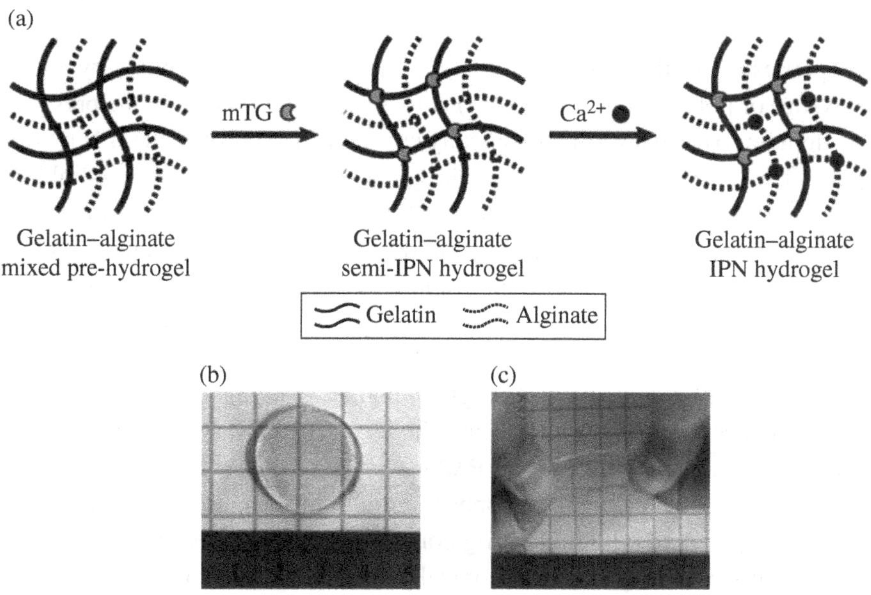

FIGURE 13.5 (a) Schematic illustration of the formation of IPN hydrogel via sequential interpenetrating network formation using enzymatic and ionic crosslinking approaches. (b) Image of typical gelatin–alginate IPN hydrogel. (c) Gelatin–alginate IPN hydrogel being stretched illustrating good tensile strength. Source: Wen *et al.* [16], pp. 504–513. Reproduced with permission of Wiley

the gelation of alginate. The double crosslinked hydrogels were referred to as interpenetrated network (IPN) hydrogels. Both semi-IPN and IPN hydrogels were rinsed thoroughly with distilled water. Then, 4.5 cm^2 disks of hydrogels were punched out from the hydrogel sheet for further characterization. The hydrogels were incubated in distilled water at room temperature for 24 h to reach its equilibrium swollen state and the aging characteristics were measured.

The aging data illustrated that the gelatin–alginate IPN hydrogels showed a significantly decreased swelling ratio in comparison with the gelatin gels. This can be attributed to the increasing crosslinking degree of the IPN hydrogel, which hindered the expansion of water in the gel. The semi-IPN hydrogels displayed a trend of slight increase in swelling ratio. This is because they were generated by intertwining alginate polymers into gelatin hydrogels without crosslinking, resulting in a relatively loose structure and a higher swelling ratio.

13.2.3.3 Covalently Crosslinked Polyacrylamide (PAAm) or Poly(acrylic Acid) (PAA)–Alginate (Alg) IPN Hydrogels IPNs made of a covalently crosslinked network and an ionically crosslinked network were produced to investigate the effective parameters that control the physical and mechanical properties of an IPN hydrogel [17]. Covalently crosslinked polyacrylamide (PAAm) or poly(acrylic acid) (PAA) networks were produced in the presence of alginate (Alg) that was then ionically crosslinked to produce the IPN hydrogels. The effect of ionic crosslinking, degree of covalent crosslinking, AAm:Alg and AA:Alg ratio on the swelling ratio, tensile properties, indentation modulus, and fracture energy of IPN hydrogels was studied. A clear difference in equilibrium swelling ratio was observed in all cases between samples prepared with and without ionic crosslinking. In ionically crosslinked PAAm–Alg and PAA–Alg IPN hydrogels, a higher Alg concentration resulted in lower swelling ratios, with slightly higher swelling occurring when a smaller concentration of chemical crosslinker (MBAA) was used. Surprisingly, the equilibrium swelling of both PAAm- and PAA-based hydrogels without ionic crosslinking showed the opposite trend with increased swelling in the hydrogels prepared with higher Alg concentrations. To assess the stability of these hydrogels, samples of both ionically crosslinked and nonionically crosslinked hydrogels were fully dried and their dry mass was determined. Samples of each hydrogel were also stored in deionized water (with daily change of water) for 1 week, then fully dried, and weighed. The difference in dry mass of the as-prepared and water-immersed samples was assumed to be due to the loss of alginate from the gel during water immersion.

13.2.3.4 Polydivinylbenzene/Polyacryldiethylenetriamine Interpenetrating Polymer Networks Liu *et al.* prepared hydrophobic–hydrophilic IPNs composed of polydivinylbenzene (PDVB) and polyacryldiethylenetriamine (PADETA), and its adsorption performance toward salicylic acid was studied from aqueous solutions [18]. Equilibrium adsorption of salicylic acid on the IPNs was performed at three different temperatures (293, 303, and 313 K, respectively) from aqueous solutions. About 0.1 g of the resins was accurately weighed and mixed with 50 ml of salicylic acid solution at different initial concentrations. The initial concentrations of salicylic acid were preset to be 200.1, 400.2, 600.3, 800.4, and 1000.5 mg/l, respectively. The mixture solution was then continuously shaken at a desired temperature for 8 h so that the adsorption reached equilibrium. After that, the IPNs were filtered and the equilibrium concentration of salicylic acid (milligram per liter) was determined by UV absorption analysis performed on a UV-2450 spectrophotometer.

Dynamic adsorption and desorption was also measured by immersing the IPNs in deionized water at room temperature for 24 h and then packed densely in a glass column (16 mm diameter). The volume of the wet resin column was 10 ml. The salicylic acid solution at an initial concentration of 650.4 mg/l was passed through the resin column at a flow rate of 7.2 BV/h and the residual concentration of salicylic acid from the effluent, C (mg/l), was dynamically recorded until it reached the initial concentration. After the dynamic adsorption, the resin column was roughly rinsed by 20 ml of deionized water and then a mixture solution composed of 0.01 mol/l of sodium hydroxide (w/v) and 40% of ethanol (v/v) was used as the desorption solvent for the dynamic desorption. Desorption solvent (200 ml) was passed through the resin column at a flow rate of 3.6 BV/h and the concentration of salicylic acid from the effluent was determined until it was about zero. The adsorption of salicylic acid on PDVB/PADETA IPNs was much superior to the hydrophobic PDVB as well as the hydrophilic PADETA, and the adsorption of PDVB/PADETA IPNs was superior to much polymeric adsorbents and comparative to the hypercrosslinked resins. The Freundlich model is shown to be more suitable for fitting the equilibrium isotherm data than the Langmuir model and the isosteric enthalpy decreased with increasing of the equilibrium capacity.

13.2.3.5 Porous Gelatin–Alginate–Polyacrylamide Interpenetrating Polymer Networks Stancu *et al.* reported a simple method for the synthesis of porous gelatin–alginate–polyacrylamide scaffolds with interpenetrating network structure [19]. The materials are obtained through a step-by-step procedure, starting with the polymerization of acrylamide and followed by the crosslinking of the natural components. Porosity was induced through freeze-drying. The aging of the IPNs was assessed in four different aqueous media including distilled deionized water (ddw), phosphate buffer saline (PBS) pH 7.4, artificial saliva (AS), and physiological serum (PhS), at 37°C. The gravimetric method was used to estimate the swelling degree (SD) at predetermined time intervals. It has been concluded that the swelling of the porous hydrogels can be fine-tuned by modifications of the composition of the initial polymer solutions. The results obtained indicated that these materials are appropriate for drug delivery, bioreactors, or even soft TE applications.

13.2.3.6 Semi-Interpenetrating Polymer Networks Composed of ι-Carrageenan and Montmorillonite Karadag *et al.* very recently reported the synthesis of a series of chemically crosslinked hybrid composite hydrogel systems containing polysaccharide/clay polyelectrolyte based on the acrylamide/sodium methacrylate (AAm/SMA) and ι-carrageenan and clay such as montmorillonite with free radical solution polymerization by using ammonium persulfate/*N,N,N′,N′*-tetramethylethylenediamine as a redox-initiating pair in the presence of poly(ethylene glycol) diacrylate as a crosslinker [20]. FTIR analysis and scanning electron microscopy (SEM) technique were applied for characterization. The hydrogels, the semi-IPNs, and the hybrid composite hydrogel systems that were synthesized in this study have shown high water absorbency. For swelling studies, chemically crosslinked dried hydrogel systems were accurately weighted and transferred into water. Water uptake with respect to time was obtained by periodically removing the samples from water, quickly blot drying, and reweighing. The measurements were conducted at 25 ± 0.1°C in a water bath.

Figure 13.6 shows that swelling increases with time up to a certain level, and then levels off. This swelling may be called as the equilibrium swelling ratio.

The semi-IPNs were sorbed from the monovalent cationic dye such as Safranin T (ST) from aqueous ST solutions. To determine the sorption characteristics of ST onto the hydrogel systems, some sorption parameters were calculated. From the results, it was concluded that the IPN systems could be used as the potential water adsorbent for the removal of the dye molecules or other big molecular structures similar to dye molecules. Consequently, semi-IPNs developed in this study may serve as a potential device for water and dye sorbents. It was also proposed that the new IPNs reported could be used to carry substances in an aquatic field for agriculture, environment, separation processes, and water purification.

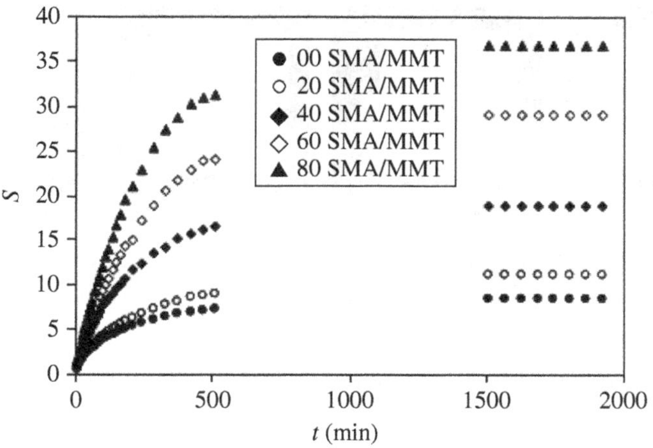

FIGURE 13.6 Swelling isotherms of AAm/SMA/MMT hydrogels. Source: Karada *et al.* [20]. Reproduced with permission of Wiley

13.2.3.7 Semi-Interpenetrating Polymer Networks of Polyacrylamide and Carboxymethyl Cellulose (CMC) Aalaie *et al.* prepared semi-IPN hydrogels by crosslinking of hydrolyzed polyacrylamide (HPAM)/CMC aqueous solutions with chromium triacetate [21]. Study of the gelation behavior using dynamic rheometer showed that the limiting storage modulus increased more than two times for the semi-IPN gels containing 50 wt% CMC. Increasing the CMC content decreased the loss factor, indicating that the elastic properties of this gelling system increase more strongly than their viscous properties. The swelling ratio of the semi-IPN gels in tap water, NaCl, $CaCl_2$, and synthetic oil reservoir water slightly decreased with increasing the CMC content of the network.

The water absorbency in NaCl solution was higher than that in $CaCl_2$ solution. This behavior was due to the reduced concentration of free ions in the gels and the ionic crosslinking by the divalent cations (Ca^{2+}) with the anionic sites of the HPAM chains, which resulted in loss of swelling. The increased CMC content of the network leads to an increase of the crosslink density of the polymeric network and hence less swelling of the resulting hydrogels. Therefore, the improved storage modulus and slightly decreased swelling capacity in oil reservoir water make these semi-IPN hydrogels potentially good candidates for excess water treatment in oil recovery applications.

13.2.3.8 Semi-interpenetrating Network of Sodium Alginate/Carboxymethylated Bacterial Cellulose Composite Very recently, Lin *et al.* prepared and studied the swelling behavior of a novel ionic crosslinking sodium alginate (SA)/carboxymethylated bacterial cellulose (CM-BC) composite with a

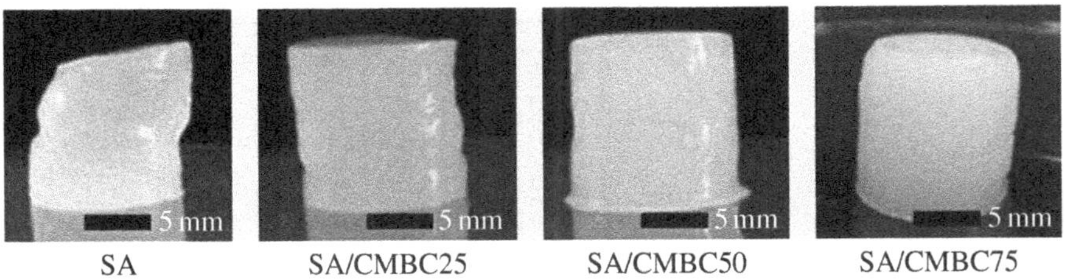

FIGURE 13.7 Photographs of SA, SA/CM-BC25, SA/CM-BC50, and SA/CM-BC75. Source: Lin *et al.* [22]. Reproduced with permission of Wiley

semi-IPN structure [22]. The composite was prepared through the blending of an SA gel with CM-BC and then crosslinking by Ca^{2+} followed by a freeze-drying process. SEM showed the composite matrix organized in a three-dimensional network of CM-BC interpenetrated against SA molecular chains with a quantity of calcium alginate microspheres upon the surface. Photographs of the prepared IPNs are given in Figure 13.7.

To measure the swelling properties of these composites, each kind of preweighed freeze-dried sample (SA, SA/CM-BC25, SA/CM-BC50, and SA/CM-BC75, cylinder height 55 mm, cylinder diameter 530 mm) was immersed in deionized water for a sufficient time to completely swell up. The sheets were then taken out of the storage container with tweezers. The samples were put in a sieve, quickly shaken twice to remove the surface water, and then weighed. The SA/CM-BC50 swelling test at pH values of 2.32, 6.91, and 10.83 (adjusted by the addition of 0.1 mol/l HCl and NaOH solutions) was also carried out in this way. The swelling behavior is given in Figure 13.8, and the swelling ratios of the composite were enhanced by 183%, 198%, and 212% with the supplementation of CM-BC weight fractions of 25%, 50%, and 75%, respectively; the swelling ratios changed with changing pH.

In addition, the tensile modulus, tensile strength, and elongation at break of SA were enhanced by 165%, 152%, and 188%, respectively, with the addition of 50 wt% CM-BC. This study demonstrated that the semi-IPN structure dramatically changed the swelling and mechanical properties of the composite, and the semi-IPN will be a promising candidate for biomedical applications such as wound dressings and skin tissue engineering.

13.2.3.9 Semi-Interpenetrating Polymer Networks Superabsorbent Resin Based on Chicken Feather Protein An interesting semi-IPNs superabsorbent resin (SAR) from chicken feather protein-g-poly(potassium acrylate)/polyvinyl alcohol (CFP-g-PKA/PVA) based on feather protein, acrylic acid (AA), and

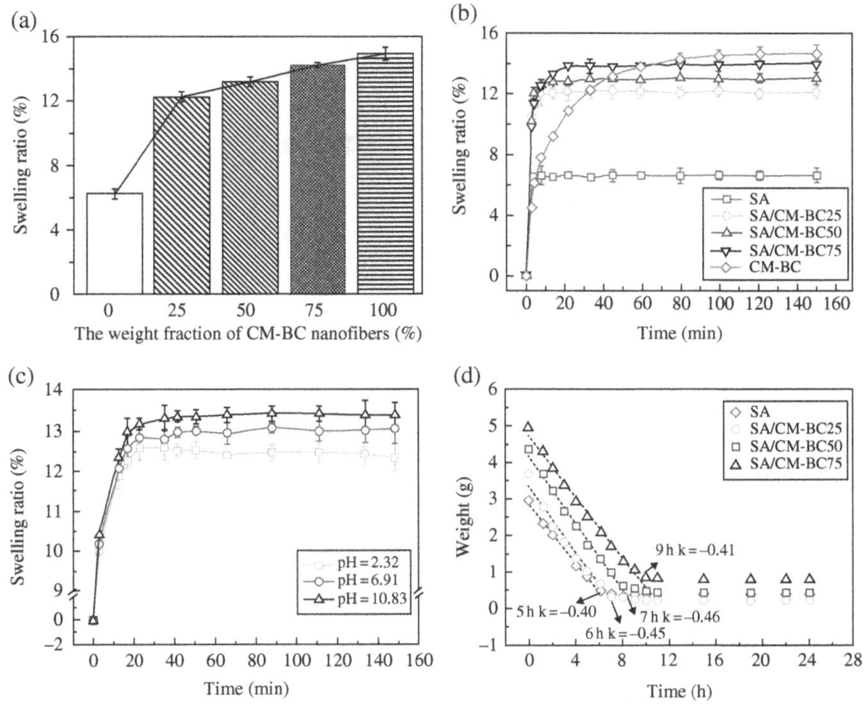

FIGURE 13.8 Swelling behaviors of the SA/CM-BC composites: (a) maximum swelling ratio, (b) swelling behaviors at pH 7.0 with different CM-BC weight fractions, (c) swelling behaviors of SA/CM-BC50 at different pHs, and (d) water loss behaviors of the SA/CM-BC composites with different CM-BC weight fractions, where *k* represents the gradient of the fitted lines. Source: Lin *et al.* [22]. Reproduced with permission of Wiley

polyvinyl alcohol (PVA) was synthesized using graft copolymerization and semi-interpenetrating technology [23]. PVA reacted with PKA during the polymerization steps and was confirmed by FTIR, SEM, and TGA analyses. The effects of AA, PVA, initiator, and crosslinker content on water absorbency of semi-IPNs SAR were studied. The swelling behavior in various pHs and saline solutions was also investigated. The water absorbency of SAR reached the maximum at a pH of 6. The effect of various salt solutions on the swelling characteristics is given in Figure 13.9.

Swelling capacity for SAR in salt solutions with the same concentration was in the order $NaCl > KCl > MgCl_2 > Na_2SO_4 > CaCl_2 > AlCl_3$. The superior performances of CFP-g-PKA/PVA semi-IPNs SAR could promote their potential applications in agriculture and horticulture fields as suggested by the authors.

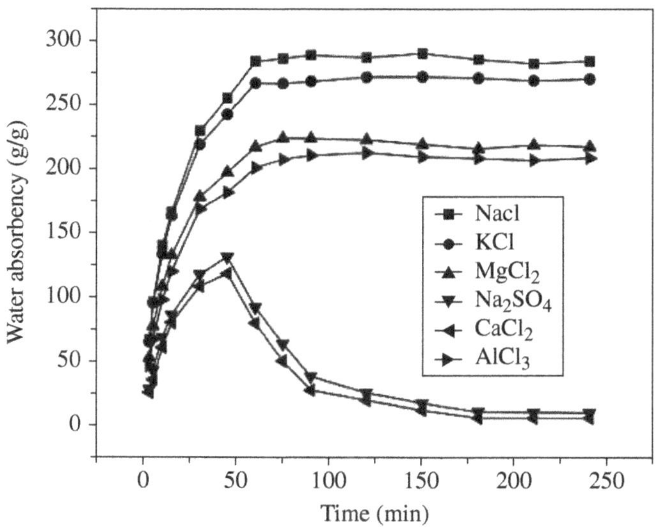

FIGURE 13.9 Effect of various salt solutions on swelling kinetics of CFP-g-PKA/PVA semi-IPNs SAR. Source: Li *et al.* [23]. Reproduced with permission of Wiley

13.2.4 Aging by Other Sources

Aging of IPNs in other mediums were also reported recently. Very recently, an interpenetrating network biohydrogel of gelatin and gellan gum was prepared by using a combination of enzymatic and ionic crosslinking approaches [24]. As previously known, gelatin is a popular substrate for cell culture applications due to its biocompatibility and biodegradability. However, the mechanical property of gelatin is not satisfactory in certain tissue engineering areas where tunable and higher mechanical strengths are required. To achieve this purpose without exposure of materials to cytotoxic chemicals or procedures, a new biohydrogel of gelatin and gellan gum with an interpenetrating network (IPN) structure was prepared using a combination of enzymatic and ionic crosslinking approaches. Rheological study of the prepared IPNs revealed the gelation procedure and thermal stability. The resulted IPN biohydrogel exhibited significantly increased and tunable mechanical strength, decreased swelling ratios, and lower degradation rate compared with pure gelatin gel.

Swelling tests of the IPNs were conducted by using disc samples that were immersed in 2 ml of PBS solution at 37°C with shaking at 80 rpm for 24 h. Then the biohydrogels were removed from the solution and the gel surfaces were gently wiped using filter paper. The IPNs were also studied by *in vitro* enzymatic degradation and *in vitro* cytotoxic evaluation with appropriate reagents. As shown in Figure 13.10, the remaining mass of the IPN biohydrogels linearly decreased to 40% after 12 h, which was similar to the behavior of

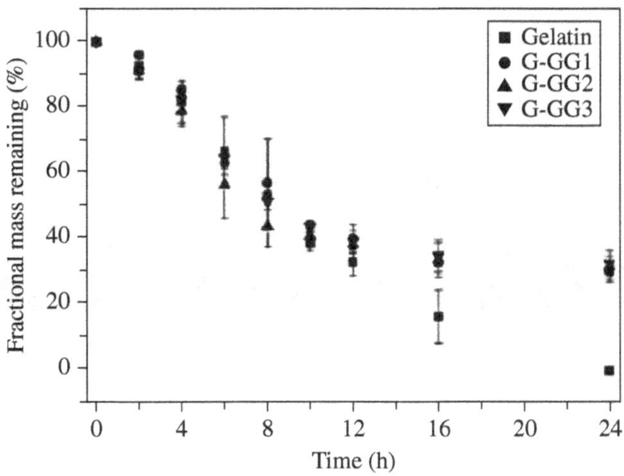

FIGURE 13.10 Degradation profile of gelatin and IPNs incubated in PBS with papain (90 U/ml) at 37°C. Source: Wen *et al.* [24], pp. 1643–1649. Reproduced with permission of Wiley

gelatin. The weight change in the gels plateaued in the degradation curve of IPN hydrogels in the following 12 h, retaining around 30% of the initial weight of the gel, while the gelatin gel degraded completely.

Ray *et al.* prepared poly(vinyl alcohol-*co*-acrylic acid)/sodium chloride hydrogel IPNs without any added crosslinker, using benzoyl peroxide as initiator and sodium chloride (NaCl) as additive [25]. The response of the hydrogels with and without NaCl was observed by studying their swelling behavior, biodegradability, and thermal stability. Biodegradation of the IPNs and quantitative estimation of CO_2 was done as given. A cultured medium was prepared by taking agar nutrient broth. The nutrient broth so prepared was sterilized maintaining at a pressure of 15 lb/in.^{2} at 121°C for 20 min. Then to 10 ml of sterilized broth 0.1 g each of the samples, that is, (PVA-*co*-PAA)/NaCl and (PVA-*co*-PAA), was taken in test tubes under aseptic conditions. To the medium, bacteria, that is, *Escherichia coli*, was inoculated. The extent of biodegradation through *E. coli* was studied by weight loss basis and the amount of CO_2 evolved during the incubation periods of 8, 15, 21, and 28 days. Also the degradation of the materials in simulated intestinal condition also studied with respect to weight loss. Materials were conditioned to minimum weight at 37 ± 1°C in an oven containing desiccant prior to being immersed into 100 ml of a simulated fed intestinal fluid (0.05 M potassium dihydrogen orthophosphate, 0.015 M sodium hydroxide, and 10 g/l. pancreatin, pH adjusted to 6.8 by sodium hydroxide). The specimens were removed at regular intervals of 4 h, being taken out of the solution, blotted on filter paper to

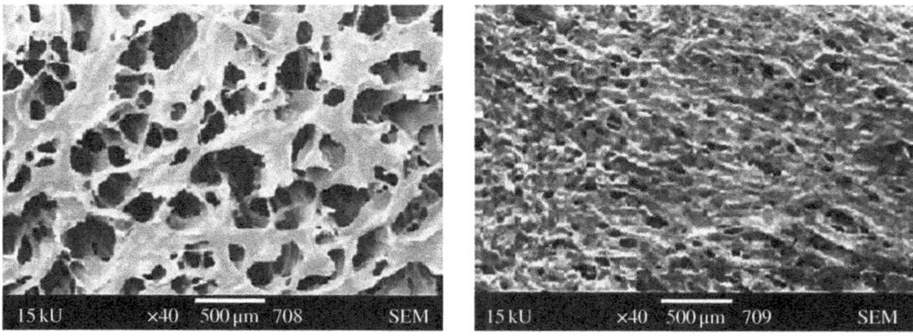

FIGURE 13.11 Scanning electron microscopy of (PVA-co-PAA)/NaCl before and after biodegradation, BPO (2.5% (w/v)), NaCl (1.0% (w/v)). Source: Ray *et al.* [25] Public Domain

remove surface solution and dried in an oven at 50°C to constant weight in order to determine eventual weight loss, taking an average of two readings.

The biodegradation study by *E. coli* showed that (PVA-*co*-PAA)/NaCl was more degradable than that without NaCl. From the pH-dependent swelling studies, IPN hydrogel showed comparatively higher swelling at intestinal pH environment than that of gastric medium and presence of NaCl in the IPN increases the swelling properties in both media. The swelling ratio of the (PVA-*co*-PAA)/NaCl IPN was found to be maximum at 1% w/v NaCl concentration. The IPN hydrogel showed moderate thermal stability, and the surface morphology by SEM revealed that the pores were mostly open in the presence of NaCl, thus making the IPN macroporous (Figure 13.11).

Essawy *et al.* explored the fabrication of novel amphiphilic interpenetrating conetwork structures that may afford a reasonable solution to the problem of oil spills [26]. The concept was developed upon compatibilization of the binary immiscible rubber blend, containing the polar acrylonitrile–butadiene rubber and the nonpolar styrene–butadiene rubber. The compatibilization process was accomplished successfully using organic–inorganic hybrid structures based on montmorillonites, priory treated with a hydrophobic cationic agent to different levels so that they acquire wide range of amphiphilicity. With achieving the goal of compatibilization using bipolar montmorillonites, the formed amphiphilic interpenetrating network structures are thought to qualify as oil absorbent materials for oils exhibiting a broad spectrum of hydrophilicity/hydrophobicity. For this purpose, toluene, which exhibits both polar and nonpolar characters, was chosen as the model to predict the oil absorption and swelling extent as a function of the clay philicity. The behavior of the swelling is given in Figure 13.12 with respect to different clay loadings and showed interesting results. Also, the capacity of oil uptake showed good correlation to the crosslinking density.

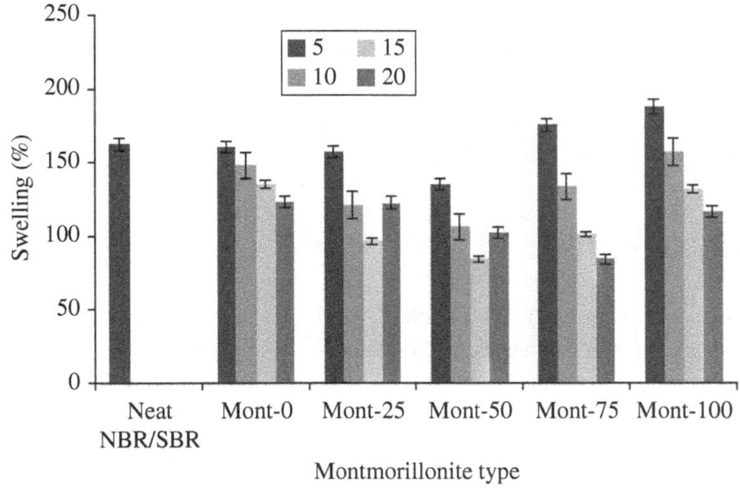

FIGURE 13.12 Swelling behavior of NBR/SBR blends based on montmorillonites bearing different organophilicity. Source: Essawy *et al.* [26]. Reproduced with permission of Wiley

Another important report came out recently from Yamak and colleagues [27]. Laccase enzyme (L) from *Trametes versicolor* was entrapped in three hydrogel structures, namely, poly(acrylamide-*N*-isopropylacrylamide), P(AAm-NIPA), and semi-interpenetrating networks of poly(-acrylamide)/alginate, P(AAm)/Alg, and poly(acrylamide-*N*-isopropylacrylamide)/alginate, P(AAm-NIPA)/Alg. The optimum temperatures for free and all immobilized systems were found to be 40°C. For free and the same immobilized systems, the enzymes retained 42, 91, 79, and 86% of their initial activities at the end of 56 days of storage. After using the mentioned immobilized systems repeatedly 10 times, they retained 77, 71, and 84% of their original activities, respectively. For free and the same immobilized systems, decolorization of acid orange 52 (AO52) in 6h was found to be 63%, 50%, 48%, and 66%, respectively. Addition of 2,2-azino-bis-(3-ethylbenzothiazoline-6-sulfonic acid), ABTS, into the assay medium increased these values up to 73%, 73%, 74%, and 75%, respectively.

13.3 CONCLUSION

IPNs as peculiar structural forms have a lot of promising applications in different fields such as drug delivery, biomedical applications, membranes, and fuel cells. The material aging has an important role to design these applications. As discussed in the previous sections, various types of aging can be thought of for IPNs. However, the numbers of publications in these areas is not much so that a generalization is difficult. Even then, a few promising results and thereby applications are proposed in this active field of research recently. New innovations and findings will help the area to grow and extraordinary materials of IPNs are expected in the near future and their aging characteristics will be very important.

REFERENCES

[1] Kim, S.C., Klempner, D., Frisch, K.S. *et al.* (1976) Polyurethane interpenetrating polymer networks. I. Synthesis and morphology of polyurethane-poly(methyl methacrylate) interpenetrating polymer networks. *Macromolecules*, **9**, 258.

[2] Murayama, S., Kuroda, S. and Osawa, Z. (1993) Hydrophobic and hydrophilic interpenetrating polymer networks (IPNs) composed of polystyrene and poly(2-hydroxyethyl methacrylate): 2. Gradient composition in the IPNs synthesized by photopolymerization. *Polymer*, **34**, 3893.

[3] Alexandratos, S.D., Grady, C.E. and Crick, D.W. (1991) Synthesis of bifunctional interpenetrating polymer networks as ion-complexing agents. *Macromolecules*, **24**, 6365.

[4] Diez-Pena, E. (2002) On the water swelling behaviour of poly(N-isopropylacrylamide) [P(N-iPAAm)], poly(methacrylicacid)[P(MAA)], their random copolymers and sequential interpenetrating polymer networks (IPNs). *Polymer*, **43**, 4341.

[5] Kim, S.J. (2003) Swelling behavior of interpenetrating polymer network. *Reactive Functional Polymers*, **55**, 53.

[6] Sperling, L.H. (1981) *Interpenetrating Polymer Networks and Related Materials*, Plenum, New York.

[7] Sperling, L.H. and Mishra, V. (1996) The current status of interpenetrating polymer networks. *Polymers for Advanced Technologies*, **7**, 197.

[8] Verge, P., Vidal, F., Aubert, P.-H. *et al.* (2008) Thermal ageing of poly(ethylene oxide)/poly(3,4-ethylenedioxythiophene) semi-IPNs. *European Polymer Journal*, **44**, 3864.

[9] Morlat, S. and Gardette, J.L. (2001) Phototransformation of water-soluble polymers. I: photo- and thermooxidation of poly(ethylene oxide) in solid state. *Polymer*, **42**, 6071.

[10] Gupta, P. and Alam, S. (2011) Studies on novel semi-2-IPNs from polyetherimide and citraconimide. *Journal of Applied Polymer Science*, **120**, 2790.

[11] Nasr, G.M., Amin, M. and Shaker, H.M. (2011) Influence of physical aging on the dielectric relaxation of ethylene-co-vinylacetate (EVA)/polypyrrole (PPy) interpenetrating polymer network composite using TSDC technique. *Journal of the Korean Physical Society*, **58**, 301.

[12] Rosu, D., Rosu, L., Mustata, F. and Varganici, C.-D. (2012) Effect of UV radiation on some semi-interpenetrating polymer networks based on polyurethane and epoxy resin. *Polymer Degradation and Stability*, **97**, 1261.

[13] Vancaeyzeele, C., Fichet, O., Laskar, J. *et al.* (2006) Polyisobutene/polystyrene interpenetrating polymer networks: effects of network formation order and composition on the IPN architecture. *Polymer*, **47**, 2046.

[14] Lin, M.-S., Wang, M.-W., Lee, C.-T. and Shao, S.-Y. (1998) Accelerated ageing behavior of compatible IPNs based on epoxy and methacrylated epoxy resins. *Polymer Degradation and Stability*, **60**, 505.

[15] Stamatopoulou, C., Klonos, P., Koutsoumpis, S. *et al.* (2014) Hydrophilic nanocomposites based on polyurethane/poly(2-hydroxyethyl methacrylate) semi-IPNs and modified/unmodified nanosilica for biomedical applications. *Journal of Polymer Science, Part B: Polymer Physics*, **52**, 397.

[16] Wen, C., Lu, L. and Li, X. (2014) Mechanically robust gelatin–alginate IPN hydrogels by a combination of enzymatic and ionic crosslinking approaches. *Macromolecular Materials Engineering*, **299**, 504.

[17] Naficy, S., Kawakami, S., Sadegholvaad, S. *et al.* (2013) Mechanical properties of interpenetrating polymer network hydrogels based on hybrid ionically and covalently crosslinked networks. *Journal of Applied Polymer Science*, **130**, 2504.

[18] Wang, X., Liang, X., Huang, J. and Liu, Y.-N. (2014) Hydrophobic–hydrophilic polydivinylbenzene/polyacryldiethylenetriamine interpenetrating polymer networks and its adsorption performance toward salicylic acid from aqueous solutions. *AIChE Journal*, **60**, 2636.

[19] Stancu, I., Lungu, A., Dragusin, D. *et al.* (2013) Porous gelatin-alginate-polyacrylamide scaffolds with interpenetrating network structure: Synthesis and characterization. *Soft Materials*, **11**, 384.

[20] Karadag, E., Hasgul, B., Kundakc, S. and Uzum, O. (2014) A study of polymer/clay hybrid composite sorbent-based AAm/SMA hydrogels and semi-IPNs composed of ι-carrageenan and montmorillonite for water and dye sorption. *Advance in Polymer Technologies*, **33**, 21432.

[21] Aalaie, J., Vasheghani-Farahani, E., Rahmatpour, A. and Semsarzadeh, M. (2013) Gelation rheology and water absorption behavior of semi-interpenetrating polymer networks of polyacrylamide and carboxymethyl cellulose. *Journal of Macromolecular Science, Part B: Physics*, **52**, 604.

[22] Lin, Q., Zheng, Y., Ren, L. *et al.* (2014) Preparation and characteristic of a sodium alginate/carboxymethylated bacterial cellulose composite with a cross-linking semi-interpenetrating network. *Journal of Applied Polymer Science*, **131**, 39848.

[23] Li, Q., Liu, J., Su, Y. *et al.* (2014) Synthesis and swelling behaviors of semi-IPNs superabsorbent resin based on chicken feather protein. *Journal of Applied Polymer Science*, **131**, 39748.

[24] Wen, C., Lu, L. and Li, X. (2014) An interpenetrating network biohydrogel of gelatin and gellan gum by using a combination of enzymatic and ionic crosslinking approaches. *Polymer International*, **63**, 1643.

[25] Ray, D., Sahoo, P.K. and Mohanta, G.P. (2008) Designing of biodegradable interpenetrating polymer network of poly(vinyl alcohol-co-acrylic acid)/sodium chloride hydrogel: an approach to drug delivery. *International Journal of Pharmaceutical Sciences and Nanotechnology*, **1**, 144.

[26] Essawy, H.A., El-Sabbagh, S.H. and Tawfik, M.E. (2015) Novel interpenetrating amphiphilic co-networks based on compatibilized NBR/SBR-Montmorillonite composites: a study on the oil absorption characteristics. *Polymer Composites*, **36**, 1494–1501.

[27] Yamak, O., Kalkan, N., Aksoy, S. *et al.* (2009) Semi-interpenetrating polymer networks (semi-IPNs) for entrapment of laccase and their use in Acid Orange 52 decolorization. *Process Biochemistry*, **44**, 440.

14

THEORETICAL MODELING AND SIMULATION OF IPNs

PRATHAB BASKAR

Materials Modeling and Product Design, Research & Development, Tata Steel, Jamshedpur, Jharkhand, India

14.1 INTRODUCTION

An interpenetrating polymer networks (IPNs) is any material containing two polymers, each in network form. These networks are prepared by polymerizing and crosslinking one polymer in the presence of a prefabricated networked polymer, resulting in a permanent mixture often referred to as a supramolecular blend in which two or more molecules are related by noncovalent forces. IPNs are also known as entanglements of polymer networks that are ideally held together only by permanent topological interactions [1, 2]. Generally, IPNs are created for the purpose of combining individual properties of two or more polymers. In some cases, entirely new properties are exhibited by the IPNs that are not observed in either of the two single networks alone [3]. Materials formed from IPNs share the properties that are characteristic of each network. A composite or an ideal IPNs of two or three different polymers would be a better choice.

In IPNs, high molecular weight polymers have extremely low mixing entropy, usually resulting in microphase separation [4]. Thus, the interfacial compatibility of IPNs incorporating two or more components plays an important role in

Micro- and Nano-structured Interpenetrating Polymer Networks: From Design to Applications, First Edition. Edited by Sabu Thomas, Daniel Grande, Uroš Cvelbar, K.V.S.N. Raju, Ramanuj Narayan, Selvin P. Thomas and Akhina H.

determining its performance. A unique feature of these systems is that the surface composition as well as structure and consequently the surface properties of the material are often different from that of the bulk due to a surface enrichment of the component that will minimize the total surface energy of the system. Understanding these surface aspects becomes important in designing high-performance IPNs required for various applications such as coatings, adhesives, and as biomaterials. Even though the study of IPN surfaces and interfaces is important in many engineering areas, until recently, their study remained largely empirical. With the advent of sophisticated instrumentation and development of new computational models, the investigation on surface characteristics of IPNs is one of the most rapidly growing research areas in theoretical chemistry. Obtaining the detailed microscopic information on the conformation of polymer chains at the surfaces would greatly enhance our understanding of the existing knowledge on many useful interpenetrating polymeric networks. In order to establish compatibility of IPN components, their surface properties in addition to chemical/physical interactions as well as mechanical stability are important. Two main classes of IPN surfaces and interfaces can be distinguished:

1. A symmetric polymer interface, where two or more or less identical polymers are physically crosslinked. Also, an asymmetric polymer interface, involving two different polymers. These are generally termed as thermoplastic apparent interpenetrating polymer networks (thermoplastic AIPNs). Thermoplastic AIPNs are hybrids between polymer blends and IPNs.
2. A composite interface composed of a new organic and inorganic hybrid materials. Usually, the nonpolymer is a solid phase such as glass, carbon, boron fibers, steel, or metal oxides. The interface of polymer with metal/metal oxides is described by a general term of "organic–inorganic composites IPN materials," which are usually the heterogeneous materials that display a wide variety of properties.

14.2 THEORETICAL SIMULATIONS

The tremendous advances in computing power at relatively low price in recent years have made molecular simulation a powerful tool for the elucidation of surface phenomena that controls the surface properties [5]. Besides, computer simulation is an excellent tool for exploring the local structure and properties at the interfacial region, especially when the experimental evidence on macromolecular organization at the surface is scarce. In this chapter, numerous computational methods were used to address the issues of IPN surfaces.

14.2.1 Quantum Mechanics

Among the quantum mechanical tools, the most rigorous is that of *ab initio* method, which is based on the first principles of density functional theory (DFT) [6]. These are based on the local-density approximation model and they rely solely on the quantum mechanical principles. The key to this approach is that the energy of the system can be described by the charge density instead of using electron wave functions. This can be generalized by the equation:

$$E[\rho]=T[\rho]+U[\rho]+E_X[\rho] \tag{14.1}$$

where $T[\rho]$ is the kinetic energy of a system of noninteracting particles of density ρ, U is the energy due to Coulombic interactions, and $E_X[\rho]$ includes the many-body contributions to the total energy. The system is solved by minimizing the total energy with respect to charge density. Calculations done with these methods can reveal the locations of the atomic nuclei at equilibrium as well as the energies and spatial distribution of the electrons. The only input to these calculations is the chemical identity of the atoms that comprise the surface or interface. These methods can also be used to determine the forces exerted on the atoms, from which one can deduce the motion of atoms using the laws of classical physics. If the forces are accurate, then calculations can yield valuable insights into the dynamic processes that occur at the surfaces. However, the *ab initio* simulations are restricted to small numbers of atoms because of the intense computational resources that are required. Hence, these are computationally exhaustive, making them difficult for simulations involving macromolecules.

14.2.2 Classical Mechanics

The next level of describing the molecular system employs the assumption that quantum effects are negligible due to the relative mass of the nuclei as compared to electron. This allows us to move from Schrödinger equation to Newton's equation of motion. A molecular dynamics (MD) simulation numerically solves the Newton's equations of motion, thus allowing structural fluctuations to be observed with respect to time. The MD methodology is founded upon the basic principles of classical mechanics and can provide a window into the microscopic dynamical behavior of the individual atoms that make up a given system. Dynamic simulation methods are widely used to obtain information on the time evolution of conformations of time scale that is otherwise inaccessible, and simulation is therefore a valuable tool in extending our understanding of model systems. Theoretical consideration of any system allows us to investigate specific contributions to a property through "computational alchemy" [7], that is, modifying the simulation in a way that is nonphysical, but nonetheless allows model's characteristics to be probed. One particular example is the artificial conversion of the energy function from that representing one system to that of another during the simulation. This is an important technique in surface free-energy calculations [8]. The MD simulations are now widely and routinely applied and especially popular in the fields of materials science and polymer chemistry [9, 10]. Thus, it is timely and useful to provide a pedagogical treatment of the theoretical and numerical aspects of MD simulation techniques that illustrate the capability of these approaches.

14.3 MOLECULAR DYNAMICS METHODS AND THEORY

14.3.1 Potential Energy Functions

Choice of an appropriate energy function for describing the intermolecular and intramolecular interactions is critical to a successful MD simulation. In the conventional MD simulations, energy function for nonbonded interactions is a simple pairwise additive function (for computational reasons) of nuclear coordinates. This use of a single nuclear coordinate to represent atoms is justified in terms of Born–Oppenheimer approximation [11]. For bonded groups of atoms, that is, those that form covalent bonds, bond angles, or dihedral angles, simple two-body, three-body, and four-body terms are used. The energy functions usually consist of a large number of parameterized terms. These parameters are chiefly obtained from the experimental and/or a quantum mechanical study on small molecules or fragments; it is assumed that such parameters may be transferred to larger molecules of interest. Set of functions along with the associated set of parameters is termed a force field. Force field development, as a fundamental issue underlying all the atomistic simulations, has drawn considerable attention in recent years. A variety of force fields have been developed for the simulation of polymers, namely, DREIDING [12], polymer consistent force field (PCFF) [13], and condensed-phase optimized molecular potentials for atomistic simulation studies (COMPASS) [14, 15].

1. DREIDING: DREIDING force field uses the general force constants and geometry parameters based on the simple hybridization considerations rather than individual force constants and geometric parameters that depend upon the particular combination of atoms involved in the bond, angle, or torsion terms. However, uncertainties in predicting the distribution of charges in molecules and in estimating van der Waals interactions are the limitations with the restricted set of parameters used in DREIDING.

2. PCFF: PCFF is an *ab initio* force field. Here, the parameters are derived based on *ab initio* data using the least squares fit technique developed by Hagler and coworkers [16]. Many of the nonbonded parameters of PCFF, which include atomic partial charges and Lennard-Jones 9-6 (LJ-9-6) parameters, were taken from the consistent valence force field (CVFF) [17]. Conversely, PCFF parameters are not suitable for MD simulations at finite temperatures. Specifically, systematic errors in pressure–volume–temperature (PVT) relationship have been observed for liquids and polymers using the MD simulations.
3. COMPASS: Basically, it is a hybrid approach consisting of both *ab initio* and empirical methods that was employed to derive the new general force field based on PCFF force field. The force field should be suitable for condensed-phase applications by modifying the nonbonded parameters and consequently valence parameters due to coupling between valence and nonbonded parameters. The outcome is a new, condensed-phase optimized *ab initio* force field called COMPASS. The combined parameterization procedure significantly improves the quality of the force field. Validation studies [18] based on a large number of isolated molecules, molecular liquids, and molecular crystals, representing large molecular classes, have shown that the present force field enables accurate and simultaneous prediction of structural, conformational, vibrational, and thermophysical properties for a broad range of molecules in isolation and in condensed phases. Like all the widely employed force fields, it contains several discrete terms. Each of these terms possesses simple functional forms; the functions used in this force field are shown in the following:

$$\begin{aligned}
V(r) = {} & \sum_{b} k_b (b-b_o)^2 + \sum_{\theta} k_\theta (\theta-\theta_o)^2 + k\varphi\big(1+\cos(n\varphi-\delta)\big) \\
& + \sum_{\chi} k_2 \chi^2 + \sum_{bb'} k_{bb'} (b-b_o)(b'-b'_o) + \sum_{b\theta} k_{b\theta} (b-b_o)(\theta-\theta_o) \\
& + \sum_{b\varphi} k_{b\varphi} (b-b_o)\big(1+\cos(n\varphi-\delta)\big) + \sum_{\theta\varphi} k_{\theta\varphi} (\theta-\theta_o)(1+\cos(n\varphi-\delta)) \\
& + \sum_{\theta\theta'} k_{\theta\theta'} (\theta-\theta_o)(\theta'-\theta') + \sum_{\varphi\theta\theta'} k_{\varphi\theta\theta'} (\theta-\theta_o)(\theta'-\theta'_o)\cos n\varphi \\
& + \sum_{ij} \varepsilon_{ij} \left[\frac{A_{ij}}{r_{ij}^{9}} - \frac{B_{ij}}{r_{ij}^{6}} \right] + \sum_{ij} \frac{q_i q_j}{r_{ij}}
\end{aligned} \tag{14.2}$$

The functions can be divided as a sum of valence terms including diagonal, cross terms, and nonbonded interaction terms. Valence terms represent the internal coordinates of bond (b), angle (θ), torsion angle (φ), and out-of-plane angle (χ), whereas cross-coupling terms include the combinations of two or three internal coordinate. The cross-coupling terms are important to predict the vibration frequencies and structural variations associated with the conformational changes. Among the cross-coupling terms given in Equation 14.2, the bond–bond, bond–angle, and bond–torsion angle are the most frequently used terms. The nonbonded interactions, which include LJ-9-6 potential function [19] for van der Waals (vdW) term and Coulomb function for electrostatic interaction, are used to understand interactions between pairs of atoms that are separated by two or more intervening atoms or those that belong to different molecules. In comparison with the most common LJ-12-6 function, which is known to be too "hard" in the repulsion region, LJ-9-6 function is somewhat softer but may be too attractive in the long separation range [20]. However, the electrostatic interaction is represented using the atomic partial charges.

For the purposes of MD, it is advantageous for the force field to have efficiently accessible first and second derivatives with respect to atomic position that correspond to physical characteristics of atomic force and force gradients, respectively. Using Equation 14.2, one can evaluate the potential energy, V(r), of the system from a single set of atomic coordinates, since the relevant distances and angles are easily determined. The energy is that of a single instance, termed a snapshot, of the system.

14.3.2 Molecular Mechanics

Molecular mechanics is the case of MD in which the temperature is 0 K or there is no kinetic energy in the system. In this methodology the potential energy of the system is minimized with respect to atomic positions. Using the aforementioned force field, energy of the system is calculated and the atomic positions are shifted until a minimum energy is achieved. The application of the force field allows the potential energy of the system to be represented by one explicit function. The negative gradient of the potential energy or forces in the system is used to guide the minimization process. The most popular mathematical algorithms [21] include those that use the derivatives of various orders, including the first-order (i.e., it utilizes the first-order derivatives) steepest descent and conjugate gradient methods as well as the second-order (i.e., utilizes second-order derivatives) Newton–Raphson method, each having their own intricacies in computational cost and stage of the minimization in which they are most effective.

1. *Steepest descent*: Steepest descent is the method most likely to converge, no matter what the function is or where it begins. It will quickly

reduce the energy of the structure during the first few iterations. However, convergence will slow down considerably as the gradient approaches zero. It should be used when the gradients are very large and configurations are far from the minimum.

2. *Conjugate gradient*: This method improves the line search direction by storing information from the previous iteration. It is the method of choice for systems that are too large for storing and manipulating a second-derivative matrix. The time per iteration is longer than found for steepest descents, but this is more than compensated for by an efficient convergence.
3. *Newton methods*: Newton–Raphson method is only recommended for systems having a maximum of 200 atoms. It has a small convergence radius, but it is very efficient near the energy minimum. Newton methods require computation and storage of second derivatives and are thus expensive in terms of computer resources. These characteristics render the basic Newton–Raphson method less useful. In general, the steepest descents are often the best optimizer to use for the first 10–100 steps, after which the conjugate gradient and/or a Newton–Raphson optimizer can be used to complete the optimization to convergence.

14.3.3 Integration of Equation of Motion

In simple terms, MD simulations involve the iterative numerical calculation of instantaneous forces present in a MM system and the consequential movements in that system. As MD calculations provide a window into the detailed motion of the individual atoms in a system, the microscopic mechanisms of energy and mass transfer can be gleaned. For an atom, i, with mass m_i, and position indicated by the 3-dimensional vector r_i, the relationship between atom's velocity and momentum, p_i, is given by

$$\frac{dr_i}{dt} = \frac{p_i}{m_i} \tag{14.3}$$

The net force, F_i, exerted on atom, i, by the remainder of the system is given by the negative gradient of the potential energy function, U, with respect to the position of the atom:

$$F_i = \frac{-dU}{dr_i} \tag{14.4}$$

Newtonian equation of motion for atom i is given as

$$\frac{dp_i}{dt} = F_i \tag{14.5}$$

Given the position with respect to a single component of vector r_i, (i.e., the position along a single dimension, x) at a specific time, t, then the position after a short and finite interval, denoted Δt, is given by the standard Taylor series:

$$x(t+\Delta t) = x(t) + \frac{dx(t)}{dt}\Delta t + \frac{d^2x(t)}{dt^2}\frac{\Delta t^2}{2} \tag{14.6}$$

The position x(t), velocity dx(t)/dt, and acceleration $d^2x(t)/dt^2$ are sufficient for the numerical solution to equations of motion if some approximation to account for higher-order terms in Taylor series can be made.

Numerous algorithms exist for integrating the equations of motion [22–24]. Many of these are finite difference methods in which integration is partitioned into small steps, each separated in time by a specific period Δt because the continuous potentials describing atomic interaction preclude an analytical solution. In MD, the most widely used time integration algorithm is probably the Verlet algorithm [22] and its variants. The simple Verlet algorithm uses atomic positions and accelerations at time t and positions from the prior step, x(t−Δt), to determine the new positions at (t+Δt). The advantage of Verlet algorithm is (i) time reversible and (ii) storage requirements are modest. The disadvantage is that the algorithm is of moderate precision and velocities are not explicitly generated. Thus,

$$x(t+\Delta t) = 2x(t) - x(t-\Delta t) + \frac{d^2x(t)}{dt^2}\frac{\Delta t^2}{2} \tag{14.7}$$

A better implementation of the same basic algorithm is the velocity Verlet algorithm [25]. With this algorithm, both positions and velocities at time, t, are calculated explicitly. Also, there is no compromise on the precision:

$$x(t+\Delta t) = x(t) + \frac{dx(t)}{dt} + \frac{d^2x(t)}{dt^2}\frac{\Delta t^2}{2} \tag{14.8}$$

$$\frac{dx(t)}{dt}(t+\Delta t) = \frac{dx(t)}{dt} + \frac{1}{2}\left[\frac{d^2x(t)}{dt^2} + \frac{d^2x(t)}{dt^2}(t+\Delta t)\Delta t\right] \tag{14.9}$$

It is the most attractive integration algorithm in use because of its numerical stability, convenience, and simplicity.

14.3.4 Statistical Ensembles

Integrating Newton's equations of motion allows us to explore the constant-energy surface of a system. However, the most natural phenomena occur under conditions where the system is exposed to external pressure and/or exchanges heat with the environment. Typically, it is important to accurately simulate experimental conditions to be replicated. Under these conditions, total energy of the system is no longer conserved and the extended forms of MD are therefore required. Various values for physical conditions, such as pressure and temperature, may be readily considered in the simulations. Several methods are available for controlling temperature and pressure, depending on which the state variables (energy, E, enthalpy, H (i.e., E+PV); number of particles, N; pressure, P; temperature, T; and volume, V), are kept constant, and different statistical ensembles can be generated. However, experiments are usually performed at the constant temperature and volume (i.e., canonical ensemble) or constant pressure and temperature (i.e., isobaric–isothermal ensemble) such that it is often desirable to simulate these conditions or instead mimic these conditions for those expected under physiological conditions. The most widely simulated ensembles are as follows.

1. *Canonical ensemble (NVT)*: This is the collection of all systems whose thermodynamic state is characterized by the fixed number of atoms, N; fixed volume, V; and fixed temperature, T. The NVT is an appropriate choice when the conformational searches of the models are carried out in vacuum without the periodic boundary conditions. The temperature control using the stochastic collision approach of Andersen and coworkers [26] was used in simulations.
2. *Isobaric–isothermal ensemble (NPT)*: This an ensemble with a fixed number of atoms, N; fixed pressure, P; and fixed temperature, T. The NPT is an ensemble of choice when correct pressure, volume, and densities are important during the simulation. Further, pressure control developed by Berendsen and coworkers [27] was used in the MD simulations.
3. *Microcanonical ensemble (NVE)*: Here, the number of atoms, N; constant energy, E; and constant volume, V ensemble is obtained by solving the standard Newton equation without any temperature and pressure control. The constant-energy surface of the conformational space is explored to elude the perturbation introduced by temperature and pressure bath coupling.

14.3.5 Simulation Environment

The size of any simulated bulk system is generally limited by the capability of available computers to handle between $N = 100$ and 10000 atoms. This occurs, since in principle, each atom can interact with every other atom, causing the time required to evaluate the forces on the atoms is proportional to N^2. Further, if we attempt to model an isolated molecular system containing such small number of atoms, clearly an appreciable fraction of atoms would lie close to the periphery of the system and significant surface effects would be evident. The standard approach is, therefore, to introduce the so-called periodic boundary conditions, in which the system is considered to be surrounded on all sides by the replicas of itself, forming an infinite macro lattice. Typically, a cubic lattice is used for replication of the central cubic box. Atoms outside the central box are simply images of the atoms simulated in that box. The so-called minimum image convention guarantees that calculating only one pairwise interaction for each pair of atoms does not include duplicate interactions between atoms. For atoms i and j, the interaction is that between the original atom i and whichever copy of atom j, original or image, is closest to atom i.

To increase the speed of calculation of nonbonded interactions in the simulations of dense molecular systems, it is customary to use cutoffs, whereby the interaction between two atoms, i and j, is ignored if the separation exceeds some specified value. The use of minimum image convention implies that the cutoff radius must be less than one half the shortest box edge lengths. The cutoffs [28] commonly encountered are atom based and group based. When atoms with large partial or formal charges are present, significant errors can be introduced if nonbonded interactions are computed using the atom-based cutoffs. In principle, it is possible to eliminate these artifacts by dispensing with the cutoffs for electrostatic interactions. In a periodic system, we might thus consider the evaluation of electrostatic interactions using the standard lattice summation methods (e.g., the Ewald method). However, one disadvantage is that such methods are computationally expensive and thus are not suited for use in generating long dynamics trajectories. However, the group-based approach is very efficient and well suited to conditions in which the system consists of molecules composed of essentially neutral fragments with dipoles and performs substantially accurate calculations in an efficient manner [29]. Thus, nonbonded interactions have been calculated using the group-based method with explicit atom sums being calculated with the distance of 9.5 Å.

14.3.6 Amorphous Cells

In case of polymers, it is important to start with a configuration in which the backbone bond conformer populations and global characteristics of the chains (e.g., mean square radius of gyration) are representative of those that would be found with a significant probability in the polymer to be studied. For short chains with a maximum of 20–30 backbone bonds, it has been demonstrated that starting with a regular crystalline arrangement and performing MD or Monte Carlo (MC) can give reasonable starting structures [30, 31]. However, this technique becomes progressively slower as the physical size of the simulated system increases and as the constituent molecules become longer and topologically more complex (e.g., with long side chains). Hence, an alternative method employed by Theodorou and Suter [32] for longer chains build molecules in two separate steps, involving the creation of an initial guess structure, followed by a relaxation of the structure to a state of minimum potential energy. Each cell may therefore be considered to comprise a portion or portions of a parent chain plus a series of image chains, which are merely displacements of the parent molecule along x, y, and z directions by some integer multiple of the cell edge length.

Theodorou–Suter chain-generation approach attempts to satisfy the requirement that the initial configuration be "typical" of one, which would be found with significant probability by taking advantage of the widely accepted idea that the configurational statistics of polymeric chains in the bulk are essentially unperturbed globally by the excluded volume effects. Accordingly, they are well represented by an interdependent rotational isomeric state (RIS) model [33], and this in principle can be used as a basis for generating polymeric chains in a bulk system. However, in a dense system, direct application of the RIS model fails to control intermolecular overlaps between the parent chain and its images. Hence, in order to avoid the excessive overlaps between chains, modified conditional probabilities were used to account for nonbonded interactions between atoms to be placed and rest of the system.

A further approach to the construction of configurations in dense systems is provided by Meirovitch's scanning method [34, 35] that is based on a method devised by Rosenbluth and Rosenbluth [36]. In the scanning method, all possible continuations of the growing polymer chain were considered while calculating the conditional probabilities. However, in practice, it was not possible to examine all continuations of the chain, and look ahead is normally restricted to a few backbone bonds. The number of look ahead bonds represents the number of forward bonds that are considered during the construction, which were subsequently refined to provide the input for further calculations.

14.4 MOLECULAR DYNAMIC STUDY OF SURFACE/ INTERFACE PROPERTIES OF THERMOPLASTIC APPARENT INTERPENETRATING POLYMER NETWORKS AND ORGANIC–INORGANIC COMPOSITE INTERPENETRATING POLYMER NETWORKS

14.4.1 Surface Energy of Thermoplastic Apparent Interpenetrating Polymer Networks

Evaluating the surface tension of hybrid polyblends such as the thermoplastic AIPNs is very important when the surface composition differs from the bulk composition. In general, this behavior observed in polyblends [37, 38] is

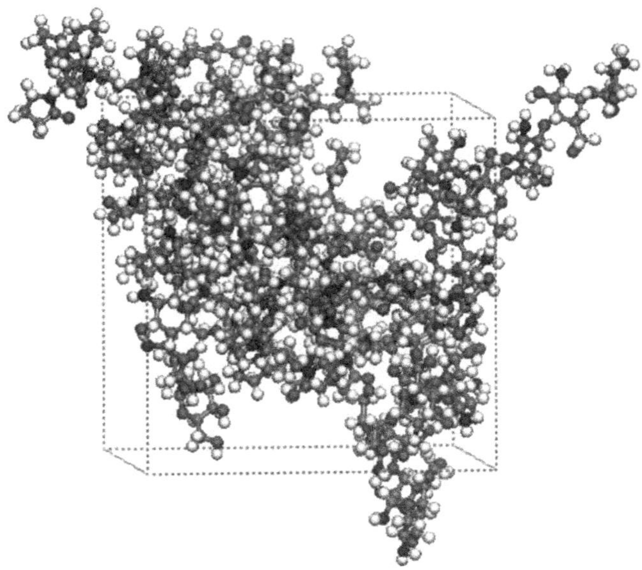

FIGURE 14.1 Snapshot of the polyblend poly(*N*-vinyl-2-pyrrolidone)/chitosan

called "surface segregation" or "surface enrichment." Surface segregation for polyblends is caused by the difference in surface tension between pure components. Model snapshot of poly(*N*-vinyl-2-pyrrolidone)/chitosan polyblend is shown in Figure 14.1. Surface energy of hybrid polyblends is computed by bulk PVT properties using equation of state approach [39]. In this context, the surface energy of thermoplastic AIPNs was evaluated using the equation [40, 41]:

$$\gamma = \gamma^* \tilde{\gamma} \tag{14.10}$$

where γ^* and $\tilde{\gamma}$ are characteristic and reduced surface tensions, respectively. According to Patterson and Rastogi [41], γ^* is related to equation of state parameters given by

$$\gamma^* = k^{1/3} P^{*2/3} T^{*1/3} \tag{14.11}$$

where k is Boltzmann constant and P* and T* are the characteristic parameters for pressure and temperature, respectively. According to Prigogine and Saraga [42], the equation for reduced surface tension is

$$\tilde{\gamma} = M\tilde{V}^{-5/3} - \frac{\tilde{V}^{1/3} - 1}{\tilde{V}_2} \ln\left[\frac{\tilde{V}^{1/3} - 0.5}{\tilde{V}^{1/3} - 1}\right] \tag{14.12}$$

where $\tilde{\gamma}$ and $\tilde{V}$ are characteristic and reduced surface tensions and molar volume; M is fractional decrease in the nearest neighbors of a cell due to migration from bulk phase to surface phase and its value varies from 0.25 to 0.29 for a closely packed cubic lattice. In the present calculations, we have used M = 0.29 to compute the PVT parameters using the Flory equation of state [43, 44]. Reduced volume was calculated from the thermal expansion coefficient, α, using the relationship

$$\tilde{V} = \frac{\left[1+(4/3)\alpha T\right]}{\left[1+\alpha T\right]^3} \tag{14.13}$$

Characteristic pressure, P*, is defined as the ratio of ε*/ν*, where ε* is total interaction energy per mer and ν* is close-packed mer volume. Thus, P* is a direct measure of the cohesiveness or strength of intermolecular interactions. Hence, P* is equal to the cohesive energy density (CED) in the close-packed state [41, 45, 46] because $CED \equiv \Delta E_{vap}/V = \varepsilon^*/\nu^* \equiv P^*$. Characteristic temperature T* was computed using

$$T^* = \frac{T\tilde{V}^{4/3}}{\tilde{V}^{1/3}-1} \tag{14.14}$$

Thermal expansion coefficient α was obtained from the slope of specific volume given by

$$\alpha = \frac{1}{\nu}\left[\frac{d\nu}{dT_p}\right] \tag{14.15}$$

In order to compute the specific volume for bulk PVT parameters, MD simulations (NPT ensemble) have been performed on the polyblends at various temperatures below 300 K. The procedure to calculate specific volume by means of a constant pressure simulation was investigated earlier by Fried and Ren [47]. The surface tension calculated using the bulk PVT parameters involved the computation of characteristic pressure, P*, which was determined from the CED. CED corresponds to cohesive energy per unit volume. If V_{mol} is molar volume of the polymer, then CED is defined as

$$CED = \left(\frac{E_{coh}}{V_{mol}}\right) \tag{14.16}$$

The Hildebrand solubility parameter δ is given as

$$\delta = \left(\frac{E_{coh}}{V_{mol}}\right)^{1/2} \tag{14.17}$$

14.4.2 Organic–Inorganic Composite Interpenetrating Polymer Network Materials

IPNs of various combinations between organic polymers and inorganic materials have been a very useful method to explore new types of materials having unique properties and desired specific performances. The role of organic–inorganic IPN composites in industrial applications has been rapidly growing due to their strength, lightweight, and chemical stability. An understanding of polymer/nonpolymer interfaces, including those of polymers and metals or metal oxides, enables the development of new materials with promising technological applications [48, 49]. The interfaces of polymers with metals and metal oxides are distinctly different from the polymer–polymer IPN interfaces. Therefore, a detailed molecular level study to understand the physical interactions and solid-state chemical reactions between polymer surface atoms and metal/metal oxide surfaces as well as near-surface atoms are important to understand the macroscopic properties of a variety of chemically bonded multilamellar dielectric and composite systems.

In this sense, the classical simulation study on polymer/nonpolymer IPN interfaces plays a vital role to investigate the miscibility and adhesion characteristics of such systems, which are otherwise difficult to study experimentally or for which experimental data are not available; even if some scanty data are available, they are of uncertain quality. From a standpoint of adhesion property, it is important to study the interaction between polymer and metal oxide surfaces, since such systems represent a new class of polymeric materials, which combine the properties of embedded inorganic particles (in terms of mechanical strength, modulus, thermal stability, etc.). Here, the related quantity of interest is the energy of adhesion, which is the energy released when two bodies of different surfaces are brought in contact with each other. The strength of adhesion holding the two materials together is a function of attractive or retaining forces at the interface between them.

Kornherr *et al.* [50] employed MD simulations to study the interaction of adsorbed silane molecules (octyl trihydroxysilane, butyl trihydroxysilane, aminopropyl trihydroxysilane, and thiolpropyl trihydroxysilane) with polar zinc oxide surface $(000\bar{1})$. Computation of adhesion energy of methyl methacrylate (MMA) with several metal oxides like corundum (Al_2O_3), hematite (Fe_2O_3), silica (SiO_2), and rutile (TiO_2) were explored [51] using MD simulation strategy. The energy of adhesion was calculated according to Equation 14.18. At first, energy (E_{total}) for the simulation box containing both oligomer and surface atoms was calculated and then the energy of oligomer (E_{Pol}) was calculated without any contribution from the surface. Finally, the surface atoms were kept and polymer was removed to calculate the energy of the surface ($E_{surface}$). The energy of adhesion of oligomer and the surface was then computed as

$$E_{adhesion} = \frac{E_{total} - \left(E_{surface} + E_{pol}\right)}{V} \tag{14.18}$$

where V is the molar volume of oligomer. The values E_{pol} and $E_{surface}$ were calculated by assembling the oligomers into the simulation box with the metal oxide surface and *c*-dimension of the box were extended to 30Å such that oligomer was at an equidistance from the metal oxide surface and it can see only one side of the surface. The MD simulation was then performed for 300 000 steps with a time step of 1*fs* at 298 K. A model structure of polymer with metal oxides is depicted in Figure 14.2.

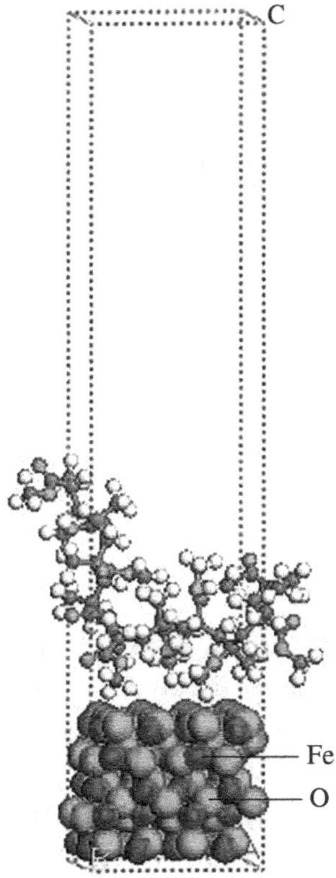

FIGURE 14.2 A typical simulated oligomer (acrylate) and its interface with metal oxide (Fe_2O_3) surface

14.5 CONCLUSIONS

Surfaces and interfaces play an important role in controlling the properties of a broad range of IPNs. Recent advances in both methodology and computational resources have led to the study of dynamics of polymer chains at the surfaces and interfaces. Hence, in this review, we have shown how MD simulations could provide valuable insights into surface energetics of the IPNs, which may not be readily available from experiments. From a methodological perspective, the use of periodic boundary conditions to enable the calculation of bulk structure and properties has become the norm for force field-based simulations in materials science and offers an efficient way to simulate bulk material and interfaces. Surface energy of polymer blends has been evaluated by the bulk PVT properties, while the surface energy through equation of state approach involves the determination of thermal expansion coefficient, which is obtained from the plots of specific volume versus temperature by performing NPT dynamics. Further, the MD simulations have been used to predict and rationalize the interfacial properties of organic–inorganic composite materials to explore the adhesion characteristics of polymers. It also contributes to a better fundamental understanding of the complicated physical interfacial process. The evaluation of interfacial chemistry between various metal oxides and polymers will provide a basis for the design and synthesis of IPN composites. The present study clearly reveals that molecular modeling and simulation tools are quite useful in computing the surface/interface properties of thermoplastic AIPNs and composite IPNs. It is also clear that the use of MD simulations will continue to grow with the increase of computer power. The outstanding challenge for the future is to integrate MD techniques to provide a more complete multiscale understanding of surface phenomena of IPNs.

REFERENCES

[1] Frisch, H.L., Klempner, D. and Frisch, K.C. (1969) A topologically interpenetrating elastomeric network. *Journal of Polymer Science. Part B*, **7**, 775.

[2] Vendamme, R., Onoue, S.Y., Nakao, A. and Kunitake, T. (2006) *Nature Materials*, **5**, 494.

[3] Myung, D., Waters, D., Wiseman, M. *et al.* (2008) Progress in the development of interpenetrating polymer network Hydrogels. *Polymers for Advanced Technologies*, **19** (6), 647–657.

[4] Sperling, L.H. and Mishra, V. (1996) *Polymers for Advanced Technologies*, **7**, 197–208.

[5] Frenkel, D. and Smit, B. (2002) *Understanding Molecular Simulation*, 2nd edn, Academic Press, San Diego, CA.

[6] Hohenberg, P. and Kohn, W. (1964) *Physics Review*, **136**, 864.

[7] Straatsma, T.P. and McCammon, J.A. (1992) *Annual Review of Physical Chemistry*, **43**, 407.

[8] Simonson, T., Archontis, G. and Karplus, M. (2002) *Accounts of Chemical Research*, **35**, 430.

[9] Kremer, K. (2003) *Macromolecular Chemistry and Physics*, **204**, 257.

[10] Rickman, J.M. and LeSar, R. (2002) *Annual Review of Materials Research*, **32**, 195.

[11] Born, M. and Oppenheimer, J.R. (1927) *Annals of Physics*, **84**, 457.

[12] Mayo, S.L., Olafson, B.D. and Goddard, W.A. (1990) *The Journal of Physical Chemistry*, **94**, 8897.

[13] Sun, H.J. (1994) *Computers and Chemistry*, **15**, 752 Macromolecules 1995,28, 701.

[14] Sun, H., Ren, P. and Fried, J.R. (1998) *Computational and Theoretical Polymer Science*, **8**, 229.

[15] Rigby, D., Sun, H. and Eichinger, B.E. (1997) *Polymer International*, **44**, 311.

[16] Maple, J.R., Dinur, U. and Hagler, A.T. (1988) *Proceedings of the National Academy of Sciences of the United States of America*, **85**, 5350.

[17] Sun, H., Mumby, S.J., Maple, J.R. and Hagler, A.T. (1994) *Journal of the American Chemical Society*, **116**, 2978.

[18] Sun, H. (1998) *The Journal of Physical Chemistry. B*, **102**, 7338.

[19] Lifson, S., Halger, A.T. and Dauber, P. (1979) *Journal of the American Chemical Society*, **101**, 5111.

[20] Halgren, T.A. (1992) *Journal of the American Chemical Society*, **114**, 7827.

[21] Leach, A. (2001) *Molecular Modeling Principles and Applications*, Prentice Hall, New York.

[22] Verlet, J. (1967) *Physics Review*, **159**, 98.

[23] Beeman, D. (1976) *Journal of Computational Physics*, **20**, 130.

[24] Gear, C.W. (1971) *Numerical Initial Value Problems in Ordinary Differential Equations*, Prentice Hall, New York.

[25] Swope, W.C., Andersen, H.C., Berens, P.H. and Wilson, K.R. (1982) *The Journal of Chemical Physics*, **76**, 637.

[26] Andersen, H.C. (1980) *The Journal of Chemical Physics*, **72**, 2384.

[27] Berendsen, H.J.C., Postma, J.P.M., van Gunsteren, W.F. *et al.* (1984) *Chemical Physics*, **81**, 3684.

[28] Brooks, C.L., Pettitt, B. and Karplus, M. (1985) *The Journal of Chemical Physics*, **83**, 5897.

[29] Eichinger, B.E., Rigby, D. and Stein, J. (2002) *Polymer*, **43**, 599.

[30] Vacatello, M., Avitabile, G., Corradini, P. and Tuzi, A. (1980) *The Journal of Chemical Physics*, **73**, 548.

[31] Boyd, R.H. (1989) *Macromolecules*, **22**, 2477.

[32] Theodorou, D.N. and Suter, U.W. (1985) *Macromolecules*, **18**, 1467.

[33] Flory, P.J. (1953) *Principles of Polymer Chemistry*, Cornell University Press, Ithaca, NY.

[34] Meirovitch, H.J. (1983) *The Journal of Chemical Physics*, **79**, 502.

[35] Meirovitch, H.J. (1985) *Macromolecules*, **18**, 569.

[36] Rosenbluth, M.N. and Rosenbluth, A.W. (1955) *The Journal of Chemical Physics*, **23**, 356.

[37] Eisenriegler, E., Kremer, K. and Binder, K. (1982) *The Journal of Chemical Physics*, **77**, 6296.

[38] Ober, R., Paz, L., Taupin, C. *et al.* (1983) *Macromolecules*, **16**, 50.

[39] Prathab, B., Subramanian, V. and Aminabhavi, T.M. (2007) *Polymer*, **48**, 417.

[40] Fox, H.W. and Zisman, W.A. (1950) *Journal of Colloid Science*, **5**, 514.

[41] Patterson, D. and Rastogi, A.K. (1970) *The Journal of Physical Chemistry*, **74**, 1067.

[42] Prigogine, I. and Saraga, L. (1952) *The Journal of Chemical Physics*, **49**, 399.

[43] Cahn, J.W. and Hilliard, J.E. (1958) *The Journal of Chemical Physics*, **28**, 258.

[44] Dill, K.A. and Flory, P.J. (1981) *Proceedings of the National Academy of Sciences of the United States of America*, **78**, 676.

[45] Sanchez, I.C. and Lacombe, R.H. (1978) *Macromolecules*, **11**, 1145.

[46] Abe, A. and Flory, P.J. (1965) *Journal of the American Chemical Society*, **87**, 1838.

[47] Fried, J.R. and Ren, P. (1999) *Computational and Theoretical Polymer Science*, **9**, 111.

[48] Allcock, H.R. (1992) *Science*, **255**, 1106.

[49] Jeevananda, T., Begum, B. and Siddaramaiah, S. (2002) *European Polymer Journal*, **37**, 1213.

[50] Kornherr, A., French, S.A., Sokol, A.A. *et al.* (2004) *Chemical Physics Letters*, **393**, 107.

[51] Prathab, B., Subramanian, V. and Aminabhavi, T.M. (2007) *Polymer*, **48**, 409.

15

APPLICATIONS OF INTERPENETRATING POLYMER NETWORKS

RADHIKA RAVEENDRAN AND CHANDRA P. SHARMA
Division of Biosurface Technology, Biomedical Technology Wing, Sree Chitra Tirunal Institute for Medical Science and Technology, Thiruvananthapuram, Kerala, India

15.1 INTRODUCTION

Interpenetrating polymer networks (IPNs) comprise of entanglements of polymeric networks that are ideally held together by topological interactions. These chemically crosslinked permanent internetwork entanglements cannot be detached from each other without fracturing any of the networks [1–3]. Various applications of these IPNs have been extensively explored in the scientific and patent literature. Research on IPNs has picked up pace and their applications have also been extrapolated to manufactured products. BioMed Sciences manufactured Silon®, an IPN based on polytetrafluoroethylene (PTFE) and silicone elastomers, which is unique for a wide range of clinical and consumer wound, scar or skin care products, as well as other medical and nonmedical applications [4]. Ion exchange resins were among

Micro- and Nano-structured Interpenetrating Polymer Networks: From Design to Applications, First Edition. Edited by Sabu Thomas, Daniel Grande, Uroš Cvelbar, K.V.S.N. Raju, Ramanuj Narayan, Selvin P. Thomas and Akhina H.

the first applications of IPNs. Anionic–cationic IPNs, prepared via suspension polymerization, are used for ion exchange resins. Rapid ion exchange is attributed to the arrangement of the positive and negative ions. Thermoplastic IPNs, based on polypropylene and ethylene propylene diene monomer (EPDM), are used as automotive bumpers, because of their high energy-absorbing ability. Kraton, Shell Chemical Company product, is another thermoplastic IPN used as an insulator for under-the-hood automotive wiring. IPNs are also efficient for use in separation techniques as well as membrane catalysis. Favorable thermoset properties coupled with a degree of flexibility, controlled phase domain sizes, and high energy absorption capabilities are the application invoking properties of IPNs. IPNs also find use as high-performance specialty polymers in electronic devices and also as sound and vibration damping materials in aircraft, automobile, and submarine technology [5]. Nowadays, there exists much interest in the development of biopolymer-based IPNs for various applications like drug delivery and tissue engineering. IPNs are coveted biomaterials due to their physical and biological characteristics such as enhanced solubility of hydrophobic drugs, excellent swelling capacity, biodegradability, biocompatibility, and weak antigenicity. IPNs are anticipated to be of value in tissue engineering and three-dimensional (3-D) cell culture applications. This chapter throws light on various applications of IPNs focusing mainly on the biomedical applications.

15.2 WHAT ARE INTERPENETRATING POLYMER NETWORKS?

IPNs are combination of two polymers in network form, polymerized or crosslinked, in the immediate vicinity of one another. Among multiphase polymeric materials, they stand out as a novel type of polymer alloys. IPN materials are differentiated into sequential IPN (SIPN), simultaneous IPN (SIN), latex IPN (LIPN), gradient IPN, thermoplastic IPN, and semi-IPN [6]. It is the route of synthesis which determines the formation of various kinds of IPNs. For example, SINs are synthesized by the simultaneous polymerization and crosslinking of addition-polymerization type and condensation-polymerization type monomers. SIPNs possess two networks that have been synthesized in sequence. In LIPNs, crosslinking and mixing of the components are carried out in latex form [7]. IPNs have also been bestowed the name "polymer catenanes" due to their permanent entanglements with covalent bonds between the polymers [8–10].

15.3 PROPERTIES OF INTERPENETRATING POLYMER NETWORKS

The interpenetration and entanglements in IPNs increase the extent of intimate mixing between the polymeric networks and also enhance the physical crosslinking density. These factors contribute to the their characteristic double network structure and exceptional synergistic behavior [11, 12]. Permanent entanglement chains restrict the free movement of the polymer segments giving rise to chemical bond-like structures that impart "forced compatibility" [8]. Therefore, compatible materials can be obtained from incompatible polymers and it's this forced compatibility that makes IPNs superior to normal blends. Absence of chemical bonds in normal blends, due to simple mixing of polymers, result in low interfacial strength. The production of blends is limited to thermoplastic polymers because the mixing takes place after the completion of polymerization. As in the case of IPNs, mixing takes place before polymerization and the two kinds of crosslinking thermosetting polymers combine freely, and the permanent entanglements in the network control the phase separation [13].

15.4 APPLICATIONS OF INTERPENETRATING POLYMER NETWORKS

IPNs are capable of providing key attributes of its constituent components or sometimes they can altogether exhibit new properties that are not observed with the individual networks [14]. Hence, they exhibit a multitude of applications in various fields (Figure 15.1). The application-inducing properties arise from a high degree of flexibility, controlled phase domain sizes, energy absorption potential, and a range of other properties in addition to thermoset properties [15].

15.4.1 Selective Transportation of Liquids and Gases

Transportation of liquids and gases is crucial where applications involving protective coatings, food and beverages packaging, membrane separation processes are concerned [16]. To design economically viable and highly efficient membranes, it is substantial to elucidate the relationship between the transport properties and molecular structure of polymers. Transport of molecules through polymeric membranes occurs via the random motion

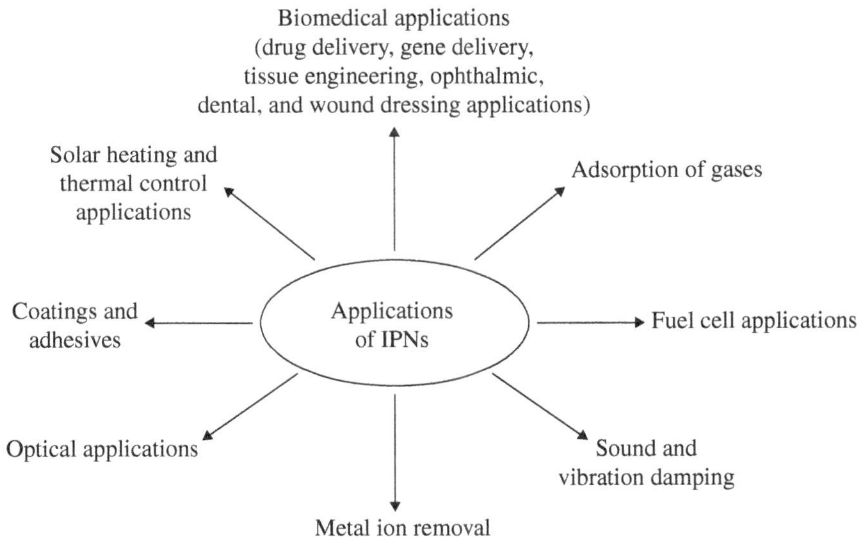

FIGURE 15.1 Applications of IPNs

and the driving force for this process arises from the difference in the concentration between the two phases [17]. IPNs exhibit different swelling characteristics compared to other crosslinked polymers. They comprise of a rubbery and a glassy component, and the behavior of these components toward solvents is different due to the difference in their molecular mobility and free volume [18]. It has been noted that the sorption behavior of IPNs depends on weight composition of components, crosslink density and hydroxyl value in case of polyols [19]. Lee *et al.* conducted a study on a series of polyurethane/polystyrene (PU/PS) IPNs, semi-IPNs, and linear blend membranes under varied composition and temperature conditions and corroborated the enhancement in permeability coefficients to the annealing effect which was dependent on the crosslinked nature of the membrane [20]. Composition change, synthesis temperature, annealing temperature, and crosslinked state, obtained by modulating the homogeneity change, are essential factors which control the gas transport characteristics in an IPN. Holdsworth *et al.* have investigated IPNs based on poly(ethyl acrylate)–poly(ethyl methacrylate) latex and poly(ethyl methacrylate)–poly(ethyl acrylate) latex for their membrane-aided transportation of gases [21]. Semi-IPNs have been developed from ionic polymers which have been utilized for pervaporation-assisted esterification to remove water from the reaction medium [22].

15.4.2 Ion Exchange Membranes

Ion exchange membranes have traversed a long way from laboratory tools to industrial commodities. They are extensively applied for treating industrial effluents, desalination of sea water, and concentration or separation of pharmaceutical and food [23]. Better energy-efficient and cleaner process is assured with the evolvement of ion exchange membranes. IPN ion exchange membranes comprise of a microporous polymeric support film interpenetrated by an ion exchange polymer constituted by the absorption and polymerization of monomers within the support film. Ion exchange ligands at the surface of the polymer induce ligand mobility to extract and transport ions across the membrane. Much research has also been carried out in the area of crosslinked proton exchange membranes (PEMs) based on SIPN and IPN architectures. A distinguishable feature which makes the SIPNs and IPNs stand out from the usual polymer network is the presence of entrapped linear polymers in their network structure [24]. Proton conductivities and methanol permeabilities for fuel cell applications have been demonstrated using semi-IPN membranes of polyvinyl alcohol/polystyrene sulfonic acid–polymethyl acrylate (PVA/PSSA–PMA) [25]. It was found that the proton conductivity was strongly dependent on the content of the linear acid polymer entrapped in PVA/PSSA–PMA semi-IPN membranes. Better performance of PVA/PSSA–PMA semi-IPN membranes over PVA/PSSA membranes was attributed to their moderate hydrolytic stabilities and stronger mechanical properties. There are also reports of semi-IPNs prepared by the combination of linear PSSA and PVA-GA polymer networks which showed high proton conductivities and high water retention capabilities for PEM applications in fuel cells [26, 27]. A PVA/PAMPS–PHEMA(poly (2-acrlamido-2-methyl-1-propane sulfonic acid–poly (2-hydroxyethyl methacrylate) IPN system has also been used recently as a PEM in combination with silica particles by taking advantage of *in situ* polymerization approach by sol–gel hydrolysis of tetraethoxysilane (TEOS) in the polymer network [28].

15.4.3 Removal of Metal Ions

Metal ion removal by membrane processes is highly effective and economical especially in the present day where heavy metal pollution has become a major cause of concern. It has superseded the conventional methods like chemical precipitation–neutralization, solvent extraction, and coagulation [29]. Designing chelating adsorbing membranes using polymers promote metal uptake in the bulk of the membranes and also at their surfaces. Immobilization in the membrane is crucial due to the use of water-soluble complexing polymer, and this could be achieved by incorporating semi-IPNs which can entrap complexing polymers in a crosslinked polymeric matrix [30]. A similar strategy has been employed by Bessboussea *et al.* by developing PVA/CP (chelating polymer) membranes by using the semi-IPN technique in view of application to the industrial wastewater treatment [31]. The performances of the membranes for the removal of Hg(II), Pb(II), Cd(II), and Cu(II) ions from aqueous solution were evaluated in this study. Qingchun and Tian have also reported the use of polyacrylonitrile/silica nanospheres with 3-D interpenetrating network structure which finds application in removal of Pb(II) from water [32].

15.4.4 Sound and Vibration Damping

Sound and vibration damping properties find applications in automotive, aircraft, construction, and appliance industries [33]. The presence of crosslinks in IPN restricts the domain size to the order of 10–20 nm which leads to the formation of a microheterogeneous morphology. The formation results in broad glass transition which induces frequency damping property [34]. Controlling certain factors like miscibility of the system, chemical structure of the polymer, glass transition temperatures, crosslinking density, and feeding sequences in polymerization can improve the damping behavior of IPNs [35]. IPNs based on polyurethanes have been widely investigated for the potential of damping property [36]. Polyurethane elastomers when combined with compatible polymers, to form IPNs, eventually result in high damping phenomenon over a broad range of frequencies and temperatures.

15.4.5 Other General Applications

Improved optical fibers have been prepared by gradient IPN technology taking advantage of a gradient refractive index which curves the light away from the surface and reduces the amount of light emitting out of the fiber [37]. IPNs also have the potential for solar heating applications due to its adjustable lower critical solution temperature and restrained phase domain growth [38]. IPNs have also been demonstrated to be promising candidates for spacecraft thermal applications [39]. Coating and adhesive applications have also been explored for IPNs [40].

15.4.6 Biomedical Applications of Interpenetrating Polymer Networks

The potentials of IPNs constituted from natural and synthetic polymers have been tapped for a host of biomedical applications.

15.4.6.1 Drug Delivery The major requisites in designing efficient drug delivery systems are the active control of site/kinetics of drug release and stability improvement prior to delivery. Delivery of drugs at a predetermined rate for a specific time period is achieved by IPN-based drug delivery. For accomplishing ideal features of an effective drug delivery systems, functionalization of polymeric excipients through the modification of their chemical structure is attracting a great deal of attention [41–43]. However, biocompatibility and toxicity concerns limit the use of new or chemically modified materials. IPNs possessing different functionalities are capable of dodging regulatory issues of toxicity and hence can provide the required performance. The difference between IPN formation and copolymerization is that the latter results in a new chemical entity and does not ensure the maintenance of the properties of each polymer, whereas the interpenetration of crosslinked polymers leads to materials that combine the features of both components in a tunable way [44]. IPN-based drug delivery systems are designed to deliver drugs in a controlled manner with minimum fluctuations. They are endowed with various physical and biological characteristics such as enhanced solubility of hydrophobic drugs, excellent swelling capacity, stable drug formulations, biodegradability, biocompatibility, weak antigenicity, and specific tissue targeting, which make them suitable for drug delivery applications.

Several factors like nature and molecular weight of the drug, density, pore size of the matrix, degree of crosslinking, and solvent type govern the release of a drug from an IPN hydrogel. A hydrogel comprises of two components, one being a strong polymer network and the other being permeation channels filled with water for diffusion of water-soluble solutes [45]. The water content of the matrix governs the release of the molecule within the hydrogel indicating that the water-sorption kinetics is one of the most important factors that influences the release kinetics of molecules from IPN hydrogels. High swelling capability of a hydrogel can facilitate better release of a solute [46]. Semi-interpenetrating network (semi-IPN) hydrogels fabricated from silk fibroin (SF) and polyacrylamide (PAAm) have been investigated as a matrix for sustained drug release of two different model molecules: trypan blue and FITC-inulin [47]. Release of various drugs and proteins has been investigated in many IPNs (Table 15.1).

Release of a drug can be favored by reduction of hydrogen bonds between the IPN components leading to a decrease in the crosslinking density and an increase in the free space within the polymeric matrix. Microgel/nanogel–hydrogel IPN composites employing acrylamide and 2-hydroxyethylmethacrylate poly(acrylonitrile)-based core–shell nanoparticles have been utilized by employing fluorescein sodium as a model drug to study the antimicrobial property of the particles against various strains of bacteria [56]. Modification

TABLE 15.1 Some IPNs Used in Drug and Protein Delivery

IPN Components	IPN Type and Property	Encapsulated Drug	References
1. Poly(acrylamide) and poly(γ-glutamic acid)	pH- and temperature-sensitive semi-IPN	Amoxicillin	48
2. Polyacrylamide grafted j-carrageenan and sodium alginate	pH-responsive IPN	Ketoprofen	49
3. Polypropylene substrates grafted with *N*-isopropylacrylamide and acrylic acid	IPN with eluting performance	Vancomycin	50
4. Poly methacrylic acid–alginic acid	Semi-IPN	Insulin	51
5. Sodium carboxymethyl cellulose and poly(vinyl alcohol)	IPN	Diclofenac sodium	52
6. Chitosan, poly (*N*-vinylpyrrolidone) and poly(acrylamide) polymers	Ternary IPN	Amoxicillin	53
7. Phthalyl chitosan–poly(ethylene oxide)	Semi-IPN	Insulin	54
8. Poly(acrylic acid-*co*-acrylamide)/*O*-carboxymethyl chitosan	Superporous hydrogel-containing IPN	Insulin	55

of the polymeric backbone and modulation of the crosslinking modality are certain factors which hold the reins for manipulating the physicochemical properties of the IPN matrix and prove beneficial in designing a material system with tailored swelling, drug release, and degradation kinetics. Semi-IPNs exhibit positive response to environmental stimuli and hence play a pivotal role in the development of controlled drug release devices.

15.4.6.2 Tissue Engineering Tissue engineering holds great potential for repair and regeneration of diseased tissues. Development of tissue engineering scaffolds has emerged as a topic of great interest in biomedical research. Over the past few years, hydrogels have evolved as promising candidates for engineered tissue scaffolds due to their biocompatibility and similarities to native extracellular matrix. However, controlling the porosity of a hydrogel is a challenging issue. The ability to control porosity and the microarchitectural features in hydrogels are vital in creating engineered tissues with structure and function similar to native tissues. The significance of porosity in cell proliferation, survival, and migration are widely reported [57, 58]. An efficient method to reinforce porous hydrogels is the creation of an IPN [59]. Hydrogels with superior mechanical properties could be obtained through the preparation of IPNs. Ferrer *et al.* fabricated a hydrophobic-hydrophilic IPN structure of porous-hybrid hydrogel of poly(ethyl methacrylate) and poly (2-hydroxyethyl acrylate) using sequential polymerization [60]. Enlarged pores can be obtained by the incorporation of IPN structure. IPNs are also extrapolated for other tissue engineering applications such as development of cartilage scaffolds and bone substitutes. Encapsulation of 3-D SMCs and 2-D culture of endothelial cells (ECs) have been demonstrated using gelatin-bonded dextran-based hydrogel with relatively high modulus [61]. These hydrogels exhibited elastic properties that were superior to PEG-based hydrogels. Adhesion of vascular endothelial cells in 2-D culture and proliferation of vascular smooth muscle cells in 3-D culture was supported for up to 14 days with the aid of these IPN hydrogels. The IPN hydrogels are capable of putting forth an increased mechanical strength beyond that of either original component and enhanced stability against enzymatic digestion or UV degradation. IPNs comprising of 1-ethyl-3-(3-dimethylaminopropyl)carbodiimide and *N*-hydroxysuccinimide crosslinked porcine atelocollagen and poly(ethylene glycol) diacrylate crosslinked 2-methacryloyloxyethyl phosphorylcholine (MPC) have been employed for making corneal substitutes [62]. These hydrogels could retain the full biointeractive, cell-friendly properties of collagen in promoting corneal cell and nerve ingrowth and regeneration (despite MPC's known antiadhesive properties). Furthermore, this hydrogel is able to interact with host tissues to promote tissue regeneration by mobilizing endogenous host stem or progenitor cells. Long-term performance of small-diameter vascular grafts made of a poly(ether)urethane–polydimethylsiloxane (PEtU–PDMS) semi-IPN was studied by Soldani *et al.* which projected the efficiency of the grafts comparison with a standard expanded polytetrafluoroethylene (ePTFE) graft [63]. The grafts exhibited ability of in vivo remodeling and negligible signs of calcification. Despite the maintenance of elasticity and enhanced strength coupled with failure stress and stiffness, synthetic and nonbiological IPN hydrogels are not preferred for cell culture due to their nondegradability [64]. Therefore, biologically derived IPNs from collagen and hyaluronic acid have been widely explored for tissue engineering applications [65]. Enzymatically degradable IPNs have also been assessed for enhancements in peri-implant bone formations [66]. Modulation of bone formation in the peri-implant region have been reported using IPNs of poly(acrylamide-*co*-ethylene glycol/acrylic acid) functionalized with an Arg-Gly-Asp-arginylglycylaspartic acid (RGD) grafted to titanium implants [67]. Natural biomaterials based IPNs possess dynamic cell-responsive mechanical properties which can provide a culture environment that supports tissue growth. Shikanov *et al.* investigated the cell-responsive mechanical properties of fibrin–alginate in the in vitro growth of ovarian follicles [68]. IPNs derived from synthetic and a natural polymer could result in materials, which combine the mechanical properties of the synthetic component with the biological properties of the natural one.

15.4.6.3 Gene Delivery Cationic polymers are most-sought-after matrices for gene delivery as they are able to mask the negative charge of DNA and help in its condensation which further leads to efficient transfection with negligible toxicity and immune response [69]. Resisting premature enzymatic degradation and aggregation, targeting specific uptake and supplying controlled release of the genetic material are the characteristics of a good delivery system. Hydrogel–hydrogel composite semi-IPNs synthesized by dispersing quaternary ammonium salt hydrogels, developed from a cationic monomer, (3-acrylamidopropyl)-trimethylammonium chloride, into neutral monomers such as acrylamide or 2-hydroxylethyl methacrylate exhibited gene delivery potential [70].

15.4.6.4 Ophthalmic Applications Artificial corneas offer an alternative treatment to patients who are blind due to corneal disease. The basis for an artificial cornea encompasses transparency, mechanical strength, oxygen permeability, permeability to glucose and other nutrient molecules, as well as cellular compatibility [71]. Glucose-permeable poly(dimethyl siloxane)–poly(*N*-isopropyl acrylamide) interpenetrating networks have been synthesized and characterized, and their potential as ophthalmic biomaterial has been proved [72]. Liu *et al.* prepared collagen–phosphorylcholine interpenetrating network hydrogels as corneal substitutes in order to retain the enzymatic resistance, hydrophilicity, and increased mechanical strength of MPC.

15.4.6.5 Dental Applications Use of IPN nanostructures is abundant in denture base polymers, fiber-reinforced composites, denture teeth, etc. The key rationale for using IPNs in dentistry lies in its toughness and mechanical interlocking to resin adhesives and also favorable handling properties [73]. Clinical success of dental treatments employing composite resin restoration is related to the extent of interlocking of resin adhesives to polymers and composites. Inadequate interlocking of materials results in the debonding of restoration subsequently leading to a failure in the treatment. Polymethyl methacrylates (PMMA) are widely used as denture base polymers. Another highlight of IPN is its use as artificial teeth. Usually, artificial teeth are made of porcelain or polymers. Porcelain is prone to chipping and this problem is overcome by the use of polymers as they are much tougher. Polymers such as crosslinked PMMA experience the drawback of being plasticized by food materials which severely decrease its mechanical properties. IPNs of PMMA can circumvent this setback as they swell less than crosslinked polymers and therefore are not easily plasticized. The artificial teeth, launched by Dentsply, are based on PMMA–PMMA homo-IPNs. These materials, under the trade names Trubyte and Bioform, are swelling resistant and are ideal for making good mouth fits as they possess excellent grindability [6]. IPNs also provide good interfacial adhesion which is an essential requirement for the success of modern dentistry.

15.4.6.6 Other Biomedical Applications The poor mechanical properties of hydrogels restrict their use to coatings on catheters and cannot be applied as urinary materials. The mechanical properties of hydrogels can be improved without affecting other key characteristics through the formation of an IPN. Jones *et al.* investigated the physicochemical and biological performance of hydrogel–poly(e-caprolactone) IPNs as urinary material [74]. Interpenetrating fibrin–alginate matrices have also shown promise in *in vitro* ovarian follicle development [68]. Recently, poly(3,4-ethylenedioxythiophene)/poly(styrenesulfonate)–poly(vinylalcohol)/poly(acrylic acid) IPNs were employed for improving optrode-neural tissue interface [75]. Investigation of IPNs for self-healing applications has been recently done by Peterson *et al.* involving an *in situ* sequential IPN system based on diglycidyl ether of bisphenol A (DGEBA) cured with 4,40-methylene biscyclohexanamine as the crosslinked phase and poly(methacrylated phenyl glycidyl ether) (pMPGE) as the linear polymer phase [76].

15.5 CONCLUSION

Interpenetrating polymer networks represent one of the fastest growing fields in polymer science and have a wide range of applications. Their functionality in adsorption processes, optical, thermal, electronic, and acoustics applications are well established. The utility of these crosslinked phase separating materials with reduced phase domain size as a biomaterial has been recently explored. They are ideal as carrier systems for the delivery of drugs, proteins, and genes. IPNs are, indeed, inevitable in revolutionizing the field of biomaterial science starting from the development of bioengineering tissues, such as bone substitutes, tissue and cartilage scaffolds to targeted drug and gene delivery.

REFERENCES

[1] Suthar, B., Xiao, H.X., Klempner, D. and Frisch, K.C. (1996) *Polymers for Advanced Technologies*, **7**, 221.

[2] Sperling, L.H. (1985) *Polymer Engineering and Science*, **25**, 517.

[3] Frisch, H.L., Klempner, D. and Frisch, K.C. (1969) *Journal of Polymer Science, Part B*, **7**, 775.

[4] Dillon, M. (1994) Silicone and Poly(tetrafluoroethylene) Interpenetrating Polymer Networks: Brief History, in *Interpenetrating Polymer Networks*, ACS Advances in Chemistry (eds D. Klempner, L.H. Sperling and L.A. Utracki), ACS Books, Washington, DC.

[5] Sperling, L.H. (1990) Basic viscoelastic definitions and concepts, in *Sound and Vibration Damping with Polymers*, ACS Symposium Series (eds R. Corsaro and L.H. Sperling), ACS Books, Washington, DC.

[6] L.H. Sperling and V. Mishra, The Current Status of Interpenetrating Polymer Networks, in S.C. Kim and L.H. Sperling, eds, *IPNs around the World: Science and Engineering*, Wiley, New York (1997).

[7] Caykara, T., Sengül, G. and Birlik, G. (2006) *Macromolecular Materials and Engineering*, **291**, 1044–1051.

[8] Frish, H.L. and Klempner, D. (1970) *Advances in Macromolecular Chemistry*, **2**, 149.

[9] Klempner, D. (1978) *Angewandte Chemie*, **17**, 97.

[10] Frisch, K.C., Frisch, H.L., Klempner, D. and Mukerjee, S.K. (1974) *Journal of Applied Polymer Science*, **18**, 689.

[11] Sperling, L.H. (1981) *Interpenetrating Polymer Networks and Related Materials*, Plenum, New York.

[12] Frisch, H.L., Klempner, D. and Frisch, K.C. (1980) *Polymer Alloys II*, Plenum, New York.

[13] Hsu, T.J. and Lee, L.J. (1985) *Polymer Composites*, **15**, 951.

[14] Myung, D., Waters, D., Wiseman, M. *et al.* (2000) *Polymers for Advanced Technologies*, **19**, 647.

[15] Sperling, L.H. and Hu, R. (2002) Interpenetrating Polymer Networks, in *Polymer Blends Handbook* (ed L.A. Utracki), Kluwer Academic Publishers, Dordrecht, p. 417.

[16] George, S.C. and Thomas, S. (2001) *Progress in Polymer Science*, **26**, 985.

[17] Chainey, M. (1989) Transport phenomena in polymer films, in *Handbook of Polymer Science and Technology, Composites and Speciality Applications* (ed N.P. Cheremisinoff), Marcel Dekker, New York, p. 499.

[18] Shanthamurthy, U.A. and Aminabhavi, T.M. (1990) *Polymer*, **31**, 1757.

[19] Ajithkumar, S., Patel, N.K. and Kansara, S.S. (2000) *European Polymer Journal*, **36**, 2387.

[20] Lee, D.S., An, J.H. and Kim, S.C. (1994) Gas Transport in Polyurethane-Polystyrene Interpenetrating Polymer Network Membranes: Effect of Cross-Linked State and Annealing, in *Interpenetrating Polymer Networks Advances in Chemistry* (eds D. Kempner and L.H. Sperling), American Chemical Society, Washington, DC, p. 463.

[21] Holdsworth, A.K. and Hourston, D.J. (1994) Gas Permeability of Latex Interpenetrating Polymer Network Films, in *Interpenetrating Polymer Networks Advances in Chemistry* (eds D. Kempner and L.H. Sperling), American Chemical Society, Washington, DC, p. 449.

[22] Nguyen, Q.T., M'Bareck, C.O., David, M.O. *et al.* (2003) *Materials Research Innovations*, **7**, 212.

[23] Xu, T. (2005) *Journal of Membrane Science*, **263**, 1.

[24] Ye, Y.S., Rick, J. and Hwang, B.J. (2012) *Polymers*, **4**, 913.

[25] Kang, M.S., Kim, J.H., Won, J. *et al.* (2005) *Journal of Membrane Science*, **247**, 127.

[26] Sahu, A.K., Selvarani, G., Bhat, S.D. *et al.* (2008) *Journal of Membrane Science*, **319**, 298.

[27] Sahu, A.K., Selvarani, G., Pitchumani, S. *et al.* (2008) *Journal of the Electrochemical Society*, **155**, 686.

[28] Fu, R.Q., Hong, L. and Lee, J.Y. (2008) *Fuel Cells*, **8**, 5261.

[29] Fu, F. and Wang, Q. (2004) *Desalination*, **164**, 105.

[30] Metayer, M. and M'Bareck, C.O. (1997) *Reactive and Functional Polymers*, **33**, 311.

[31] Bessboussea, H., Vercherea, J.F. and Lebruna, L. (2012) *Chemical Engineering Journal*, **187**, 16.

[32] Qingchun, Z. and Tian, C. (2012) *Industrial and Engineering Chemistry Research*, **51**, 4952.

[33] Aklonis, J.J. and MacKnoght, W.J. (1983) *Introduction to Polymer Viscoelasticity*, 2nd edn, Wiley-Interscience, New York.

[34] Fay, J.J., Murphy, C.J., Thomas, D.A. and Sperling, L.H. (1991) *Polymer Engineering and Science*, **31**, 1731.

[35] Utracki, L.A. (2002) *Polymer Blends Handbook*, Kluwer Academic Publishers, Dordrecht.

[36] Tang, D., Liangsheng, Q., Jin, Z. and Zhao, L. (2005) *Journal of Materials Science and Technology*, **21**, 585.

[37] Bukbinder, T.L. and Kosjakov, V.I. (1996) *Vysokomolekulyarnye Soedineniya, Seriya*, **28**, 625.

[38] Eck, W., Wilson, H.R. and Cantow, H.J. (1995) *Advanced Materials*, **7**, 800.

[39] Anne, T., Jean, P.D., Pierre, H.A. *et al.* (2012) *Solar Energy Materials & Solar Cells*, **99**, 116.

[40] Boyd, G.T., DeVoe, R.J., Gorodisher, I., and Ylitalo, D.A. (1997) Shaped Adhesives. U.S. Patent 6,054,007, April 10, 1997 (to 3M Innovative Properties Company).

[41] Bernkop, A.S. (2000) *Expert Opinion on Therapeutic Patents*, **10**, 1357.

[42] Signori, F., Solaro, R., Chiellini, E. *et al.* (2003) *Macromolecular Symposia*, **197**, 289.

[43] Twaites, B., Heras Alarcón, C. and Alexander, C. (2005) *Journal of Materials Chemistry*, **15**, 441.

[44] Gupta, N. and Srivastava, A.K. (1994) *Polymer International*, **35**, 109.

[45] Graham, N.B. and McNeil, M.E. (1984) *Biomaterials*, **5**, 27.

[46] Lee, K.Y. and Mooney, D.J. (2001) *Chemical Reviews*, **101**, 1869.

[47] Mandal, B., Kapoor, S. and Kundu, S.C. (2009) *Biomaterials*, **30**, 2826.

[48] Rodríguez-Félix, D.E., Pérez-Martínez, C.J., Castillo-Ortega, M.M. *et al.* (2012) *Polymer Bulletin*, **68**, 197.

[49] Kulkarni, R.V., Boppana, R., Krishna Mohan, G. *et al.* (2012) *Journal of Colloid and Interface Science*, **367**, 509.

[50] Muñoz-Muñoz, F., Ruiz, J.C., Alvarez-Lorenzo, C. *et al.* (2012) *Radiation Physics and Chemistry*, **81**, 531.

[51] Sajeesh, S. and Sharma, C.P. (2004) *Journal of Biomaterials Applications*, **19**, 35.

[52] Banerjee, S., Siddiqui, L., Bhattacharya, S.S. *et al.* (2012) *International Journal of Biological Macromolecules*, **50**, 198.

[53] Ekici, S. and Saraydin, D. (2004) *Drug Delivery*, **11**, 381.

[54] Rekha, M.R. and Sharma, C.P. (2008) *Journal of Applied Polymer Science*, **110**, 2787.

[55] Lichen, Y., Jieying, D., Jing, Z. *et al.* (2010) *Biomaterials*, **31**, 3347.

[56] Silan, C., Akcali, A., Otkun, M.T. *et al.* (2012) *Colloids and Surfaces, B: Biointerfaces*, **89**, 248.

[57] Mandal, B. and Kundu, S. (2009) *Biomaterials*, **30**, 2956.

[58] Lien, S.M., Ko, L.Y. and Huang, T.J. (2009) *Acta Biomaterialia*, **5**, 670.

[59] Annabi, N., Nichol, J.W., Zhong, X. *et al.* (2010) *Tissue Engineering, Part B: Reviews*, **16**, 371.

[60] Ferrer, G.G., Melia, J.M.S., Canales, J.H. *et al.* (2005) *Colloid & Polymer Science*, **283**, 681.

[61] Liu, Y. and Chan-Park, M.B. (2009) *Biomaterials*, **30**, 196.

[62] Liu, W., Deng, C., McLaughlin, C.R. *et al.* (2009) *Biomaterials*, **30**, 1551.

[63] Soldani, G., Losi, P., Bernabei, M. *et al.* (2010) *Biomaterials*, **31**, 2592.

[64] Weng, L., Gouldstone, A., Wu, Y. and Chen, W. (2008) *Biomaterials*, **29**, 2153.

[65] Brigham, M.D., Bick, A., Lo, E. *et al.* (2009) *Tissue Engineering Part A*, **15**, 1645.

[66] Takaoka, K., Nakahara, H., Yoshikawa, H. *et al.* (1988) *Clinical Orthopaedics and Related Research*, **234**, 250.

[67] Ho, J.E., Barber, T.A., Virdi, A.S. *et al.* (2007) *Journal of Biomedical Materials Research, Part A*, **81**, 720.

[68] Shikanov, A., Xu, M., Woodruff, T.K. and Shea, L.D. (2009) *Biomaterials*, **30**, 5476.

[69] Smedt, S.C., Demeester, J. and Hennink, W.E. (2000) *Pharmaceutical Research*, **17**, 113.

[70] Sahiner, N., Godbey, W.T., McPherson, G.L. and John, V.T. (2006) *Colloid & Polymer Science*, **284**, 1121.

[71] Chirila, T.V., Hicks, C.R., Dalton, P.D. *et al.* (1998) *Progress in Polymer Science*, **23**, 447.

[72] Liu, L. and Sheardown, H. (2005) *Biomaterials*, **26**, 233.

[73] Vallittu, P.K. (2009) *Journal of Adhesion Science and Technology*, **23**, 961.

[74] Jones, D.S., McLaughlin, D.W.J., McCoy, C.P. and Gorman, S.P. (2005) *Biomaterials*, **26**, 1761.

[75] Lu, Y., Li, Y., Pan, J. *et al.* (2012) *Biomaterials*, **33**, 378.

[76] Peterson, A.M., Kotthapalli, H., Aflal, M. *et al.* (2012) *Composites Science and Technology*, **72**, 330.

INDEX

Micro- and Nano-structured Interpenetrating Polymer Networks: From Design to Applications, First Edition. Edited by Sabu Thomas, Daniel Grande, Uroš Cvelbar, K.V.S.N. Raju, Ramanuj Narayan, Selvin P. Thomas and Akhina H.